D1547962

THE MATHEMATICAL THEORY
OF THE DYNAMICS OF
BIOLOGICAL POPULATIONS

THE MATHEMATICAL THEORY
OF THE DYNAMICS OF
BIOLOGICAL POPULATIONS

Based on a Conference organised by the
Institute of Mathematics and its Applications
in association with the Institute of Biology

Edited by

M. S. BARTLETT

R. W. HIORNS

*Department of Biomathematics,
University of Oxford, Oxford, England*

ACADEMIC PRESS · LONDON AND NEW YORK

ACADEMIC PRESS INC. (LONDON) LTD.
24/28 Oval Road
London NW1

United States Edition published by
ACADEMIC PRESS INC.
111 Fifth Avenue
New York, New York 10003

Library of Congress Catalog Card Number: 73-1468
0-12-079850-6

PRINTED IN GREAT BRITAIN BY
J. W. ARROWSMITH LTD., BRISTOL, ENGLAND

CONTRIBUTORS

N. T. J. Bailey, *Division of Strengthening of Health Services, World Health Organization, 1211 Geneva 27, Switzerland.*

M. S. Bartlett, F.R.S., *Department of Biomathematics, University of Oxford, Pusey Street, Oxford OX1 2JZ, England*

J. A. Bishop, *Department of Zoology, University of Liverpool, Brownlow Street, P.O. Box 147, Liverpool L69 3BX, England.*

W. F. Bodmer, *Genetics Laboratory, Department of Biochemistry, University of Oxford, South Parks Road, Oxford OX1 3QU, England.*

R. M. Cormack, *Department of Statistics, University of Edinburgh, Edinburgh, Scotland.* (Present address: *Department of Statistics, University of St. Andrews, St. Andrews, Fife, Scotland*).

J. Gani, *Department of Probability and Statistics, The University, Sheffield, S3 7RH, England.*

W. B. Hall, *Department of Agriculture and Fisheries for Scotland, Marine Laboratory, P.O. Box 101, Victoria Road, Aberdeen AB9 8DB, Scotland.*

M. P. Hassell, *Imperial College Field Station, Silwood Park, Ascot, Berkshire, England.*

R. W. Hiorns, *Department of Biomathematics, University of Oxford, Pusey Street, Oxford, OX1 2JZ, England.*

R. Jones, *Department of Agriculture and Fisheries for Scotland, Marine Laboratory, P.O. Box 101, Victoria Road, Aberdeen AB9 8DB, Scotland.*

S. Karlin, *Department of Theoretical Mathematics, The Weizmann Institute of Science, Rehovot, Israel, and Stanford University, Stanford, California, U.S.A.*

E. D. Le Cren, *The River Laboratory, East Stoke, Wareham, Dorset BH20 6BB, England.* (Present address: *Freshwater Biological Association, Windermere Laboratory, The Ferry House, Ambleside, Westmorland, England*).

P. D. M. Macdonald, *Department of Applied Mathematics, McMaster University, Hamilton, Ontario, Canada.*

M. Mountford, *The Nature Conservancy, 19 Belgrave Square, London, SW1X 8PY, England.*

G. Murdie, *Imperial College Field Station, Silwood Park, Ascot, Berkshire, England.*

R. A. Parker, *Department of Zoology, Washington State University, Pullman, Washington 99163, U.S.A.*

E. C. Pielou, *Department of Biology, Dalhouse University, Halifax, Nova Scotia, Canada.*

J. G. Pope, *Fisheries Research Laboratories, Pakefield Road, Lowestoft, Suffolk, England.*

A. Robertson, F.R.S., *Department of Genetics, Edinburgh University, Old College, South Bridge, Edinburgh EH8 9YL, Scotland.*

P. M. Sheppard, F.R.S., *Department of Zoology, University of Liverpool, Brownlow Street, P.O. Box 147, Liverpool L69 3BX, England.*

J. G. Skellam, *The Nature Conservancy, 19 Belgrave Square, London, SW1X 8PY, England.*

M. Williamson, *Department of Biology, University of York, York, YO1 5DD, England.*

Preface

These Proceedings have been assembled from the papers presented at a conference at Oxford in September, 1972 on the *Mathematical Theory of the Dynamics of Biological Populations*, arranged by the Institute of Mathematics and its Applications, in association with the Institute of Biology. The idea of such a conference arose from some discussion in the first instance between Professor Sir James Lighthill, F.R.S. and Mr. R. Beverton of the National Environment Research Council, following which an Organizing Committee was set up consisting of Professor M. S. Bartlett, F.R.S. (Chairman) (Department of Biomathematics, Oxford), Professor P. Armitage (Department of Statistics, London School of Hygiene), Mr. R. Beverton (N.E.R.C., Alhambra House), Professor W. Bodmer (Department of Genetics, Oxford), Professor J. Gani, F.I.M.A. (Department of Statistics, Sheffield), Dr. J. Skellam (Nature Conservancy, London), Professor T. R. E. Southwood (Department of Zoology, Imperial College, London), Dr. E. T. Goodwin, F.I.M.A. (National Physical Laboratory).

The committee was chosen to represent both mathematical and biological interests, and envisaged the subject-matter of the conference "as broadly based on the mathematical theory of time-change in animal populations, whether these populations refer to land animals (including humans), fishes or insects, and whether they refer to actual migration in space or to changes in size and structure. Problems such as gene flow, or epidemiological spread in infection, for which the mathematical equations are often similar, are to be included. Statistical and computing aspects arising from the essential comparison and inter-relation of theoretical models with observed data are also to be considered. While emphasis will be placed on the relevant and often common theoretical techniques and methods, these are to be discussed in close relation with problems of current and future importance in the various biological areas."

In addition to the wide spectrum of subject-matter, it was realized that different aspects could be emphasized, for example, the I.M.A. has always regarded the educational and expository aspect as important, and there were also, of course, research aspects; there was the obvious possible divergence between mathematics and biology. The committee, bearing in mind the comparatively short duration of the conference and the impossibility of a complete coverage of the subject-matter, selected a representative group of speakers in the hope of attaining a representative group of papers; and it must be left to those attending the conference or to the reader to judge how far this object was achieved. At any rate, we may speak for the I.M.A. in thanking the rest of the committee members and speakers for their co-operation in this venture. We are also grateful to Professor C. A. Coulson for his stimulating and very apposite opening remarks, which have been included here.

With regard to the individual papers, these have been left in the order they were given at the conference with, of course, the warning made at the time that the broad subject-divisions were not to be regarded too rigidly. No "post-mortem" on the papers was attempted at the conference, nor will it be held now; but we have very briefly summarized below the contributions made by the various papers to the five sessions and which now constitute the five parts of this volume.

Mathematical models for biological populations in time and space were considered in the first two sessions each starting with a theoretical paper on time changes and on space diffusion respectively, and continuing with various specific biological problems classified as falling, at least in part, in these two areas. The first paper, in addition to its general introductory remarks, attempted to include a brief but comprehensive outline of the relevant mathematical formulation of population change, aided by its mathematical appendix and bibliography, and emphasized common theoretical ground for different biological problems by referring in more detail to three applications in epidemiological, genetical and demographic fields, viz. a model for measles, the temporal history of a gene mutation, and age distributions. Pope examined the theory of fisheries management; and in particular the effects of variable rates of exploitation and of mobility of fishing effort. Jones and Hall developed one- and two-species models for studying the population dynamics of fish species such as haddock.

In the second session Skellam's general paper on diffusion put special emphasis on the relation between the mathematical equations and their biological interpretation. Murdie and Hassell discussed predator–prey models, referring to the effects of heterogeneous food or prey distribution in space and of searching behaviour by the predators. Pielou described an investigation into the geographical ranges of aphid infestation on *Solidago* species in eastern N. America. Le Cren concluded the first day with an account of the dynamics of salmonid fish, including migration effects, and territorial behaviour as a cause of density-dependent mortality.

The session on population genetics was introduced by Bodmer with a discussion of deterministic models. He referred not only to the simplest models of genetic changes assuming the Hardy–Weinberg law, but also to such complications as incompatibility selection, and linked loci with selection. Karlin presented new findings on a two locus haploid model subject to selection and unidirectional mutation, the main problem considered being the effect of recombination on the speed of selection of the double mutant when most fit. Robertson emphasized the importance of stochastic effects in artificial selection, and described computer simulation to investigate the dominating characteristics when a selected character is affected by very many loci. The fourth paper, whose title at the conference has been abbreviated, dealt with a few models involving migration and selection effects in human populations and referred to some of the difficulties of analysing population data.

The fourth session comprised a further group of papers dealing with estimation and simulation problems, both of which are, of course, a familiar feature of practical population models. The first two papers, by Cormack and by Bishop and Sheppard, discussed capture-recapture estimation, the first simplifying the theoretical basis of such estimation, and the second comparing the performance of two different estimation techniques proposed by Fisher and Ford and by Jolly by means of computer simulation and sampling of a fictitious population. Bailey examined estimation problems in the epidemiological field, referring in particular to detailed parametric models for diseases such as measles, infectious hepatitis and smallpox. In the final paper in this session, Parker described a deterministic model of a large Canadian lake subject to phosphate influx, its computer simulation and problems of parameter adjustment to obtain a "best fit".

The papers in the final session had been classified as relating in one way or another to population distribution and community structure. Gani emphasized the distinction between a *current* life table and a *cohort* life table, especially when survival rates were time-dependent; and also discussed further the topic of age distributions referred to in the first paper of the conference. Macdonald in his paper on the estimation of parameters for population models of cell proliferation also presented general formulae for "age" and related distributions for growing cell populations. Mountford investigated the role of variable clutch-size in the survival of a species in a heterogeneous environment. Finally, Williamson presented a paper on species diversity, discussing measures of diversity and drawing attention to the need for a theory of population dynamics for a whole community.

The conference papers were often followed by relevant and stimulating discussion, which somewhat reluctantly it was decided would not be feasible to record. However, it seems probable that some participants, and perhaps some readers of this volume, will be inspired to take up the more challenging topics raised in these papers for further study. The wide variety of experience and backgrounds of the participants at the Conference inevitably meant that many new encounters and acquaintances were formed and we consequently hope there should result an enhanced understanding of the theories and methodologies, whether similar or contrasting, which apply in the different problem areas.

Finally, acknowledgements must be made, in addition to those mentioned earlier, to the Secretary of the I.M.A., Mr. Norman Clarke, to the Deputy Secretary, Miss Catherine Richards and her assistants, in particular, Miss Barbara Mayne, Miss Maureen Downie and Mrs. Lesley Whittaker for their efforts in planning and executing the detailed arrangements for the conference, to Magdalen College for the hospitality and accommodation provided for participants and to the Department of Zoology in the University of Oxford for the use of an excellent lecture theatre and its facilities.

M. S. BARTLETT
R. W. HIORNS

Department of Biomathematics
University of Oxford
April, 1973

Contents

Part IV   Estimation and Simulation Problems

Part V   Population Distribution and Community Structure

Mathematics and Biology

Introductory remarks by Professor C. A. Coulson (President of the Institute of Mathematics and its Applications)

In welcoming the members of this Conference on the Dynamics of Biological Populations, I wish to remind you that for the first time in the history of the I.M.A., there has been co-sponsorship by the I.M.A. and the Institute of Biology. This is all to the good: for there is need of mathematics in studying many biological problems; and there is much to be gained by using it.

First, then, the question of need. Everyone knows about Darwin's theory of evolution, but not everyone realizes that, from the earliest days, it was clouded in uncertainty. Thus Darwin believed in the accumulation of small variations and Huxley in the advent of large ones. Here indeed the thought-forms of the mathematician are necessary if any clear-cut decision is to be made. Darwin himself was no mathematician ("stuck five fathoms deep in mathematics", as he once put it in a letter to a friend, "and likely to stay there".) This was a pity, because quite simple mathematical types of reasoning could have revealed the error of his view that, in bisexual reproduction, each offspring inherited the mean of the characteristics of his two parents. This failure to think mathematically delayed our understanding of biological processes. So there was a gap of more than seventy years between the *Origin of Species* in 1848 and the work of R. A. Fisher and J. B. S. Haldane in Britain (1910–1931) on the mathematical principles of population dynamics.

There is advantage in this coming together. I am thinking of the interplay of experiment and theory, often to the benefit of mathematics. There is good mathematics involved in understanding variations in population. We must learn to deal both with continuous and discrete time situations. The first leads most naturally to the use of differential and integral equations, the second to some of the exciting developments in Markov processes that have come in the last two decades. Moreover (as I learned to my cost some years ago when studying the growth of a population in which occasional mutation could occur), there are some most peculiar statistical distributions to be found, and studied.

In view of this cross-fertilization it is good that the I.M.A. is involved in this conference. But there is a word of caution to be said; not only is it desirable, it is absolutely essential that both the experimental biologist and the professional mathematician should establish a close relationship. For biological material does not exhibit the nice regular reproducible behaviour ideally anticipated by the simple-minded mathematician. Instead it is almost infinitely variable. I recall very vividly how this impressed itself on me more than 35 years ago, at a time when I had turned away from the physical sciences to become for a while an experimental bacteriologist. Bacteria of a given kind were clearly not all the same, in size, or activity. So the concepts to be used in studying them could not be the same sorts of concepts as would be appropriate in the physical sciences. We could summarize by saying that the mathematician needs the restraining hand of the experimental biologist, and the biologist needs the rigorous competence and analytic power of the mathematician. Without this mutual interaction we have all too often bastard mathematics or "sloppy" biology. Nowhere is this more clear than in the study of population dynamics: nowhere is it more necessary to establish an effective liaison.

In this conference our use of the word "population" means that we study all sorts of creatures. This includes not only the human population explosion recently dealt with in the Club of Rome's Report, but also of equally topical interest, the most efficient long-term policy for fishing in the North Sea and the North Atlantic; or the migrations of birds, and the growth of cells, as well as the spread of disease and the distribution of food supplies. A mature society must be one that correctly evaluates the work of Thomas Malthus, Charles Darwin and the Doomsday watchers. None of this is possible without the establishment of a proper mathematical background, within which to interpret the results of biological experiment. Our present Conference marks a significant step towards this end.

Part I

Population Processes in Time

1. Equations and Models of Population Change

M. S. BARTLETT

Department of Biomathematics, University of Oxford, Oxford, England.

1. Remarks on Theoretical Model-Building

The mathematician and the biologist have an almost inevitable confrontation in the formulation of a model for any biological situation, for, whereas the biologist will probably begin with the actual and complex set of facts that he has observed or otherwise become aware of, the mathematician is likely to be looking for an idealized model capable of theoretical investigation. This kind of dilemma has become more obvious as the use of mathematics has moved from the physical to the biological (and also to the social) sciences, and I do not believe that there is a facile answer. It is quite natural for an experimental or observational scientist to list a large collection of facts that should be taken into account in any comprehensive theoretical formulation; but the purpose of the latter is to understand and interpret the facts, and the mathematician is sometimes able to give insight by a study of idealized and over-simplified models, which are more complicated in their behaviour than is often realized. Until we can understand the properties of these simple models, how can we begin to understand the more complicated real situations? There is plenty of scope for variety of approach, but I would urge that critical criteria of success be employed wherever possible. These include the extent to which:

(i) known facts are accounted for;

(ii) greater insight and understanding are achieved of the biological situation being studied;

(iii) the theory or model can correctly predict the future pattern, even under different conditions from those pertaining to the current observed data.

Over-simplified models are obvious targets for criticism, but model-builders who are more ambitious in their attention to detail must bear in mind two often relevant points:

(a) If the simplified model has retained the basic features of the real situation, there may well be "diminishing returns" as further parameters and features are introduced;

5

(b) The more parameters, the more problems of estimation, and this can seriously add to the difficulties of building what may be in principle a more realistic model.

2. Equations of Population Change

Let me try next to survey some of the basic equations of population change, and see how they fit into the historical picture. (They are applicable to physical or chemical situations as well as biological but I shall have the biological in mind.) It is convenient and feasible to formulate these equations in a general probabilistic or *stochastic* framework, without implying by this that the so-called *deterministic* framework is not adequate in many contexts.

I shall denote a set of population sizes at time t of various interacting species or types (including if necessary different classes by age-groups or sex or location or genetic composition of the same species) by the mathematical vector notation N_t where t may be discrete (generation number) or continuous. For definiteness we shall mainly consider the latter case in this introductory presentation, though analogous formulations in discrete time have been familiar from the work of mathematical geneticists (in relation to successive *generations*), or in population theory from the work of P. H. Leslie (1945), leading to matrix recurrence equations which are in particular convenient for simulation calculations.

In most contexts it is possible to classify the process as what is called *Markovian*, by which is meant that the future evolution of the process can be calculated in principle from a knowledge of the population vector N_t at any given time t. This is obviously not valid in general if N_t omits specification by, say, age and sex, but may be valid if such classifications are included. An example which appears to be an exception is that of human populations increasing under a social custom of family planning, but even such examples may, at least in principle, be included by appropriate redefinition, such as counting married couples as units, classified by their family size. Even at this stage it is evident how the mathematical formulation can increase in complexity with the realism of the specification, but not necessarily with a change in the basic mathematical framework. (There is an additional technical complication in the exact handling of age increase or spatial movement, owing to the more complicated nature of N_t when it involves subdivision by ancillary *continuous* variables, but I will come back to that later.)

In continuous time, the probabilities of changes are reasonably assumed proportional to the time interval $(t, t + \Delta t)$ for small enough Δt (thereby ensuring a non-zero probability of change in a non-zero interval $t, t + \tau$, and a zero probability of change in a zero interval). Thus there is a transition probability function

$$R\{\mathbf{N}_{t+\Delta t}|\mathbf{N}_t\}\Delta t \text{ of order } \Delta t \text{ for } \mathbf{N}_{t+\Delta t} \neq \mathbf{N}_t,$$

by means of which the probabilistic differential equation (known as the Kolmogorov "forward" equation) governing the changes in $\mathbf{N}_t$ from t to $t + \Delta t$ may be written down.

It is often mathematically convenient, as I pointed out many years ago (e.g. Bartlett, 1949), to formulate these changes in one "omnibus" equation for what is called the probability-generating (or some equivalent) function of $\mathbf{N}_t$, as shown in more detail in the Appendix.

This equation (for, say, the probability-generating function $\Pi_t(\mathbf{z})$) may be written

$$\frac{\partial \Pi_t(\mathbf{z})}{\partial t} = H_t \Pi_t(\mathbf{z}),$$

where H_t is an *operator* (see Appendix I) acting on $\Pi_t(\mathbf{z})$. In general it will involve t explicitly, as in seasonal variation of birth- or death-rates, but if it is what we term *homogeneous* in time, then the above equation has the simple formal solution

$$\Pi_t(\mathbf{z}) = e^{Ht} \Pi_0(\mathbf{z}),$$

though this equation must not mislead us into thinking that even some of the most elementary natural examples of it, especially those involving more than one type of individual, have other than complicated or intractable solutions.

In the time-homogeneous case there may be an equilibrium situation, which will then satisfy the equation

$$H\Pi(\mathbf{z}) = 0.$$

Sometimes it is illuminating to formulate this general model in direct stochastic terms. We have postulated a stochastic change in time Δt, $\Delta \mathbf{N}_t = \mathbf{N}_{t+\Delta t} - \mathbf{N}_t$, with probability $R\{\mathbf{N}_{t+\Delta t}|\mathbf{N}_t\}\Delta t$, and this specification is equivalent to some equation

$$d\mathbf{N}_t = f(\mathbf{N}_t)\,dt + d\mathbf{Z}_t,$$

where $d\mathbf{Z}_t$ has zero mean and known distributional properties. This equation has its *deterministic* analogue

$$\frac{d\mathbf{n}_t}{dt} = f(\mathbf{n}_t),$$

which neglects $d\mathbf{Z}_t$.

When the solutions of these equations are too intractable, as is in any case likely when further practical details are inserted, it will often be more feasible and instructive to compute numerical *realizations* of the process. In the deterministic version this is in fact identical with the numerical solution (for given starting conditions); but note that in the stochastic version several simulations must be computed in order to have any idea of the complete probabilistic situation, and the known complete solutions in some relatively simple cases have warned us that these can sometimes be rather complex. It is important in more realistic formulations to judge whether stochastic effects are likely to be important, and this judgement in turn can be assisted by theoretical studies of some of the simpler cases.

The remainder of this paper refers to three examples in different biological fields: epidemiological, genetical and demographic. Firstly, some of the properties of the stochastic model for recurrent epidemics in a population with a steady influx of susceptible individuals, which I have previously employed (see, for example, Bartlett, 1960) as a model for measles epidemics, are recalled, and also the *predicted* changes in these properties following the current programme of *partial* vaccination against measles infection, recently investigated by D. A. Griffiths (1971).

3. The Recurrent Epidemics Model for Measles

A theoretical model for measles epidemics goes back to the (deterministic) formulation by W. Hamer in 1906, and has since been discussed extensively by various writers, so that full details and background should be sought in the previous literature. To recapitulate briefly, however, on the model, this represents a population of susceptible individuals in homogeneous contact with a number of infected and infective individuals, the latter being removed (or permanently recovering) analogously to a mortality, and a steady influx of susceptibles represented in the measles application by real births in the community. An influx of infective individuals may also be included if relevant. When the community is closed to such infection from outside, the only parameters are the removal or "mortality" rate μ for the infectives, depending on the incubation period, the influx rate v for susceptibles, depending on the community population size and the birth-rate, and the infectivity coefficient λ, or equivalently the quantity $\rho = \lambda v/\mu$. The value of the latter was originally estimated by Hamer for London as $1/68 \cdot 2$, though Griffiths (*loc. cit.*) has suggested that a smaller value for ρ may be more appropriate (even prior to the vaccination era).

My reasons for referring to this model here (for a summary of its mathematical formulation and properties, see Appendix III) are partly because it is an example of a model which, in spite of its simplified character as formulated above (this neglects seasonal periodicity, the discrete aspects of the incubation period for infection, and spatial heterogeneity, all of which can of course be introduced, especially into simulation studies) has the three features referred to earlier which make it a useful and illuminating model. The earlier deterministic versions of the model gave the oscillatory tendency correctly, but greater insight came from a realization of the importance of "fade-out" of infection associated with the stochastic model, leading to quantitative studies of the concept of a critical population size (around a quarter of a million), and of the inverse relation of epidemic periodicity with community size for communities below this critical size.

This success with the model justifies moreover the claim that it should be relevant even when the recent introduction of vaccine changes the conditions drastically. D. A. Griffiths (*loc. cit*; see also Appendix III), on examining this problem has drawn some conclusions which seem rather disturbing, namely, that a *partial* vaccination programme, which reduces but does not eliminate the susceptible influx, will merely increase the intra-epidemic period without lessening the intensity of epidemics; and, moreover, such an increase in period will if anything put up the average age of onset for those who become infected, with a consequent increased risk to them of complications arising from the infection.

4. The Temporal History of a Gene Mutation

The second example I wish to refer to represents a rather small, but nevertheless very vital, part of the theoretical problems of population genetics: the fate of a gene mutation. Here my purpose is again to show that simple models, if correctly used, can be informative, perhaps more so than has always been appreciated. A useful account of this problem will be found in Chapter 1 of Kimura and Ohta (1971); my remarks are intended by way of further commentary.

In addition to the usual simplifications such as random mating (in the case of sexual reproduction for diploids) there is, as Kimura and Ohta note, a choice of mathematical technique. Either we may concentrate on the early stages of the mutant gene, when its stochastic history may be dealt with by the theory of branching processes (see Appendix I, example (a)) neglecting the finite size of the population; or we may consider the later stages, when the frequency ratio of the number of mutant genes at that locus in the population to the total number of genes at that locus, is more relevant, and we treat it as effectively a continuous variable changing according to a diffusion equation (Appendix I, example (b)).

In fact, it seems advisable to make use of both techniques, so that any limitations of one are high-lighted by the other, and our conclusions are unlikely to be so restricted by the theoretical techniques used. To be more specific, let us consider a neutral mutant gene in a finite population of fixed size N (total gene number $2N$ for a diploid). If the initial number of such mutant genes is not too small, the diffusion equation may be used to obtain not only the probability of extinction or fixation of the mutant (the latter probability being $n/(2N)$ when n is the initial number of mutant genes) but also the distribution of the time to extinction or fixation. Formulae for the mean and variance of this distribution have been given by Kimura and Ohta (cf. also Ewens, 1963); in particular, the mean time conditional on fixation is approximately $4N$ (independent of n), compared with $2n \log (2N/n)$ to extinction. (In the Appendix a rather more general formulation is given.)

However, let us now consider the branching process approach, even at the cost of making N infinite. We start with just one mutant ($n = 1$), but realize that the number of mutant genes in the progeny from this mutant is very relevant. Indeed, we get a different distribution with different assumptions, the probability of extinction at the first generation being e^{-1} if the number of mutant genes has a Poisson distribution with mean one, but $\frac{1}{2}$ if the number of progeny from a mating is restricted to two. The complete distribution of the time to extinction may be investigated (cf. Nei, 1971a), but in view of these difficulties let us simplify further by taking time continuous (as in the diffusion approximation), and represent the process by a simple birth-and-death process (with $\lambda = \mu$ for the neutral case). The value of the parameter λ determines the time scale and may be adjusted to give any desired probability of extinction at $t = 1$ (one generation), consistently with the arbitrariness already referred to.

My justifications for this somewhat naïve suggestion are:

(i) The resulting distribution agrees well with simulation results reported by Kimura and Ohta (1969b), a little better for $\lambda = e^{-1}$ than $\lambda = \frac{1}{2}$, but in neither case is χ^2 significant.

(ii) The distribution is known to have infinite moments, and, while this is in effect a consequence of N infinite, it is sufficient to warn us that an appeal to means and variances may be most misleading for some of the extremely J-shaped distributions arising in this context.

I would in fact suggest that Kimura's earlier (1964) solution for the cumulative distribution (see Appendix II) is more relevant, and that percentiles should be used in this context in preference to moments. Another problem where this suggestion is relevant is that of the total number of new mutant individuals before the mutant type finally dies out (as it does eventually if N is infinite); here again the theoretical mean is infinite in the neutral case. Even in the case of a deleterious mutant (for example, the interesting application by Nei, (1971b), to the haemoglobin mutant Hb M_{IWATE}) the distributions of extinction time and total mutant numbers may be extremely skew. Incidentally the continuous time birth-and-death adaptation to this problem must keep track of new mutant individuals as new "births", but this is technically feasible (Appendix II).

5. Auxiliary Variables such as Age

Coming finally to a few remarks on relevant equations in the general area of demography, I will first of all recall that equations of population growth were classically formulated, for example, by the American actuary A. J. Lotka (see Keyfitz, 1968) in deterministic terms, the growth and change of the population being developed theoretically or numerically from its detailed age composition and its mortality and birth-rates for individuals of different ages. (The complication of the two sexes is a further technical difficulty.) This formulation is sufficient for most actuarial and demographic purposes; but it is as well to realize the theoretical need for the more comprehensive stochastic formulation by noticing that any finite population can merely consist of a number of individuals of specific ages (and sex), and asking what the relation is of such a population with theoretical concepts such as stable age distributions and ultimate exponential growth.

The answer for many of the more straightforward models (in fact, "linear" models) of population growth is that these theoretical results refer to *expected* values, and fluctuations from these expected values may, if desired, also be investigated. The complete stochastic formulation is a matter of some sophistication (see D. G. Kendall, 1949), but may often be relevant if only to ensure that when we refer to, say, an age distribution we are speaking about a property of our model which makes theoretical sense. Sometimes, however, this wider approach enables us to ask relevant questions which cannot even be formulated in a deterministic framework. This has obviously been true in the epidemiological and genetic examples already discussed, but if we refer to the derivation (see Bartlett, 1970, or Appendix IV) of stable age distributions not only in the standard population case but also in the more recent application to proliferating cell populations (Cleaver, 1967) we may notice several queries concerning the stochastic independence of successive phase-times of a dividing cell, the stochastic independence of subsequent daughter cells from the same parent, and so on, some or all of which may affect the analysis and interpretation of particular experimental results (cf. Macdonald, 1970).

Mathematical Appendix

I. BASIC EQUATIONS

Population number N_t at time t; or, more generally, vector $\mathbf{N}_t$, or even $\mathbf{N}_t(u)$, where u may be age or position.

Changes in N_t from 0 to t assumed governed by matrix

$$\prod_{i=1}^{t} \mathbf{Q}_i$$

where $\mathbf{Q}_i$ matrix of transition probabilities between times i and $i + 1$. In time-homogeneous case, this is $\mathbf{Q}^t$, or equivalently $e^{\mathbf{R}t}$ in continuous time case.

If we define (in latter case)

$$\pi_t(\mathbf{z}) = M_t(\boldsymbol{\theta}) = E\{\exp \boldsymbol{\theta}' \mathbf{N}_t\}, \qquad (\boldsymbol{\theta} = \log \mathbf{z}),$$

then (Bartlett, 1949)

$$\frac{\partial M_t(\boldsymbol{\theta})}{\partial t} = H\left(\boldsymbol{\theta}, t, \frac{\partial}{\partial \boldsymbol{\theta}}\right) M_t(\boldsymbol{\theta}), \tag{1}$$

where

$$H(\boldsymbol{\theta}, t, \mathbf{n}) = \lim_{\Delta t \to 0} E\left\{\frac{\exp(\boldsymbol{\theta}' \Delta \mathbf{N}_t) - 1}{\Delta t} \middle| \mathbf{N}_t = \mathbf{n}\right\},$$

or equivalently in terms of $\pi_t(\mathbf{z})$.

In time-homogeneous case, equilibrium solution (if it exists) satisfies

$$H\left(\boldsymbol{\theta}, \frac{\partial}{\partial \boldsymbol{\theta}}\right) M(\boldsymbol{\theta}) = 0. \tag{2}$$

Direct stochastic representation, say

$$d\mathbf{N}_t = \mathbf{f}(\mathbf{N}_t)\, dt + d\mathbf{Z}_t,$$

where $E\{d\mathbf{Z}_t\} = 0$, with deterministic approximation

$$\frac{d\mathbf{N}_t}{dt} = \mathbf{f}(\mathbf{N}_t). \tag{3}$$

Equilibrium points in (3) determined by

$$\mathbf{f}(\mathbf{N}_t) = 0. \tag{4}$$

Examples

(a) Multiplicative or branching processes
Discrete time:

$$\mathbf{z} \to \mathbf{G}(\mathbf{z})$$

$$\pi_{t+1}(\mathbf{z}) = \pi_t(\mathbf{G}(\mathbf{z})),$$

or equivalently ("backward" equation, with $\pi_0(\mathbf{z}) = \mathbf{z}$)

$$\pi_{t+1}(\mathbf{z}) = \mathbf{G}(\pi_t(\mathbf{z})). \tag{5}$$

In continuous time, let

$$\mathbf{G}(\mathbf{z}) \to \mathbf{z} + \mathbf{g}(\mathbf{z})\,dt,$$

then (5) becomes

$$\frac{\partial \pi_t(\mathbf{z})}{\partial t} = \mathbf{g}(\pi_t(\mathbf{z})). \tag{6}$$

(b) Normal diffusion approximation (one variable X_t)

$$H(\theta, t, X_t) \sim \mu(t, X_t)\theta + \tfrac{1}{2} v(t, X_t)\theta^2.$$

For probability density $f_t(x)$,

$$\frac{\partial f_t(x)}{\partial t} + \frac{\partial}{\partial x}[\mu(t, x)f_t(x)] = \tfrac{1}{2}\frac{\partial^2}{\partial x^2}[v(t, x)f_t(x)]. \tag{7}$$

From (7), in equilibrium (time-homogeneous case),

$$f(x) = Ah(x) \qquad \text{(Sewall Wright formula)} \tag{8}$$

where

$$h(x) = 1/[v(x)Q(x)], \quad Q(x) = \exp\left\{-2 \int_0^x \mu(u)\,du/v(u)\right\}$$

II. First Passage (e.g. extinction) Probabilities and Times

For absorption (fixation) problems in time-homogeneous case, consider "backward" equation equivalent to (7) viz if $f_t(x|x_0)$ is written $f_t(x_0)$ for brevity,

$$\frac{\partial f_t(x_0)}{\partial t} - \mu(x_0)\frac{\partial f_t(x_0)}{\partial x_0} - \tfrac{1}{2}v(x_0)\frac{\partial^2 f_t(x_0)}{\partial x_0^2} = 0 \tag{9}$$

Write

$$F_t(x) = \int_{-\infty}^x f_t(x), \qquad P_0(t) = F_t(0), \qquad P_1(t) = 1 - F_t(1),$$

$$P(t) = P_0(t) + P_1(t),$$

$$L(\psi) = L_0(\psi) + L_1(\psi) = \int_0^\infty e^{-\psi u}\left[\frac{\partial P(u)}{\partial u}\right] du,$$

then distribution of first passage time to 0 or 1 determined by (cf. Cox and Miller, 1965, § 5.10)

$$\tfrac{1}{2}v(x_0)\frac{\partial^2 L_i}{\partial x_0^2} + \mu(x_0)\frac{\partial L_i}{\partial x_0} = \psi L_i, \qquad (L_i(x_0 = j) = \delta_i^j). \tag{10}$$

Solution of (10) well-known (by other methods) if v & μ constant. In more general case, one method of solution (when moments exist) is to write

$$L_i = \sum_{r=0}^{\infty} (-\psi)^r L_i^{(r)}(x_0)/r!,$$

then

$$L_i^{(0)} = P_i, \qquad P_1 = \int_0^{x_0} Q(u)\,du \Big/ \int_0^1 Q(u)\,du, \qquad P_0 = 1 - P_1,$$

and

$$L_i^{(r)} = \left[\int_0^1 Q(u)\,du\right]\left[P_1 \int_{x_0}^1 2rP_0 h L_i^{(r-1)}\,du + P_0 \int_0^{x_0} 2rP_1 h L_i^{(r-1)}\,du\right] \tag{11}$$

In particular,

$$L_i^{(1)} = \left[\int_0^1 Q(u)\,du\right]\left[P_1 \int_{x_0}^1 2P_i P_0 h\,du + P_0 \int_0^{x_0} 2P_i P_1 h\,du\right] \tag{12}$$

(cf. Ewens, 1963; Kimura and Ohta, 1969a, b).

Alternatively, we may solve for $P_i(t)$ directly. For example, in case $\mu(x) = 0$, $v(x) = \kappa x(1 - x)$,

$$P_0(t) = 1 - x_0 + \frac{1}{2}\sum_{i=1}^{\infty} (-1)^i[L_{i-1}(2x_0 - 1) - L_{i+1}(2x_0 - 1)]\,e^{-\frac{1}{2}\kappa i(i+1)t} \tag{13}$$

where $L_i(x)$ are Legendre polynomials (Kimura, 1964), $P_1(t|x_0) = P_0(t|1 - x_0)$. Asymptotically,

$$P_0(t) \sim 1 - x_0 - 3x_0(1 - x_0)\,e^{-\kappa t}. \tag{14}$$

From (5) or (6), for $G(z)$ or $g(z)$ (one type of individual), extinction probability $\pi_t(0)$. (In discrete time, if $G(z) = \exp(z - 1)$, $\pi_1(0) = e^{-1}$; if $G(z) = \tfrac{1}{2}(1 - z^2)$, $\pi_1(0) = \tfrac{1}{2}$).

For simple birth-and-death process in continuous time,

$$g(z) = (\lambda z - \mu)(z - 1)$$

and

$$\pi_t(0) = \frac{\mu(T-1)}{\lambda T - \mu}, \qquad T = e^{(\lambda-\mu)t}. \tag{15}$$

($\lambda < \mu$; or, $\lambda > \mu$, *given extinction*, interchange λ and μ). For this process, distribution of time to extinction given by

$$\left[\frac{\partial \pi_t(0)}{\partial t}\right] dt = \frac{\mu(\mu - \lambda)\, dT}{(\lambda T - \mu)^2}, \qquad (T = 0 \text{ to } 1). \tag{16}$$

In particular, if $\lambda = \mu$,

$$\left[\frac{\partial \pi_t(0)}{\partial t}\right] dt = \frac{\lambda\, dt}{(1 + \lambda t)^2}, \qquad (t = 0 \text{ to } \infty). \tag{17}$$

From (16),

$$E\{t\} = -\frac{1}{\lambda} \log\left(1 - \frac{\lambda}{\mu}\right). \tag{18}$$

Number (N) of new individuals per ancestor
In discrete time,

$$\pi(z) = G(z\pi(z)), \tag{19}$$

whence

$$E\{N\} = \frac{m_1}{1 - m_1}, \qquad \sigma^2\{N\} = \frac{v_1}{(1 - m_1)^3}, \tag{20}$$

where m_1 and v_1 mean and variance from $G(z)$ (cf. Nei, 1971b).
  In continuous time,

$$\pi(z) = p + qz\pi^2(z),$$

where

$$p = \mu/(\lambda + \mu) \qquad (>\tfrac{1}{2}),$$

whence

$$E\{N\} = q/(p - q), \qquad \sigma^2\{N\} = pq/(p - q)^3 \tag{21}$$

(cf. Bartlett, 1966 § 5.22).

Use of (17) *as an approximation in population genetics*

TABLE I. Distribution of time to extinction (generations) of a neutral mutant

t	Simulation[a]	e^{-1}	λ	$\frac{1}{2}$
1	84	73·6	100·0
2	25	24·9	25·0
3	23	16·7	15·0
4	10	12·0	10·0
5	5	9·0	7·1
6	6	7·0	5·4
7	3	5·7	4·2
8	6	4·6	3·3
9–10	4	7·1	5·0
11–20	14	17·1	11·4
21–30	4	6·7	4·2
31–50	6	5·9	3·6
51–100	5	4·7	2·9
over 100	5	5·0	2·9
Total	200	200	200
χ^2 (13 d.f.)		10·9	15·6

[a] Kimura and Ohta (1969b).

If alternatively (16) were used (to correspond more closely with finite population for which $P_0 < 1$), we could take $\mu/\lambda = P_0 = 1 - \frac{1}{2}N$ in neutral case.

III. RECURRENT EPIDEMICS MODEL AND ITS APPLICATION TO MEASLES

In (1), let N_t denote number S_t susceptibles and I_t infectives, then postulated epidemic model may be written

$$\frac{\partial \pi_t}{\partial t} = \lambda(w^2 - zw)\frac{\partial^2 \pi_t}{\partial z \partial w} + \mu(1 - w)\frac{\partial \pi_t}{\partial w} + v(z - 1)\pi_t + \varepsilon(w - 1)\pi_t \quad (22)$$

(cf. McKendrick, 1926; Bartlett, 1949).

As (pre-vaccination) model for measles, take $\varepsilon = 0$ (no immigration of infection, $\mu = \frac{1}{2}$ (incubation period about two weeks), $\rho = v\lambda/\mu \sim 1/68\cdot2$.

From (22), when $\varepsilon = 0$,

$$\frac{dx}{dt} = \lambda E\{IS\} - \mu x, \qquad \frac{dy}{dt} = -\lambda E\{IS\} + v, \quad (23)$$

where $x = E\{I\}$, $y = E\{S\}$.

Deterministic approximation: $E\{IS\} \sim xy$.

Linearizing approximation: $x = m(1 + \mu)$, $y = n(1 + v)$, where $m = v/\mu$, $n = \mu/\lambda$,

$$\frac{du}{dt} = \mu v(1 + u) \sim \mu v, \qquad \frac{dv}{dt} = -\rho(u + v + uv) \sim -\rho(u + v). \quad (24)$$

Period $2\pi\{\rho\mu - \frac{1}{2}\rho^2\}^{-\frac{1}{2}} = 73 \cdot 7$ weeks.
Damping coefficient proportional to ρ.
Expected time to "fade-out" of infection (Bartlett, 1960)

$$T \sim \sqrt{(2\pi n)} \, e^{\frac{1}{2}(m + n/m)^2/n}/(\mu m), \qquad (n \gg m). \quad (25)$$

Below critical size ($\varepsilon > 0$), period $\tau \sim 1/\rho + \sqrt{1/\rho\varepsilon}$.
Post-vaccination model (Griffiths, 1971):

$$v' = vf, \, (v \text{ influx of susceptibles}, f > 0),$$

$$m' = fm, n' = n, \rho' = f\rho.$$

Period $\sim 2\pi/\sqrt{\rho'\mu} = [2\pi/\sqrt{\rho\mu}]/\sqrt{f}$,

$$T' = T(fm, n) \sim T(m, n/f^2).$$

Critical community size multiplied by factor $1/f^2$.
Below critical size, $\tau' \sim \tau/f$ if $\varepsilon' = f\varepsilon$.
(Broadly speaking, reduction in susceptibles spreads out epidemics, but does not lessen intensity; Griffiths also notes that average age of onset will be increased.)

IV. AGE DISTRIBUTIONS

(i) For a uni-sex population $N_t(u)$ with birth- and death-rates $\lambda(u)$ and $\mu(u)$ dependent on age u, the deterministic age density $f_1(u, t)$ is the expected age-density $E\{dN(u, t)\}/dt$ in the complete stochastic formulation. For f_1 we have the continuity equation

$$\frac{\partial f_1}{\partial t} + \frac{\partial f_1}{\partial u} = -\mu(u)f_1, \qquad (0 < u < t), \quad (26)$$

together with the "renewal" equation at $u = 0$,

$$f_1(0, t) = \int_0^\infty \lambda(u)f_1(u, t) \, du. \quad (27)$$

These equations determine $f_1(u, t)$ from given initial conditions. In particular,

$$f_1(u, t) = f_1(0, t - u) \exp\left[-\int_0^u \mu(v)\,dv\right], \qquad (0 < u < t); \tag{28}$$

and if the population is not intrinsically decreasing, $f_1(0, t)$ has in general the asymptotic solution $C\,e^{\kappa t}$, where $\kappa\ (\geq 1)$ is the dominant root of the equation in s,

$$\int_0^\infty e^{-st}\lambda(t)\,e^{-\int_0^t \mu(v)\,dv}\,dt = 1. \tag{29}$$

(For the more complicated case of two sexes, see, for example, Goodman, 1968, Bartlett, 1970.)

(ii) In the case of proliferating cell populations, denote the four mitotic phases G_1 (resting), S (DNA synthesis), G_2 (resting) and M (mitosis) by $i = 1, 2, 3, 4$ respectively. Then we may wish to consider either

$$N_i(u, t), \qquad u \text{ age from beginning of cycle,}$$

or

$$N_i(u_i', t), \qquad u_i' \text{ age from beginning of phase.}$$

For simplicity consider the case of zero death-rate. Then, if $r_1(u)$ are transition rates from phase i,

$$Df_1 = -r_1(u)f_1, \qquad (u > 0),$$
$$Df_i = -r_i(u)f_i + r_{i-1}(u)f_{i-1} \qquad (i \neq 1), \tag{30}$$

where

$$D \equiv \frac{\partial}{\partial t} + \frac{\partial}{\partial u}.$$

Asymptotically, if $f_i(u, t) \sim g_i(u)\,e^{\kappa t}$,

$$g_1 = C_1\,e^{-\kappa u - \int_0^u r_1(v)\,dv},$$

etc. If we conveniently define

$$r_{1,i}(u) = r_i(u)g_i(u)/[g(u) + \ldots g_i(u)], \qquad i \neq 1,$$

then

$$g_1(u) + \ldots g_i(u) = C_1\,e^{-\kappa u - \int_0^u r_{1,i}(v)\,dv}$$
$$= C_1\,e^{-\kappa u}\{1 - \Phi_{1,i}(u)\}, \tag{31}$$

say, where $C_1 = 2\kappa$ from relation

$$\int_0^\infty 2\,e^{-\kappa u}\,d\Phi_{1,4}(u) = 1.$$

Alternatively, for u_i',

$$D_i f_i' = -r_i(u_i')f_i', \qquad (u_i' > 0),$$

and if

$$f_i'(u_i', t) \sim g_i'(u_i')\,e^{\kappa t},$$
$$g_i'(u_i') = C_i'\,e^{-\kappa u_i'}\{1 - \Phi_i(u_i')\}. \qquad (32)$$

If

$$L_i(s) = \int_0^\infty e^{-su}\,d\Phi_i(u), \qquad C_i' = \kappa/[1 - L_i(\kappa)],$$

and fraction of $g(u)$ in phase i is

$$\int_0^\infty g_i(u)\,du = 2\{-L_{1,i}(\kappa) + L_{1,i-1}(\kappa)\}.$$

(If phase times are *independent*, then

$$L_{1,i}(s) = L_1(s)L_2(s)\ldots L_i(s).)$$

REFERENCES AND SHORT BIBLIOGRAPHY

I. General

BARTLETT, M. S. (1949). Some evolutionary stochastic processes. *J. R. Statist. Soc.* B **11**, 211–29.

BARTLETT, M. S. (1966). "An Introduction to Stochastic Processes." (2nd Edn.) Cambridge Univ. Press, London.

COX, D. R. and MILLER, H. D. (1965). "The Theory of Stochastic Processes." Methuen, London.

KOLMOGOROV, A. (1931). Über die analytische Methoden in der Wahrscheinlich-keitsrechnung. *Math. Ann.* **104**, 415–58.

LEVINS, R. (1966). The strategy of model building in population biology. *Amer. Scientist* **54**, 421–31.

MAYNARD SMITH, J. (1968). "Mathematical Ideas in Biology." Cambridge Univ. Press, London.

*WILLIAMSON, M. (1972). "The Analysis of Biological Populations." Edward Arnold, London.

WIT, C. T. de (1970). Dynamic concepts in biology. *In* "The use of Models in Agricultural and Biological Research." (Ed. J. G. W. Jones), 9–15.

II. Ecology and epidemiology

*BAILEY, N. T. J. (1957). "The Mathematical Theory of Epidemics." Griffin, London.

BARTLETT, M. S. (1960). "Stochastic Population Models in Ecology and Epidemiology." Methuen, London.

GAUSE, G. F. (1934)."The Struggle for Existence." Dover Publications, New York.

GRIFFITHS, D. A. (1971). Epidemic models (D. Phil. Thesis, University of Oxford).

LOTKA, A. J. (1926). "Elements of Physical Biology." Dover Publications, New York.

MACARTHUR, R. H. (1970). Species packing and competitive equilibrium for many species.*Theor. Pop. Biol.* **1**, 1–11.

*MACARTHUR, R. H. and WILSON, E. O. (1967). "The Theory of Island Biogeography." Princeton Univ. Press, U.S.A.

MCKENDRICK, A. G. (1926). Applications of mathematics to medical problems. *Proc. Edin. Math. Soc.* **44**, 98–130.

*PATIL, G. P., PIELOU, E. C. and WATERS, W. E. (1971). "Statistical Ecology." (3 vols.) Penn. State Univ. Press, U.S.A.

*PIELOU, E. C. (1969). "An Introduction to Mathematical Ecology." Wiley, New York.

VOLTERRA, V. (1926). Variazioni e fluttuazioni del numero d'individui in specie animali conviventi. *Mem. Acad. Lineci Roma*, **2**, 31–113.

III. Population genetics

*CAVALLI-SFORZA, L. L. and BODMER, W. F. (1971). "The Genetics of Human Populations." Freeman, San Francisco.

EWENS, W. J. (1963). Diffusion equation and pseudo-distribution in genetics. *J. R. Statist. Soc.* B **25**, 405–12.

*EWENS, W. J. (1969). "Population Genetics." Methuen, London.

FISHER, R. A. (1930). "The Genetical Theory of Natural Selection." Oxford Univ. Press, London.

HALDANE, J. B. S. (1924). A mathematical theory of natural and artificial selection. *Trans. Camb. Phil. Soc.* **23**, 19–41.

*KARLIN, S. (1969). "Equilibrium behaviour of Population Genetic Models with Non-random Mating." Gordon and Breach, New York.

KIMURA, M. (1964). Diffusion models in population genetics. *J. Appl. Prob.* **1**, 177–232.

KIMURA, M. and OHTA, T. (1969a). The average number of generations until fixation of a mutant gene in a finite population. *Genetics* **61**, 763–71.

KIMURA, M. and OHTA, T. (1969b). The average number of generations until extinction of an individual mutant gene in a finite population. *Genetics* **63**, 701–9.

*KIMURA, M. and OHTA, T. (1971). "Theoretical Aspects of Population Genetics." Princeton Univ. Press.

MORAN, P. A. P. (1962). "Statistical Processes of Evolutionary Theory." Oxford Univ. Press, London.

NEI, M. (1971a). Extinction time of deleterious mutant genes in large populations. *Theor. Pop. Biol.* **2**, 419–25.

NEI, M. (1971b). Total number of individuals affected by a single deleterious mutation in large populations. *Theor. Pop. Biol.* **2**, 426–30.

WRIGHT, S. (1931). Evolution in Mendelian populations. *Genetics* **16**, 97–159.

IV. Demography, including cell kinetics

BARRETT, J. C. (1966). A mathematical model of the mitotic cycle and its application to the interpretation of percentage labelled mitoses data. *J. Natn. Cancer Inst.*, **37**, 443–50.

BARTLETT, M. S. (1969). Distributions associated with cell populations. *Biometrika* **56**, 391–400.

BARTLETT, M. S. (1970). Age distributions. *Biometrics* **26**, 377–85.

*CLEAVER, J. E. (1967). "Thymidine Metabolism and Cell Kinetics." Elsevier, Amsterdam.

GOODMAN, L. A. (1968). Stochastic models for the population growth of the sexes. *Biometrika* **55**, 469–87.

KENDALL, D. G. (1949). Stochastic processes and population growth. *J. R. Statist. Soc.* B **11**, 230–64.

*KEYFITZ, N. (1968). "Introduction to the Mathematics of Population." Addison-Wesley, New York.

LESLIE, P. H. (1945). On the use of matrices in certain population mathematics. *Biometrika* **34**, 183–212.

*MACDONALD, P. D. M. (1970). Statistical inference from the fraction labelled mitoses curve. *Biometrika* **57**, 489–503.

POWELL, E. O. (1956). Growth rate and generation time of bacteria, with special reference to continuous culture. *J. Gen. Microbiol.* **15**, 492–511.

SISKEN, J. E. and MORASCA, L. (1965). Intrapopulation kinetics of the mitotic cycle. *J. Cell. Biol.* **25**(2), 179–89.

* *Authors with useful further bibliographies.*

2. An Investigation into the Effects of Variable Rates of the Exploitation of Fishery Resources

J. G. Pope

M.A.F.F. Fisheries Laboratory, Lowestoft, England.

1. Introduction

The underlying theory of fisheries population dynamics is simple in comparison with many other biological systems, but as with most other systems the practical applications rapidly lead to considerable complexities. The basic theory evokes optimization procedures based on the exploitation of resources in equilibrium, a state which is seldom realized in practice. This paper is in the nature of a summary of the progress that has been made at Lowestoft in investigating the theoretical effects of exploitation in a group of fishery resources where the fishing intensity on each stock fluctuates in response to changes in the relative abundance of fish, a system which bears a greater similarity to actual fisheries.

2. The Simple Theory of Fisheries and the Current Situation

The simple population dynamics of fisheries are based on a death process. The population of an age cohort decreases exponentially at a constant rate due to fishing mortality (F) and natural mortalities (M), so that after a time t the initial population $P(o)$ is reduced to

$$P(t) = P(o)\exp\left\{-(F + M)t\right\};$$

the catch in numbers is that proportion of the total deaths due to fishing:

$$C(t) = \frac{F}{F + M}P(o)(1 - \exp\left\{-(F + M)t\right\}).$$

It is possible to calculate the catch in weight (the yield) if the weight at each age of life is known.

This leads to two important concepts. If a group of 1000 fish is considered, then if they are not fished their biomass would increase with age to some

23

maximum and then decline. Figure 1a shows such a function for a cod stock. This is composed of an increase due to growth and a decrease due to natural mortality. The maximum of this curve places an absolute upper limit on the catch that could be made from these fish. However, realization of this maximum requires that the cohort could be exploited:

(1) independently of adjacent aged cohorts, and

(2) with an infinite fishing mortality,

and since neither of these is generally possible, the theoretical limit is not generally attainable.

Yields that can be achieved are usually considered in terms of various constant rates of fishing mortality. Figure 1b shows the yield of the cod stock

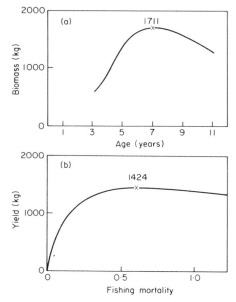

FIG. 1. The biomass at each age and the yield for various fishing mortalities for a cod stock. The biomass and the yields are those from a cohort of fish that numbered 1000 at the third birthday.

at various levels of exploitation. It is a function of the growth and natural mortality, already mentioned, and the selection proportions, i.e., the proportion of the maximum fishing mortality that acts at each age of life. The maximum of this curve is termed the maximum sustainable yield. The shape of the yield curve and its maximum may be altered by changing the selection curve (see Beverton and Holt, 1957). This would be achieved in trawl fisheries, such as those for the North Atlantic cod, by changing the mesh size of nets. The scientific management of fisheries has, in the past, been concerned with the problem of assessing the optimum levels of exploitation for particular stocks, and in general such investigations have been made in terms of constant rates of exploitation.

The comparative simplicity of this theory of fisheries is due in part to considering only the effect of the predator on the prey and not the effect of the prey on the predator. The predator in this case is the fishing fleet. Despite most scientific work concentrating on constant rates of exploitation it is an observable fact that the rates of exploitation of most fisheries vary, sometimes considerably from year to year. This is certainly the case for the North Atlantic code (I.C.E.S., 1973), which is not surprising since a considerable part of the fishing mortality on many stocks is generated by highly mobile fishing vessels. These vessels naturally tend to fish most where catches are currently most favorable and to change the stock they are exploiting in response to changes in the relative abundance of fish, a pattern of behaviour particularly pronounced since the development of the freezer trawler. This mobility of fishing effort has had the effect of making it difficult to predict catches from the various fisheries, due to the unpredictability of the rate of exploitation in future years. It has also had the effect of making the scientific management of fisheries difficult since, in practice, it is not possible to regulate one fishery without the risk of affecting others.

More seriously, the fluctuating distribution of fishing has generated an observed tendency for fleets to aggregate on the most abundant year-classes of fish almost as soon as they become available, with the result that they are fished out at a low average weight. In terms of the basic theory this implies a non-optimum form of exploitation of the individual resources which has contributed to current political problems in the management of North Atlantic fisheries. At the same time it has raised the question of optimizing the yield from a complex of several stocks.

3. The Simulation of a Complex of Fish Stocks

Because of the problems and possibilities mentioned in the previous section, it has proved worth while to investigate the behaviour of fishing fleets, particularly their distribution in relation to the complex of the north Atlantic cod stocks. In this context the most important factors are those that affect the size and disposition of fishing fleets. Factors which influence the size of the fishing fleets are naturally interesting but since they include consideration of national policy and economics they are difficult to explain simply. However, fishing vessels have a life of 20 years or more, and changes in size of fleets usually occur at a comparatively slow rate in established fisheries, and we are here more concerned with the effect of the varying disposition of a particular level of effort. The factor that most affects the disposition of fishing fleets is essentially the catch rate of each fishery. The problem of examining this over a group of stocks has been approached using a computer simulation. This was first developed by Clayden (1972) and extended by Garrod and Pope (1972).

Figure 2 shows a flow diagram which outlines the workings of the simulation, this being normally made on a monthly basis. Each month the catch per unit effort is calculated for each stock from a knowledge of the number

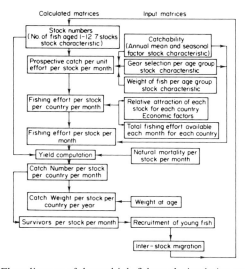

Fig. 2. Flow diagram of the multiple fish stock simulation program.

at each age, the weight at age, the selection ogive and a catchability coefficient for each stock. The fishing effort of each national fleet or fishing fleet type is then assigned to the various stocks in proportion to the catch per unit of effort available, multiplied by a weighting factor. This weighting factor allows national preferences for particular fish stocks to be taken into account and also allows for some fishing effort to be tied to particular stocks as inshore non-mobile effort. Having calculated the fishing effort the fishing mortality can be found and hence the catch. Revised populations for each stock can then be calculated. At the end of each year simulated the computer program provides statistics of effort, catch and stock size for each stock and nation. The simulation can also handle migrations of fish both within and between stocks.

This program has made it possible to simulate catches from the North Atlantic cod stocks in a reasonably satisfactory fashion (see Garrod and Pope, 1972). It has been used extensively in investigation of these stocks in the last two years. Some of the uses have been to project the most likely catches and catch rates in the various cod stocks for the next five years and to examine the effect of particular conservation measures on various stocks. Most recently its use was explored by an international body, the I.C.E.S./ I.C.N.A.F. North Atlantic Cod Working Group, which used it to determine the effect of different future levels of fishing (I.C.E.S., 1973).

4. The Yield of Fisheries Exploited by Mobile Fishing Fleets

The nature of applications of the simulation program essentially required answers to practical questions, and in these various applications the need for the simulation approach was often underlined by the amount by which some simulated results differed from preconceived notions of what the answer should be. One of the more surprising results was that the long-term yields of management schemes, based on a mixture of mobile and immobile fishing effort, were in general greater than those obtained from equivalent schemes using only non-mobile fishing effort. This result would seem to contradict, at least for the cod stocks, the fear that mobile fishing effort would lead to non-optimum forms of exploitation of fishery resources— indeed, it seemed to suggest that the optimum exploitation of a group of fisheries might be achieved by fluctuating (as opposed to constant) rates of exploitation. This suggestion leads to two important questions: firstly what is the optimum exploitation strategy for a particular stock, given the growth, natural mortality and selection proportions, and, secondly, how could this optimum be most closely approached in a practical fisheries situation? This section examines the first question and the next section is concerned with optimization in the practical fisheries situation.

A preliminary investigation of these problems has been made based on two different stocks, the first of cod and the second of haddock. The cod stock has an absolute upper limit to the catch and maximum sustainable yields which differ by 20% (see Fig. 1), while the haddock stock has an equivalent difference of 40%. Thus, particularly for the haddock, there is considerable room for improvement in yield. A procedure for establishing optimum exploitation of a stock has yet to be developed, but it is possible to indicate that fluctuations in the rate of exploitation can generate greater yields. The following examples show some gains that might be made from extreme cases of fluctuating rates of exploitation.

The most obvious way to use fluctuations of exploitation rates to increase the yield is to take advantage of fluctuations in year-class strength, or, put more simply, to tune fishing to the numbers of fish. Thus, if the recruitment pattern for the haddock stock were that shown in Fig. 3a, exploiting all these year-classes at the maximum sustainable yield level at an F of 0·25

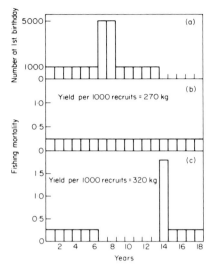

FIG.3. Recruitment patterns and patterns of exploitation for a haddock stock.

would yield in all 270 kg per 1000 recruits (Fig. 3b). If, however, the stock was left unfished until the largest year-classes were six and seven years old and fished for one year at an F of 2·0 (Fig. 3c) and thereafter at 0·25, the average yield from those year-classes would be 320 kg per 1000 recruits, an increase of 18 %. A similar pattern of recruitment and exploitation for the cod would yield an increase of 4 %.

Although the patterns of recruitment shown in this previous example might be considered to be rather artificial, the recruitment of actual stocks of fish can vary considerably, for example in the north-east Arctic cod. The recruitments shown for this stock (Fig. 4) might well have been exploitable

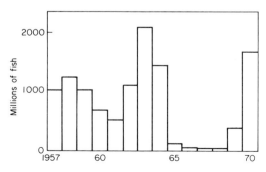

FIG. 4. Recruitment of three year old fish to the Arcto-Norwegian stock of cod.

in a similar manner to the example and indeed, it is possible that they were. Certainly it is difficult to show that constant exploitation rates would have resulted in a greater overall yield from this fishery than was in fact achieved with fluctuating rates.

It is fairly apparent that it might be possible to take advantage of the best year-classes to gain an enhanced yield, but it is rather less apparent that with fluctuating effort gains might be made from some stocks even when the recruitments are constant.

As an example, consider the haddock stock to have had a constant recruitment each year; let us assume that the number of recruits was a thousand of these fish each year. If this stock were fished with a fishing mortality of 0·25 each year (a rate which would cause 21 % of mature fish to be caught each year) the yield for each thousand recruits would be 270 kg—the maximum sustainable yield. If, instead of doing this, it is assumed that the fishing fleets which were able to generate an annual fishing mortality of 0·33 per year (a rate which would cause 26 % of mature fish to be caught each year) were diverted to other stocks for five years, and in the sixth year the stock was fished with a fishing mortality of 2·0 (a rate which would cause 82 % of mature fish to be caught each year)—in other words, using the same amount of fishing on average—then the yield would be 284 kg, an increase of 5 %. To understand why this occurs it is necessary to consider the exploitation of six consecutive year-classes under this fishing regime. Table 1 shows the numbers at each age for each year-class. If these are averaged, a mean value for the fishing mortality at each age can be calculated. It can be seen that this has been less on the younger ages and in general, greater on the older ages than an equivalent constant rate of exploitation would have been. Thus, the use of a fluctuating exploitation rate has had the effect of changing the apparent selection factors. It will be remembered (this was pointed out in the section on simple theory) that changing the selection brings about changes in the yield curve and the maximum sustainable yield.

It is an important feature of the concept of fluctuating fishing mortalities that it can be used to modify the effective selection of fish at age in a manner analogous to the conventional manipulation of selection to the fishing gear by control of mesh size. This observation that the long-term average yield might be increased beyond the maximum sustainable yield by deliberate manipulation of fluctuating fishing mortality prompts the development of techniques to optimize serial variation in the fishing mortality in relation to the known biological characteristics of the stock, given the age composition and possibly some future recruits. Some progress can be made towards discovering the optimum series of F_i by optimizing F_1 for some assumed values of $F_2, F_3 \ldots$, the fishing mortalities in future years. The most obvious choice for F_2, F_3 etc. is to adopt the constant level of fishing mortality

TABLE 1. The yield and number at age from six consecutive year-classes of haddock fished with a fishing mortality of 2·0 every sixth year. Average fishing mortalities from these year-classes are contrasted with the equivalent constant fishing mortalities. The italic values indicate where fishing occurred

Age (years)	Year-class						Average	Average	Comparable constant F
	1960	1961	1962	1963	1964	1965			
Numbers at each birthday								F	F
1	1000	*1000*	1000	1000	1000	1000	1000	0·09	0·13
2	861	403	*861*	861	861	861	785	0·14	0·21
3	741	346	210	*741*	741	741	587	0·18	0·26
4	638	298	181	134	*638*	638	421	0·25	0·33
5	549	257	156	115	74	*549*	283	0·34	0·33
6	*472*	221	134	99	64	64	176	0·50	0·33
7	55	*190*	115	85	55	55	93	0·36	0·33
8	47	22	*99*	74	47	47	56	0·30	0·33
9	41	19	12	*63*	41	41	36	0·30	0·33
10	35	16	10	8	*35*	35	23	0·25	0·33
11	30	14	9	6	4	*30*	16	0·34	0·33
12	*26*	12	7	5	3	4	10	0·50	0·33
Yield in kg									
	365	193	227	256	306	355	284		

compatible with the maximum sustainable yield. In that case, the optimum choice for F_1 must give a long-term average yield at least as great as the maximum sustainable yield. Table 2 gives the weight of catch that a haddock which survived to the next age would yield, on average, if subsequent fishing mortalities were at the maximum sustainable yield level. It can be seen, by comparing the average weight of catch, that survivors from each age would yield, with the weight of fish at each age, that it is better to allow the younger fish to survive. However, it is also better to catch fish aged five or more rather than to allow them to survive.

Given a value for survivors it is simple to calculate the yield from a particular age group. Thus if in the haddock stock there were P_5 fish aged five, if these were fished at a fishing mortality of F_1 in the next year (the maximum sustainable yield level) then the catch in the year would have a weight of

$$680P_5\frac{F_1}{F_1 + M}(1 - \exp\{-(F_1 + M)\})\text{ grammes}$$

TABLE 2. The weight of fish at each age (g) and the average weight (g) of the ultimate catch from the survivors of each age, for the haddock stock

Age (y)	Weight (g)	Average weight (g) of the ultimate yield from each fish surviving to the next age
1	120	332
2	250	405
3	350	490
4	500	578
5	680	653
6	820	722
7	980	775
8	1110	816
9	1220	842
10	1300	856

and the ultimate catch of the survivors would weigh

$$653P_5 \exp\{-(F_1 + M)\} \text{ grammes.}$$

Similar equations could be written for each age, bearing in mind that the fishing mortality for each age is given by F_1 multiplied by the selection proportion for age a (Sa).

Thus the sum of all these equations gives a function for the total yield of all the fish present in terms of F_1, the fishing mortality acting in the next year. This can be solved numerically for the maximum fairly simply, particularly since only solutions in the range $0 \le F_1 \le 2.0$ are likely to be of any interest in a fisheries context. It is however informative to use the approximation

$$(1 - \exp\{-F\})\exp\{-M/2\} \simeq \frac{F}{F + M}(1 - \exp\{-(F + M)\})$$

(see Pope, 1972) and to differentiate the function with respect to F. The resulting first differential of the yield function $Y(F_1)$ for the haddock is of the form

$$Y'(F_1) = -174P_1R_1 - 115P_2R_2 - 97P_3R_3 - 34P_4R_4 + 69P_5R_5$$
$$+ 139P_6R_6 \ldots + 469P_{10}R_{10}$$

where

$$R_a = \exp\{-S_aF_1\}.$$

Hence, since for the haddock for $F_1 \geq 0$

$$R_1 > R_2 > R_3 > R_4 = R_5 \ldots = R_{10},$$

then if $Y'(o) < 0$, $Y'(F)$ will be negative for all positive values of F and hence the maximum yield under these conditions will be achieved if no fishing takes place in the next year. The condition $Y'(o) \leq 0$ is likely to occur when a large year-class of haddock enters the fishery.

Although this technique does not lead to the optimum yield if repeated year after year, it does perhaps indicate one possible means of approaching it. This is important because a method for choosing the strategy of varying fishing to optimize the yield of fisheries would be most valuable, either to use directly or as an upper limit with which other strategies could be compared. The development of such a method would be just one way in which mathematicians might perhaps be able to help fisheries scientists. In the meantime the technique outlined above does indicate that for some fish stocks, at least, the total yield of the fish present in the stock could be increased beyond the maximum sustainable yield by a suitable fluctuation in the rate of exploitation.

5. Mobile Fishing and the Practical Management of Fisheries

The examples of the previous section indicate that fluctuating rates of exploitation can in some cases increase the total yield of fisheries, but the extreme form of fishing effort fluctuation examined in the examples are hardly likely to prove attractive to practical fishermen. It is therefore interesting to examine how the optimum yield could be most closely approached in a practical situation. It is particularly interesting to see if any of these gains in total yield could be made as a result of the forms of fluctuations likely to be generated by mobile fleets. To examine this possibility a series of simulations was made using the computer program described previously. These were made to compare the action of mobile fishing fleets with the action of non-mobile fishing fleets generating constant rates of exploitation. The simulations were made over seven stocks of fish with identical biological parameters. The numbers of fish entering each stock each year were drawn from a rectangular distribution, the same series being used for simulations of mobile and of non-mobile fishing. Two sets of simulations were made, based on the cod and haddock stocks of previous examples.

The results from these simulations showed that the overall yields from mobile and non-mobile fishing fleets were virtually identical, all differences being slight and unsystematic. Table 3 shows the percentage differences at one such comparison between the action of mobile and non-mobile fishing

TABLE 3. Percentage difference between the yield from seven stocks of haddock fished by a mobile fleet with an average fishing mortality of 0·25, and the yield of the stocks from a constant fishing mortality of 0·25. Results are based on 10-year sums

Year of simulation	Stock No.							
	1	2	3	4	5	6	7	Total
1–10	−4	− 3	+15	−1	−1	−13	+5	+1
11–20	0	− 1	−14	+2	+3	+ 8	−5	−1
21–30	+3	−13	+ 6	+4	−2	+ 2	+1	+1
Total	+1	− 5	0	+2	0	+ 2	−1	0

fleets, based on the haddock stocks. The percentage differences between the yields in each ten-year period simulated are given for each stock. This shows the close similarity of the yield to be expected from the two different forms of fishing fleet for one particular average level of fishing effort. In this example the mobile fishing fleet generated a mean fishing mortality of 0·25 with a standard deviation of 0·04. It is perhaps disappointing that these simulations of the action of mobile fishing fleets fail to show any of the increases in yield found in the examples in Section 4, or even of those shown in the more realistic context of the North Atlantic cod, but the fact that the figures show such similarity to the results which would be obtained using constant levels of exploitation is in itself very valuable. This result means that it would be just as effective to manage a group of stocks by applying some overall limitation on fishing effort as to manage each stock separately at some constant exploitation rate.

So far the action of mobile fishing fleets has only been studied in a very simple fashion and many of the problems associated with mobile fishing deserve further attention. Some of the more pressing problems are to examine the effect of mobile fishing fleets on any relationships between the number of adult fish and the number of young fish recruited, and also the effects of mobile fishing in multi-species fisheries as opposed to the multiple stocks of one species already partly explored.

This greater understanding of the action of mobile fishing fleets is most desirable, since it could very well lead to methods of management which have previously not been considered.

6. Summary

The interactions between fish and fishing vessels commonly subject fisheries to fluctuating rates of exploitation because of the mobility of some fishing fleets. The questions posed for the prediction and management of fisheries have been partly answered using computer simulations. In addition, by investigating some extreme situations, it is possible to show for some species that increased yield might be generated as a result of fluctuating rates of exploitation, but the strategy of mobile fishing that generates the optimum yield in particular circumstances has not yet been developed. However, it can be shown that in the more practical area of fisheries management, a knowledge of the action of mobile fishing fleets can generate new schemes of management that may very well be most valuable.

REFERENCES

BEVERTON, R. J. H. and HOLT, S. J. (1957). On the dynamics of exploited fish populations. *Fishery Invest., Lond.,* Ser. 2, **19**, 533.
CLAYDEN, A. D. (1972). Simulation of changes in abundance of the cod (*Gadus morhua* L.) and the distribution of fishing in the North Atlantic. *Fishery Invest., Lond.,* Ser. 2, **27**, 58.
GARROD, D. J. and POPE, J. G. (1972). The assessment of complex fishery resources. *In* "Economic aspects of fish production." O.E.C.D. Symposium on Fisheries Economics. 480.
I.C.E.S. (1973). Report of the I.C.E.S./I.C.N.A.F. Working Group on Cod Stocks in the North Atlantic. *Coop. Res. Rep. int. Coun. Explor. Sea,* **33**, 52.
POPE, J. G. (1972). An investigation of the accuracy of virtual population analysis using cohort analysis. *I.C.N.A.F. Res. Bull.,* No. 9, 65.

3. A Simulation Model for Studying the Population Dynamics of some Fish Species

R. JONES AND W. B. HALL

DAFS, Marine Laboratory, Aberdeen, Scotland

1. Introduction

In a recent paper (Jones, 1973) a model was described for dealing with the population dynamics of the larval stage in species such as haddock and cod. These are characterized by a high fecundity and by a pelagic larval stage in which growth rates and mortality rates are very high, being initially about 12 and 10% per day respectively. During its first year of life a haddock, for example, increases its weight by a factor of about 10^5 and has a probability of survival of about 10^{-4}. To account for a growth rate of this magnitude it is necessary to suppose that the rate of food intake in the larval stage must also increase at a comparable rate. In the case of haddock and cod, it seems that this is achieved by eating progressively larger organisms, rather than by eating progressively larger numbers of a particular size of organism. In fact, haddock and cod larvae normally hatch during the spring at the time of the spring zooplankton outburst and it seems likely that the situation is one in which successive cohorts of larvae grow up simultaneously with cohorts of the food species on which they feed. Until the larvae are large enough to eat something else, it is, therefore, suggested that there might be a critical period during which a cohort of larvae is to all intents and purposes, restricted to its own private food supply. Therefore if there were initially more than a certain number of larvae, it would be theoretically possible for them to graze down their particular cohort of food organisms to some threshold density below which they could no longer sustain their normal growth rate. It is assumed that they would then be liable to die, although the sequence of events leading to their death is subject to speculation. For example, it could be assumed that a haddock larva is unlikely ever to become an adult haddock if its growth rate drops below some critical value during the very rapid growth phase in the larval stage. A reduction in food density during this initial period may therefore simply cause the growth rate of the larvae to fall behind to such an extent that recovery may not be possible. Alternatively, the growth rate during this "critical" period might be genetically determined, in which case a larva that failed to get enough food would gradually lose condition and presumably weaken.

In either event, death would ultimately be due to starvation or, because of a weakened state of the larvae, to predation. The important point however is that whatever the final cause of death, the numbers dying ought to be a function of the number of food organisms and the numbers of larvae feeding, rather than a function of, for example, the numbers of predators.

This theory is biologically attractive for a number of reasons. For example it provides at least one explanation of how population size in the common gadoids might be controlled. This could be important, as there is so far no evidence of an adequate population control mechanism in any other part of the life history of these species. Furthermore, because of the tendency for the major fish species in any area to spawn successively, it is suggested that this process might operate for each species independently. Even if there were some overlapping there might still be sufficient independence to prevent an excess production of eggs by one species from necessarily causing a decline in the numbers of any other species. Further points in favour of this theory emerged when the theory was quantified and some calculations were made. These calculations showed that larval mortality rates of the order of 10% per day could be accounted for, using biologically reasonable values for the density of the food organisms and the searching capacity of the larvae. They also showed that the maximum numbers of survivors occurred when the initial numbers of larvae were between 10^8 and 10^9 per $3\,km^2$. The significance of this is that the density of eggs produced by the North Sea haddock population is estimated to fall within this range.

There remains one feature of the model, the implications of which were not fully considered in the earlier paper (Jones, 1973), and that is that larval survival was found to be particularly sensitive to variations in the initial density of food organisms. This suggests that annual variations in larval food may account for the often very large variations that are known to occur in gadoid year class strengths. The object of this paper is, therefore, to investigate this hypothesis by determining what variations in larval food are actually needed to account for the observed variations in the year class strengths of species like the haddock. To do this it has been necessary to incorporate the larval model in a model representing a complete gadoid-type life history. This has been done, and the first application of the model has been to investigate what variations in larval food supply were needed in order to account for year class fluctuations of the magnitude actually observed in North Sea haddock and whiting.

2. The Larval Model

The principle of the larval model has already been described (Jones, 1973). By assuming that the distribution of food organisms in the sea was exponential, it was possible to show that the probability of a larva encountering X_t or more organisms t days after feeding commenced could be given by the expression:

$$\exp - (X_t)/(N_{Ft}S_t)$$

where N_{Ft} is the density of food organisms t days after the larvae have commenced feeding and S_t is a measure of the volume effectively "searched" for food per day on that day. The coefficient S_t is so defined that the product $N_{Ft}S_t$ is equivalent to the mean number of food organisms encountered during any one day, i.e. S_t is equivalent to the proportion of the unit volume effectively searched for food per day. If X_t is made to refer to some critical number of food organisms, such that only larvae expecting to encounter this number, or more, on day t survive, this expression is equivalent to the probability of a larva of age t surviving until age $t + 1$. The number of larvae alive on successive days can therefore be related by the expression

$$N_{L.t+1} = N_{Lt} \exp - (X_t)/(N_{Ft}S_t) \tag{1}$$

where N_{Lt} is the number of larvae alive t days after feeding commences and $N_{L.t+1}$ is the number of larvae alive after $t + 1$ days.

For calculating the rate of consumption, and hence the rate of decline of the food organisms, one of several assumptions may be made. In the previous paper it was argued that on any one day, each larva would encounter $N_{Ft}S_t$ food organisms. All N_{Lt} larvae would therefore encounter a total of $N_{Lt}N_{Ft}S_t$ food organisms on day t. This leads to the expression

$$N_{F.t+1} = N_{Ft} - N_{Lt}N_{Ft}S_t. \tag{2}$$

Given the initial values of N_L and N_F (i.e. N_{Lo} and N_{Fo}) as well as values for S_t and X_t at each age, equations (1) and (2) can be used simultaneously to calculate the numbers of larvae and the numbers of food organisms on successive days.

An alternative assumption that might have been made about the rate of food consumption is that each larva eats a specified number of food organisms per day, the actual number being determined by the daily number of food organisms required to account for the observed larval growth rate. If the number eaten on day t is represented by U_t, one arrives at an alternative to expression (2) given by:

$$N_{F.t+1} = N_{Ft} - N_{Lt}U_t. \tag{3}$$

For the purposes of this paper, calculations were made using equations (1) and (2), (1) and (3), and also (1) and the greater of (2) and (3) without affecting the conclusions in any way.

In an earlier paper it was estimated that the "critical" period might be as long as 45 days. Here, calculations have been made using both this value and also one of 20 days.

Some examples of the relationship between the initial numbers of larvae and the numbers of larvae still alive after 45 days are given in Fig. 1. By

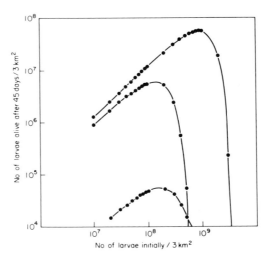

FIG. 1. Some of the family of curves relating numbers of larvae at the beginning and end of the "critical" period.

varying the values of N_{Fo}, the values of X_t, S_t and U_t, the rates at which these three values change with the age of the larvae, and also the length of the critical period, a family of curves, can be obtained as shown in Fig. 1. It is not intended to describe these in any detail in this paper. For application to the model described here a relationship between egg production and numbers of recruits at an arbitrary age, such as one year, is required, i.e. it is necessary to multiply the number of larvae alive at the end of the "critical" period by some coefficient P to represent the proportion subsequently surviving to an age of 1 year. In the case of haddock it is known that landings from the North Sea, for a considerable period averaged 1 ton/3 km². An estimate of the number of 1 year old recruits which would be required annually to account for this figure has been calculated as about 5000/3 km². This estimate was obtained using the principles of Beverton and Holt (1957) and details are given in Table 1. Other estimates could be obtained using

TABLE 1. Calculation of the numbers of North Sea haddock recruits

---

Mean annual haddock landings from the North Sea from 1923–1963 (excluding the war years) was 97,000 tons (from *Bulletin Statistique* for these years).

If it is assumed that about 10% of the fish caught are rejected at sea, the mean annual catch for this period becomes $97,000/0.9 = 108,000$ tons.

The yield per recruit of North Sea haddock has been calculated using the principles of Beverton and Holt (1957) for the following values of the basic parameters:

$$
\begin{aligned}
F &= 0.9 \\
M &= 0.2 \\
W_\infty &= 1340 \\
K &= 0.26 \\
t_0 &= -0.75 \\
t_c &= 2.0 \text{ years} \\
t_r &= 1.0 \text{ year} \\
t_\lambda &= 20 \text{ years}
\end{aligned}
\left.\begin{aligned} \\ \\ \\ \\ \end{aligned}\right\} \text{estimated from Scottish data}
$$

Using these values, the yield per one year old recruit was calculated as 210 g. This leads to an estimate of the average annual number of one year old recruits in the North Sea of:

$$\frac{108 \times 10^9}{210} = 5.1 \times 10^8 \text{ fish}$$

The area of the North Sea occupied by haddock is approximately 300,000 km². This leads to an estimate of the number of one year old recruits per 3 km²

$$\frac{3 \times 5.1 \times 10^8}{3 \times 10^5} = 5100 \text{ fish}$$

i.e. annual haddock recruitment in the North Sea has been about 5000 fish/year/3 km².

---

other values for the basic stock parameters and by allowing different values for rejection of fish at sea, but these are not critical to the present investigation. It can be further estimated (see Table 2) that the average annual egg production ought to have been about 2.5×10^8 eggs/3 km² in a stock with this level of recruitment and subject to the growth and mortality rates known to be applicable to North Sea haddock. If, therefore, recruitment is to be plotted against egg production, the resultant curve should pass through a point with these co-ordinates, or through a point related to these co-ordinates.

Whether, for North Sea haddock, co-ordinates of $R \simeq 5000$ and $E \simeq 2.5 \times 10^8$ are appropriate, depends on some of the assumptions made about the feeding behaviour of the larvae. Since haddock larvae appear to feed on the young growing stages of their food species, the actual number of food organisms available to any one larva ought to depend on the age range of the food organisms being taken at any one time. There would therefore appear to be a number of possible feeding strategies. For example,

TABLE 2. North Sea Haddock. Calculation of egg production per recruit

Age (years)	Relative[a] frequencies	Fecundity[b] $(\times 10^{-5})$
1	10,000	—
2	8187	0·7
3	2725	1·4
4	907	2·2
5	302	3·0
6	101	3·8
7	33	4·4
8	11	5·0
9	4	5·4
10	1	5·8
11	0	—

Egg production per one year old female recruit if maturity is at 2 years of age

$$= \frac{1}{10,000}[(8187 \times 0·7) + (2725 \times 1·4) + \ldots (1 \times 5·8)] \times 10^5$$

$$= 1·31 \times 10^5$$

Egg production per one year old female recruit if maturity is at 3 years of age

$$= \frac{1}{10,000}[(2725 \times 1·4) + \ldots (1 \times 5·8)] \times 10^5$$

$$= 0·73 \times 10^5$$

If the annual recruitment is equivalent to 5000 one year old recruits/3 km² of both sexes (see Table 1) the annual egg production ought to be about 1·8 or 3·3 × 10⁸ egg/ 3 km² depending on whether first maturity is at 3 or 2 years of age. Since North Sea haddock tends to mature at either 2 or 3 years of age, the intermediate value of 2·5 × 10⁸ eggs/3 km² seems an appropriate one to use.

[a] Calculated starting with an arbitrary number of 10,000 one year old fish and assuming $Z = 0·2$ until 2 years of age and $Z = 1·1$ thereafter.
[b] Fecundity $= 196W^{1·14}$ (from Raitt 1932)
and Weight $(W) = 1340 (1 - e^{-0·26(t + 0·75)})^3$
so that fecundity $= 7·20 (1 - e^{-0·26(t + 0·75)})^{3·42} \times 10^5$

restriction of the diet to the largest organisms that could be eaten would necessarily minimize the number available to be eaten. Conversely an increase in the number of organisms acceptable as food could be achieved, but only at the expense of a reduction in mean particle size and hence an increase in the daily number required. In practice, feeding strategy is likely to be a compromise between these alternatives.

To allow for this it has been assumed that the eggs produced in any one season can be divided into a number (D) of independent cohorts, each with its own cohort of food organisms.

In Fig. 2 some examples of the theoretical relationships between egg production and recruitment are shown for one larval cohort in conditions where 2 larval cohorts exist together; i.e., the curves are similar to those in Fig. 1, but have been adjusted, by simple proportion, to pass through a point with co-ordinates $R = 2500$ and $E = 1.25 \times 10^8$.

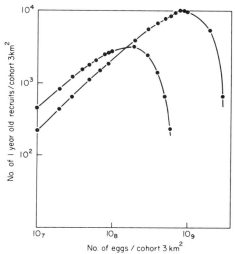

FIG. 2. The theoretical relationships between egg production and subsequent recruitment for one larval cohort of North Sea haddock.

By fixing one point on the curve, a considerable restriction is placed on the number of ways in which a curve relating recruitment to egg production can be drawn, and it should be noted that this is equivalent to placing a restriction on the possible values of P. For any given adult stock, there has to be a particular value of P for each of the curves of the type shown in Fig. 1. Values of P, although unknown in practice, are therefore determined by the properties of the larval model and by the characteristics of the adult stock. For any particular fish stock within the limits of this restriction, variations in the shapes and positions of the curves that could be drawn were found to have no effect on the conclusions drawn here.

3. The Principle of the Life History Model

Once one has a means of relating egg production to the subsequent recruitment of one year old fish, it is a relatively straightforward process to construct a model for simulating the entire life history.

The principle of this model is illustrated diagrammatically in Fig. 3. The composition of the stock in any one year is represented by two sets of

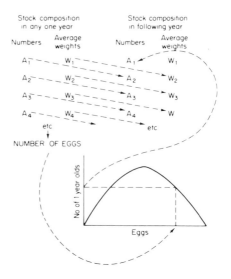

FIG. 3. Principle of the life history model.

numbers. One set represents the numbers of fish aged 1, 2, 3 years etc. in a unit of area occupied by the stock in year j. A second set, W_j, etc. represents the average weights of these fish at ages 1, 2, 3 years, etc. in year j. Given values of A_j and W_j for any one year, it is possible to calculate values for A_{j+1} and W_{j+1} in the following way:

The numbers of 2 year old fish for example are defined as the survivors of the numbers of 1 year old fish, i.e.

$$A_{2.j+1} = A_{1j} \exp - (F_1 + M_1)$$

where A_{2j+1} is the number of 2 year old fish in year $j + 1$;

    A_{1j} is the number of 1 year old fish in year j;

    F_1 is the instantaneous rate of fishing mortality experienced by the 1 year old fish;

    M_1 is the instantaneous rate of natural mortality experienced by the 1 year old fish.

In general, for fish aged i years

$$A_{i+1.j+1} = A_{ij} \exp - (F_i + M_i). \qquad (4)$$

The number of 1 year old fish in year $j + 1$, is determined from the number of 1 year old fish expected from the eggs produced in year j using equations (1) and (2) (or alternatively equations (1) and (3)). This means that initially an estimate of egg production in year j is needed and this can be determined by noting that the fecundity of a single fish of weight W_{ij} can be a relationship of the form: fecundity $= pW_{ij}^q$. Therefore the total eggs (E_j) produced by all the fish during year j can be represented by the expression:

$$E_j = 0\cdot5p \sum_{i=m}^{t_\lambda} A_{ij}W_{ij}^q$$

where m is the age of first maturity and t_λ is the age of the oldest spawning age group. A value of $0\cdot5$ has been used since the sex ratio of North Sea haddock is about $1:1$. Where the sex ratio is not unity some other proportion would be more applicable.

To allow for mortality during the egg stage, a coefficient p' can be introduced such that $p'E_j$ represents the initial number of larvae at the commencement of feeding. Finally, if D independent cohorts of larvae are assumed, the value of N_{Lo} for use in equation (1) is given by

$$N_{Lo} = p'E_j/D$$

The output from equations (1) and (2) (or equations (1) and (3)) multiplied by D and by an appropriate value of P determines A_{1j+1}, the number of 1 year old fish in year $j + 1$. This part of the operation is shown diagrammatically, at the bottom of Fig. 3.

The estimation of the average weights (W_{j+1}) proceeds in a manner that is analogous to the determination of the values of A_{j+1} from the values of A_j. For example, the mean weight of a 3 year old fish in year $j + 1$ is determined from the mean weight of a 2 year old fish in year j, plus the estimated growth, between ages 1 and 2. The most useful growth expression for subsequent manipulation in the model seems to be one in which there is an explicit term for food density. An examination of the growth curves of different stocks of haddock and cod throughout the North Atlantic suggests that food intake is proportional to a fractional power of body weight where the value of the exponent is about $0\cdot6$ or $0\cdot7$. This relationship holds approximately over a reasonable range of body weights for each species and will be used here as a first approximation. This leads to the general expression

$$W_{s+1.j} = W_{sj} + (\alpha_1 W_{sj}^r - \alpha_2 W_{sj}^{0\cdot8} - \alpha_3 W_{sj}). \tag{5}$$

The expression for growth during some specified interval of time is given in brackets and is proportional to

(energy intake—energy for maintenance—energy for reproduction).

The exponent of $0\cdot8$ in the term for maintenance energy comes from Winberg (1956).

An exponent of unity is used in the term for reproductive loss, on the grounds that ovary weight in haddock and cod and whiting is approximately proportional to body weight. A value of α_2 has been selected on the basis of physiological considerations. A value of α_3 has been selected on the basis of the estimated energy cost of reproduction. Given α_2 and α_3, a value of α_1 can be determined empirically, using equation (5), to fit the observed growth curve of the species under investigation.

For reasonably realistic results, it is appropriate to adopt fairly small intervals of time, and for this reason the incremental period has been taken as 1/100th of a year, and the values of α_s have been adjusted accordingly.

Given a value of $W_{sj} = W_{ij}$, the average weight of a fish aged i years in year j, 100 successive applications of equation (5) then leads to an estimate of $W_{s+100.j} = W_{i+1.j+1}$, the average weight of a fish aged $i + 1$ years in year $j + 1$.

For fish of mature ages, α_3 has been made equal to zero for the first 99 of these steps, and then given an appropriate annual value at the hundredth step. This simulates a loss of body weight, due to spawning, at a particular moment in time and seems more realistic than adopting a constant value of α_3 throughout a year, which would be equivalent to a uniform shedding of eggs throughout the 12 months. For the immature ages, α_3 has, of course, been made zero at all times of the year. For the investigations described here the parameter α_1 was kept constant during any series of computations. However, because this coefficient is directly proportional to food consumed, it could be made to vary with time by allowing it to become a function of the food consumed by the population in the preceding year. This enables the effect of population size on growth to be investigated and this will be considered in the Appendix.

There remains the determination of W_{1j+1}, the average weight of 1 year old fish in year $j + 1$. For the purpose of this paper, this was simply made a constant for all years.

4. Year Class Fluctuations

A characteristic of the common gadoids, is that year class fluctuations are relatively large, and empirically uncorrelated with the size of the parent stock. One test to which the model can be submitted therefore is to determine if year class variations of the appropriate order of magnitude can be generated using reasonable variations in the appropriate parameters. A biologically meaningful way of generating year class fluctuations in the model is by varying the parameter representing food density at the time when the larvae commence to feed. This was done by specifying maximum and minimum values of initial larval food density (N_{F_o} max and N_{F_o} min) and determining an actual value of N_{F_o} at random between these limits each year. An example of the year class strengths obtained over a 40-year period

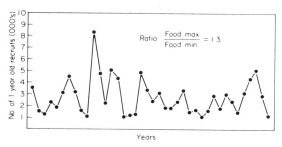

FIG. 4. Simulation of successive year class strengths using a haddock model.

from the model in this way is shown in Fig. 4. This particular sequence was obtained using a ratio of maximum larval food density to minimum larval food density of 1·3:1. The important feature of this result is that the year class strengths obtained in this way were found to be empirically un-correlated or only poorly correlated with the numbers of eggs from which they had been derived, as is found to be the case in practice.

Naturally, by increasing the ratio of maximum larval food to minimum larval food, the variability of the fluctuations in year class strength can be increased. One way of measuring this is by calculating the standard deviation of the individual values of year class strength. This has been done, and, in order to arrive at values independent of the mean, the results have been expressed as coefficients of variation.

Actual values of the coefficient of variation have been plotted against the ratio of maximum larval food/minimum larval food in Fig. 5. These were found to increase from 0–1·33 as the ratio food max/food min increased from 1–2. To assess the significance of these values, it is appropriate to compare them with the actual values of the coefficient of variation for species such as North Sea haddock and whiting for which a reasonably long series of records of year class strengths are available. For example, for the haddock year classes from 1918–1938 and 1944–1970, mean annual research vessel catches varied from 7 to 20,000 fish per 10 hours fishing in their second year of life. Two year classes, those of 1962 and 1967, were exceptionally large, however, and if these are ignored, the remainder fall into the range 7 to 4000 fish per 10 hours fishing. For these 46 year classes, the mean was 1004 and the standard deviation was 951, giving a coefficient of variation of 0·9. Reference to Fig. 5 shows that year class strengths with this coefficient of variation could be generated using a ratio of maximum larval food/minimum larval food of 1·7:1 (see Table 3).

For North Sea whiting, mean year class strengths for the years 1931–1938 and 1945–1970 was 467. The coefficient of variation was 1·3. From Fig. 5 this ratio corresponds to a ratio of food max/food min of about 2·0 (Table 3).

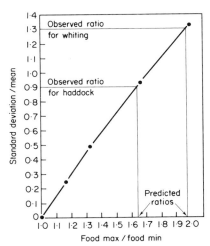

FIG. 5. Diagram to show how the Ratio of Standard Deviation Mean Increases as the Ratio Food Max./Food Min. Increases (North Sea haddock).

Strictly speaking, this result would be valid only if there were one larval cohort, or if there were several larval cohorts that were subject to identical larval food densities within any one year. However it is quite probable that in any one year there are several larval cohorts, and that each is dependent on a different larval food density. In that event the variance of the subsequent number of recruits would not be as great as would be expected if there were only one larval cohort. To account for a given level of variation in recruitment, it would therefore be necessary to assume a greater variation in larval food than that deduced above. For example, suppose that recruitment consists of R 1 year old fish, with a variance V. The coefficient of variation will be $\sqrt{V}/R$. Now, if there are D larval cohorts, each cohort will contribute R/D recruits with a variance V'.

The total recruitment from all D larval cohorts combined will still be R fish, and the variance will be $DV' = V$. Therefore $V' = V/D$, and the coefficient of variance of the number of recruits produced from any one cohort will be $(\sqrt{V/D})(D/R) = \sqrt{D}(\sqrt{V}/R)$ and this equals $\sqrt{D}$ times the coefficient of variance of the total annual recruitment. Consider for example the effect on the calculations for North Sea haddock of assuming three independent larval cohorts. To account for an overall coefficient of variation of 0·9 it is necessary to suppose that the coefficient of variation of the number of recruits produced from any one larval cohort is equal to $0·9\sqrt{3} = 1·56$. To account for this requires a ratio of food max/food min of 2·3. Similarly, in the case of North Sea whiting the ratio becomes 3·1.

5. Frequency Distribution of Year Class Strengths

An interesting feature of the simulation was that the frequency distributions of year class strengths generated by the model tended to be asymmetrical. This was also found to be a feature of the observed year class strengths, as is shown in Table 3 for haddock and whiting, and in Fig. 6a for haddock.

TABLE 3. North Sea haddock and whiting year class strengths showing the observed frequency distributions of the numbers/10 hours' fishing of 1+ fish

Haddock 1918–1938 and 1944–1970			Whiting 1931–1938 and 1945–1970		
No./10 hours	Observed frequency	% frequency	No./10 hours	Observed frequency	% frequency
0–400	18	39	0–300	19	55
401–800	5	11	301–600	8	24
801–1200	9	20	601–900	2	6
1201–1600	4	9	901–1200	2	6
1601–2000	3	7	1201–1500	1	3
2001–2400	3	6	1501–1800	0	—
2401–2800	0	0	1801–2100	0	—
2801–3200	2	4	2101–2400	1	3
3201–3600	1	2	2401–2700	0	—
3601–4000	1	2	2701–3000	1	3
	46	100		34	100

	Haddock	Whiting
Mean	1004	467
Standard deviation	951	613
Standard deviation/mean	0·9	1·3

Figure 6a, for example, shows the percentage frequency distribution of the 46 North Sea haddock year class strengths (excluding those for 1962 and 1967). It is seen that the distribution is asymmetrical, with by far the greater proportion of year classes (39%) falling into the smallest size category. At the other extreme, only 2% of year classes fell into the largest size category.

For comparison, some theoretical distributions based on 1 larval cohort are shown in Table 4 for values of food max/food min ranging from 1·17 to 2·00. It is seen that these distributions are all asymmetrical and that the

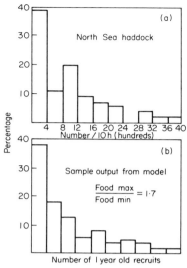

FIG. 6. Frequency distribution of the relative occurrence of different year class strengths.

percentage falling into the smallest year class category increases as the ratio of food max/food min increases. An example of a theoretical distribution obtained using a ratio of food max/food min of 1·7; 1 is shown in Fig. 6b.

This part of the investigation therefore shows that annual variations in larval food density of 1·7–3·1 can account for variations in year class strength

TABLE 4. Some theoretical percentage frequency distributions of relative year class strengths

Relative year class strength				
0–1	11	16	32	57
1–2	12	16	22	10
2–3	10	14	11	8
3–4	18	14	12	6
4–5	11	8	3	4
5–6	10	9	4	2
6–7	9	7	3	4
7–8	7	7	9	4
8–9	7	5	2	3
9–10	5	4	2	2
Food max./Food min.	1·17	1·33	1·67	2·00
Standard deviation/mean	0·25	0·49	0·93	1·33

comparable with those observed for North Sea haddock and whiting. It also shows that the model is able to account for the asymmetrical distribution of year class strengths which is observed.

The data so far available on annual variations in the abundance of organisms that might constitute the food of gadoid larvae is inadequate for testing these results. It should be noted however, that the adoption of random values of larval food density (N_{F_o}) between two limits is equivalent to the adoption of a rectangular distribution for representing annual variation in the overall mean density of the larval food. In practice, annual variations in larval food density are more likely to conform to a normal or log normal distribution. The ratio of the maximum to the minimum values observed will therefore depend on the number of observations made, and is likely to be greater than the ratio of the upper and lower limits of the best fitting rectangular distribution. This point must be taken into account when data are available for testing the conclusions reached here.

Equilibrium in the Long Term

In the special case when there is no variation in annual recruitment the mean values of recruitment and egg production can be readily inferred from the intersection of stock and recruitment curves of the kind described by Beverton and Holt (1957). Figure 7a, for example, shows the expected numbers of 1 year old recruits (R) for any given value of egg production (E).

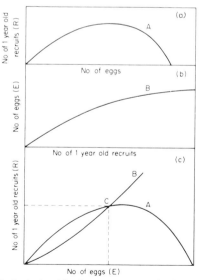

FIG. 7. Hypothetical relationships between egg production and recruitment.

This relationship depends on the population dynamics of the first year of life. Figure 7b shows the expected numbers of eggs for any given number of 1 year old recruits, and is the outcome of the demersal part of the life history. These two curves can be combined into one graph as in Fig. 7c. Curve A shows the expected value of R for a given value of E. The E axis therefore represents the independent variable while the R axis represents the dependent variable. Curve B shows the relationship depicted in Fig. 7b, but with the axes interchanged, i.e. it shows the expected value of E for a given value of R. For this curve, the R axis represents the independent variable and the E axis represents the dependent variable. The position of intersection of the curves, at point C, gives the expected equilibrium values of R and E when there is no annual variation in recruitment.

Since gadoids are the product of a fairly long evolutionary process, a necessary feature of any biologically meaningful simulation model is that it has, or at least appears to have, stability in the long term. To satisfy this requirement, the co-ordinates of R and E ought to describe points that converge towards some positive finite value (e.g. C in Fig. 7c). Failing this, the points should at least vary about a fixed position without increasing amplitude.

When stability was investigated using the model, it was found that the pattern of variations in year class strength depended to some extent on the position of intersection of the two egg/recruitment curves (e.g. on the position of point C in Fig. 7c). The model appeared to be most stable when the intersection of the curves was on the ascending limb, or on the top of curve A. Under these conditions, if larval food was kept constant, the numbers of eggs and recruits converged rapidly to constant values. Alternatively, if variations in year class strength were generated by varying the number of larval food organisms each year, results such as those depicted in Fig. 4 were obtained. The numbers of recruits varied but without any tendency for the degree of variation to increase with time. Under these conditions, the frequency distribution of relative year class strength was asymmetrical as depicted in Fig. 6.

When the intersection of the curves was on the right hand descending limb of curve A (Fig. 7c), a somewhat different result was obtained. Under these conditions, even if larval food was kept constant, the numbers of recruits produced annually was observed to oscillate. An example is shown in Fig. 8, also based on North Sea haddock. Year class strengths alternated regularly between good and poor but without any tendency for the amplitude of the oscillation to change with time. Even when the degree of variation was increased by superimposing variations due to changes in the number of larval food organisms, there was no tendency for the model to become unstable. An example is given in Fig. 9a. The alternation of good and poor

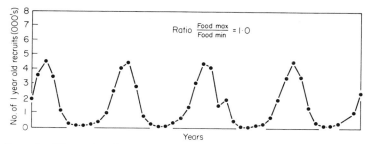

FIG. 8. Example of fluctuations in year class strength when point C (Ref. Fig. 7c) is on the right hand descending limb of curve A.

year classes is similar to that in Fig. 8, but less regular. Also shown (in Fig. 9b) are the alternations in the biomass of the stock corresponding with the results in Fig. 9a. When egg production was so high that egg numbers lay to the extreme right of curve A (Fig. 7c) recruitment was poor. After a sequence of poor year classes, however, the spawning stock started to decline until the egg number approached the middle of curve A, and recruitment became good again. The oscillations continued with no apparent tendency for their amplitude to increase. This conclusion is similar to that reached by Ricker (1954), and also confirms his conclusion that the presence of several age groups in the spawning stock confers stability for all positions of point C. Another feature of the results obtained when the equilibrium was on the right hand descending limb of curve A (Fig. 7c), was that the frequency distribution of year class strengths was not the same as that

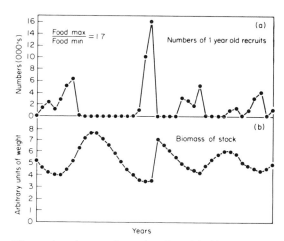

FIG. 9. Example of fluctuations in year class strength and in biomass of stock when point C (Ref. Fig. 7c) is on the right hand descending limb of curve A.

shown in Fig. 6b. An example is shown in Fig. 10, based on the results depicted in Fig. 9a. Under these conditions, the year class strength tended to be either good or bad with few, if any, intermediate values, so that the distribution tended to be U shaped.

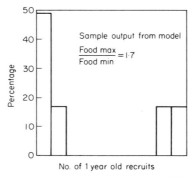

FIG. 10. Frequency distribution of the relative occurrence of different year class strengths.

So long as curves A and B intersect each other, it appears likely that the model is potentially stable. Once initiated such a system might be expected to persist, unless of course the basic parameters changed to such an extent that curves A and B no longer intersected, i.e. a decline in egg production per recruit, such as might occur with overfishing, might move curve B so far to the left (see Fig. 7c) that is no longer intersected curve A. Alternatively, an increase in mortality during the first year of life, due possibly to a reduction in larval food or to other natural causes, might move curve A downwards until again there was no point of intersection. Under either of these conditions, the population would become extinct very rapidly.

External agencies apart, however, it can be argued that natural selection is more likely to confer, rather than detract from stability in the long run. For example, other things being equal, it seems reasonable to suppose that a genotype that happened to favour survival during the first year of life ought to succeed over less favourable genotypes. Thus, one can assume that selection pressure would tend to operate upwards on curve A (with reference to the axes in Fig. 7c). Similarly, other things being equal, a genotype that favoured the production of a large number of eggs/recruit ought to succeed over genotypes that led to smaller numbers, i.e. selection pressure might be expected to operate upwards on curve B with reference to Fig. 7b (and to the right on curve B with reference to Fig. 7c). Clearly selection pressure would operate in this upward direction only to a certain point in each instance. However, the most probable way in which selection would work is upwards on curve A and to the right on curve B (Fig. 7c) moving the equilibrium point to the right, which is the direction of greater stability.

6. A General Model

For a more rigorous consideration of evolutionary trends, it seems appropriate to consider an extension of the life history model to a more general model based on several co-existing genotypes. A suggested model for this purpose is described in the Appendix.

Fundamentally, the model is based on the same principles as those depicted in Fig. 3, except that instead of one genotype A, there is assumed to be a number of genotypes A_k. It is assumed that only a proportion π of the eggs from each genotype become larvae of that genotype. The remaining proportion $(1 - \pi)$ of the eggs is assumed to yield larvae of the remaining $k - 1$ genotypes in equal proportions (see Appendix, eqn. (3)). By making $\pi = 1$, the genotypes become independent and the model can be used to simulate co-existence between a number of species. In this model, a term Z_L has been added to the daily larval mortality rate, to allow for the possibility that some larval mortality may be directly due to factors other than starvation (see eqn. (4)).

For some species, the coefficient of variation of year class strength is much lower than that of haddock or whiting. For North Sea plaice, for example, the value is 0·4. For this species it seems likely, as suggested by Steele and Edwards (1970), that an upper limit to recruitment is imposed by a limitation of food at the time when the fish first take to the bottom. To allow for a stage such as this in the life history, an upper limit A_k max has been placed on A_{1jk}, the numbers of 1 year old fish of genotype k in any year j. If this restriction is not required it can be removed by making A_k max extremely large.

It has been assumed that from age 1 year onwards, the food consists of one type until the fish reach some critical weight W_L and that from then onwards the fish are considered large enough to eat a different class of food organisms. Within each of these classes it is assumed that there may be more than one food type and that the genotypes consume these with different relative efficiencies. The food consumption coefficients α_{uvkj} are subject to alteration each year using equation (9). This has the effect of adjusting the growth rate of each genotype until the actual quantity of each food type consumed is equal to some specified quantity (ψ_{uv} max) of that food type. In the situation where there is no variation in larval food and hence in recruitment, it should be possible, by specifying the total quantity of each kind of food, to predict a growth curve for each genotype. It is hoped this might provide insight into why fish grow at the rates they do.

For example differences in growth rates between species could be investigated, or alternatively, differences within species could be simulated. Provision is made in the model for different genotypes to mature at different ages. It is hoped that the model will provide some insight into why, in practice, different species of genotypes mature at such different ages. In the situation where variations in year class strength are simulated, the application of equation (9) should generate annual fluctuations in growth rate. For this purpose, however, some of the equations in the model may be based on oversimplifications, and a more general application of the model will be to try out different assumptions about the fundamental underlying relationships between growth, natural mortality, and food density.

7. Summary

In an earlier paper, a model was proposed to account for the mortality during the larval stage, of fish such as haddock and cod. These species are characterized by a high fecundity and by a pelagic larval stage during which the growth and mortality rates are extremely high. The basis of the model is that food density is critical for survival during the larval stage. One feature of the model is that larval survival was found to be extremely susceptible to variations in the annual mean density of the food organisms and a primary object of this paper was to investigate what variations in larval food density were needed to account for the observed variations in the year class strengths of haddock and whiting.

To do this, it was first necessary to incorporate the larval model into a life history model for simulating the complete life history of the species, and part of the paper is concerned with a description of this model. It was then possible to simulate variations in year class strength by varying the parameter for larval food density at random. For simplicity, a rectangular distribution of larval food density was assumed, and variations in food density were expressed as the ratio of the upper to the lower limits of that distribution. In the event of there being one larval cohort the results showed that a ratio of 1·7 : 1 in larval food density was able to account for the observed variations in year class strength of North Sea haddock. For North Sea whiting a ratio of 2 : 1 was required. With three independent larval cohorts, the ratio of the upper to the lower limit of the larval food distribution was 2·3 for haddock and 3·1 for whiting. An interesting result was that the simulated year class strengths obtained using the model were asymmetrically distributed, and some of these were superficially very similar to those observed in practice. Also investigated was the long term stability of the life history model. It was argued that the model ought to be potentially stable for a wide range of values of the basic parameters.

Finally, a general model is described for simulating the co-existence of a number of genotypes with provision for varying the growth rates as functions of the quantities of food available.

Appendix

DESCRIPTION OF THE GENERAL MODEL

Basic definitions

Let

A_{jk}	$= \{A_{ijk}\}$ the set of numbers of fish at each age (i years) during year j of genotype k.
A_k max	= maximum permissible value of A_{1jk}.
W_{jk}	$= \{W_{ijk}\}$ the set of average weights of fish at each age (i years) during year j of genotype k.
W_{1k}	= average weight of 1 year old fish of genotype k.
F_{ik}	= the instantaneous annual rate of fishing mortality of fish aged i years of genotype k.
M_{ik}	= the instantaneous annual rate of natural mortality of fish aged i years of genotype k.
m_k	= age of first maturity of genotype k.
N_{kt}	= the number of larvae of genotype k alive t days after feeding commences.
N_{Ft}	= number of food organisms alive t days after feeding commences.
N_{Fo} and σ^2	it is assumed that the initial number of food organisms each year is lognormally distributed with a mean N_{Fo} and a variance of the logarithms $= \sigma^2$.
N_{ko}	= number of larvae alive of genotype k at the commencement of feeding.
p'	= proportion of the number of eggs produced that survive until feeding commences.
P_k	= proportion of the larvae of genotype k surviving from the end of the larval phase to age 1 year.
D	= number of larval cohorts.
E_{jk}	= total number of eggs produced in year j by genotype k.
π	= proportion of eggs produced by each genotype with the characteristics of that genotype.
p_k	= fecundity coefficient for genotype k.
q	= fecundity exponent for all genotypes.
Z_{Lt}	= daily instantaneous rate of mortality of larvae due to causes other than shortage of food.
X_{kt}	= minimum food density for survival for one day, of larva aged t days of genotype k (in units of number of food organisms encountered per day).

S_{kt} = searching coefficient of larva aged t days of genotype k.

U_{kt} = daily food requirement of larva aged t days of genotype k.

$\left.\begin{array}{l}X_{ko}, S_{ko}, \\ U_{ko}\end{array}\right\}$ = values of X, S and U when $t = 0$.

GX_{kt} = exponent for rate of change of X_{kt} with t for genotype k.

GS_{kt} = exponent for rate of change of S_{kt} with t for genotype k.

GU_{kt} = exponent for rate of change of U_{kt} with t for genotype k.

t_{λ} = oldest age group (for all genotypes).

c = length of critical period for larvae (days).

ψ_{uvj} = total quantity of food species v, consumed in year j, by fish of size group u.
There are two values of u. $u = 1$ refers to "small" fish and $u = 2$ refers to "large" fish.

W_L = value of W_{ijk} at which a fish is considered large enough to transfer from food suitable for small fish ($u = 1$) to food suitable for large fish ($u = 2$).

ψ_{uv} max = maximum sustainable value of ψ_{uvj}.

α_{uvkj} = food consumption coefficient for one hundredth of a year for a fish of size group u, belonging to genotype k, feeding on food species v, in year j.

r = food consumption exponent.

α_2 = maintenance energy coefficient for one hundredth of a year.

α_3 = coefficient for loss of weight due to reproduction for an entire year.

Basic operations

Determination of the set $A_{j+1.k}$ from the set A_{jk}:

$$A_{jk} = \{A_{1jk}, \quad A_{2jk}, \quad \cdots \quad A_{t_\lambda-1.jk}, \quad A_{t_\lambda jk}\}$$

$$A_{j+1.k} = \{A_{1j+1.k}, \quad A_{2j+1.k}, \quad A_{3j+1.k}, \quad \cdots \quad A_{t_\lambda j+1.k}\}$$

in general

$$A_{i+1.j+1.k} = A_{ijk} \exp - (F_{ik} + M_{ik}) \tag{1}$$

or $A_{i+1.j+1.k} = $ zero if $W_{s+1.jk} < W_{sjk}$ for the first step in equation (8).

Determination of $A_{1j+1.k}$:

$A_{1j+1.k}$ is calculated in a series of steps as follows:

Determination of the number of eggs:

Let

$$E_{jk} = 0.5 p_k \sum_{i=m_k}^{t_\lambda} A_{ijk} W_{ijk}^q \tag{2}$$

and

$$N_{ko} = p'/D\left[\pi E_{jk} + \frac{(1-\pi)}{k-1} \sum_{\substack{s \\ s \neq k}} E_{js}\right],$$ (3)

using the Larval model:

$$N_{kt+1} = N_{kt} \exp - [(X_{kt})/(N_{Ft}S_{kt}) + Z_{Lt}]$$ (4)

$$N_{Ft+1} = N_{Ft}\left(1 - \sum_k N_{kt}S_{kt}\right)$$ (5)

or

$$N_{Ft+1} = N_{Ft} - \sum_k N_{kt}U_t.$$ (6)

But, if N_{Ft} becomes negative then all N_{kt} become zero.

Note: there is an equation corresponding to (4) for each genotype. The model gives a choice of either equation (5) or equation (6) or the one that happens to give the greater value of N_{Ft+1}. The selected option is then used in conjunction with the set of equations (4) to calculate successive values of N_{kt} and N_{Ft}, with the ultimate object of determining a set of values of N_{kc} after the specified number (c) of days.

The values of N_{ko} obtained in (3) are used as the initial value of N_{kt} in these equations.

To generate variations in year class strength, a value of N_{Foj} is chosen at random, assuming a lognormal distribution of food organisms with mean N_{Fo} and variance σ^2.

Number of 1 year old recruits in year $j + 1$ ($A_{1j+1.k}$):

$$A_{1j+1.k} = DP_kN_{kc}$$ (7)

or A_k max whichever is the least.

Determination of the set $W_{j+1.k}$ from the set W_{jk}:

$$W_{jk} = \{W_{ijk}, \quad W_{2jk}, \quad \cdots \quad W_{t_\lambda-1.jk}, \quad W_{t_\lambda jk}\}$$

$$W_{j+1.k} = \{W_{1j+1.k}, \quad W_{2j+1.k}, \quad W_{3j+1.k}, \quad \cdots \quad W_{t_\lambda j+1.k}\}$$

$W_{1jk} = W_{1j+1.k}$ for all values of j.

Determination of $W_{i+1.j+1.k}$ from W_{ijk}:

Let

$$W_{s+1.jk} = W_{sjk} + \left[\sum_v \alpha_{uvkj}W_{sjk}^r - \alpha_2 W_{sjk}^{0.8} - \alpha_3 W_{sjk}\right]$$ (8)

$\alpha_{uvkj} = \alpha_{1vkj}$ for $W_{sjk} < W_L$ and $\alpha_{uvkj} = \alpha_{2vjk}$ for $W_{sjk} \geq W_L$.

The initial value of W_{sjk} is equal to W_{ijk}. Then: the value of $W_{s+100.jk}$ after 100 steps is the value of $W_{i+1.j+1k}$.

For the different steps, n:

$$\text{if} \quad n < 100 \qquad \alpha_3 = 0$$

$$n = 100 \qquad \alpha_3 = 0 \quad \text{for} \quad i < m_k - 1$$

$$n = 100 \qquad \alpha_3 = \alpha_3 \quad \text{for} \quad i \geq m_k - 1.$$

This simulates a loss of weight from the age of maturity onwards due to the liberation of the spawning products on just one day each year.

Determination of $\alpha_{uvk_{j+1}}$ from α_{uvkj}:

$$\alpha_{uvk_{j+1}} = \frac{\alpha_{uvkj}\psi_{uv} \max}{\psi_{uvj}}. \tag{9}$$

Catch (in numbers) in year j:

$$\sum_i \sum_k \frac{A_{ijk}F_{ik}}{F_{ik} + M_{ik}} [1 - \exp - (F_{ik} + M_{ik})]. \tag{10}$$

Catch in weight in year j:

$$\sum_i \sum_k \frac{A_{ijk}F_{ik}}{101} \sum_{n=0}^{100} W_{s+n.jk} \exp - \left(\frac{F_{ik} + M_{ik}}{100}\right)n. \tag{11}$$

Numerical size of stock in year j:

$$\sum_i \sum_k \frac{A_{ijk}}{F_{ik} + M_{ik}} [1 - \exp - (F_{ik} + M_{ik})]. \tag{12}$$

Biomass in year j:

$$\sum_i \sum_k \frac{A_{ijk}}{101} \sum_{n=0}^{100} W_{s+n.jk} \exp - \left(\frac{F_{ik} + M_{ik}}{100}\right)n. \tag{13}$$

Mean weight of fish in catch in year j:

$$\frac{\text{Catch in weight}}{\text{Catch in number}}. \tag{14}$$

Mean weight of fish in stock in year j:

$$\frac{\text{Biomass}}{\text{Numerical size of stock}}. \tag{15}$$

Food consumed in year j of category uv:

$$\psi_{uvj} = \sum_i \sum_k \frac{100\alpha_{uvkj}A_{ijk}}{101(0\cdot25)} \sum_{n=0}^{100} W^r_{s+n.j}\exp - \left(\frac{F_{ik} + M_{ik}}{100}\right)n. \tag{16}$$

Where for $u = 1$, the summation with regard to i is for those ages for which

$$W_{ijk} < W_L$$

and, for $u = 2$, the summation with regard to i is for those ages for which

$$W_{ijk} \geq W_L.$$

Note: 0·25 is a factor for growth efficiency.

REFERENCES

BEVERTON, R. J. H. AND HOLT, S. J. (1957). On the dynamics of exploited fish populations. *Fishery Invest., Lond.*, Ser. 2, **19**, 533 pp.

JONES, R. (1973). Density dependent regulation of the numbers of cod and haddock. *ICNAF/ICES/FAO Symposium on Stock and Recruitment, Aarhus, Denmark, 7–10 July* 1970. (In press).

RAITT, D. S. (1932). The fecundity of the haddock. *Scient. Invest. Fishery Bd Scotl.* 1932 (1) 42 pp.

RICKER, W. E. (1954). Stock and recruitment. *J. Fish. Res. Bd Can.* **11**, 559–623.

STEELE, J. H. AND EDWARDS, R. R. C. (1970). The ecology of O-group plaice and common dabs in Loch Ewe. IV. Dynamics of the plaice and dab populations. *J. exp. mar. Biol. Ecol.* **4**, 174–187.

WINBERG, G. G. (1956). [Rate of metabolism and food requirements of fishes] *Nauchnye Trudy Belorisskovo Gosudarstvennovo Universiteta Imeni V.I. Lenina, Minsk.* 253 pp. and *Fish. Res. Bd Can. Trans. Ser.* (194), 1960, 202 pp.

Part II

Population Processes in Space

4. The Formulation and Interpretation of Mathematical Models of Diffusionary Processes in Population Biology

J. G. SKELLAM

The Nature Conservancy, Belgrave Square, London, England

1. Introduction

Between the observation of natural phenomena on the one hand and the sophisticated mathematical exercises which may eventually ensue on the other, there lies a broad spectrum of theorization.

The term "model" will be employed here in accordance with its time-honoured usage in natural science to mean the whole of the conceptual construct, and not to mean a system of mathematical relations which can be extracted from the model and studied in isolation. Some regard the mathematical features as part of the descriptive clothing of the model, whilst others look on them as the bare bones. It is enough, however, to make the worthwhile distinction between the scientist's model in all its conceptual richness and those highly abstract systems which form part of its logical structure.

The results which are implicit in the logical structure are often revealing and enlightening, but tautologies are not strictly informative. They emerge because, either intentionally or unwittingly, we have already planted the seeds from which they grow. Furthermore these results are without empirical meaning unless they are brought once again into proper conjunction with the scientific concepts which give the model substance. For example, a mathematically derived statement about the probability of ultimate extinction of a theoretical population is empty unless we know whether the time-scale envisaged is ecological, geological or astronomical.

Roughly speaking, a model is a peculiar blend of fact and fantasy, of truth, half-truth and falsehood. In some ways a model may be reliable, in other ways only helpful and at times and in some respects thoroughly misleading. The fashionable dogma that hypothetical schemes can be tested

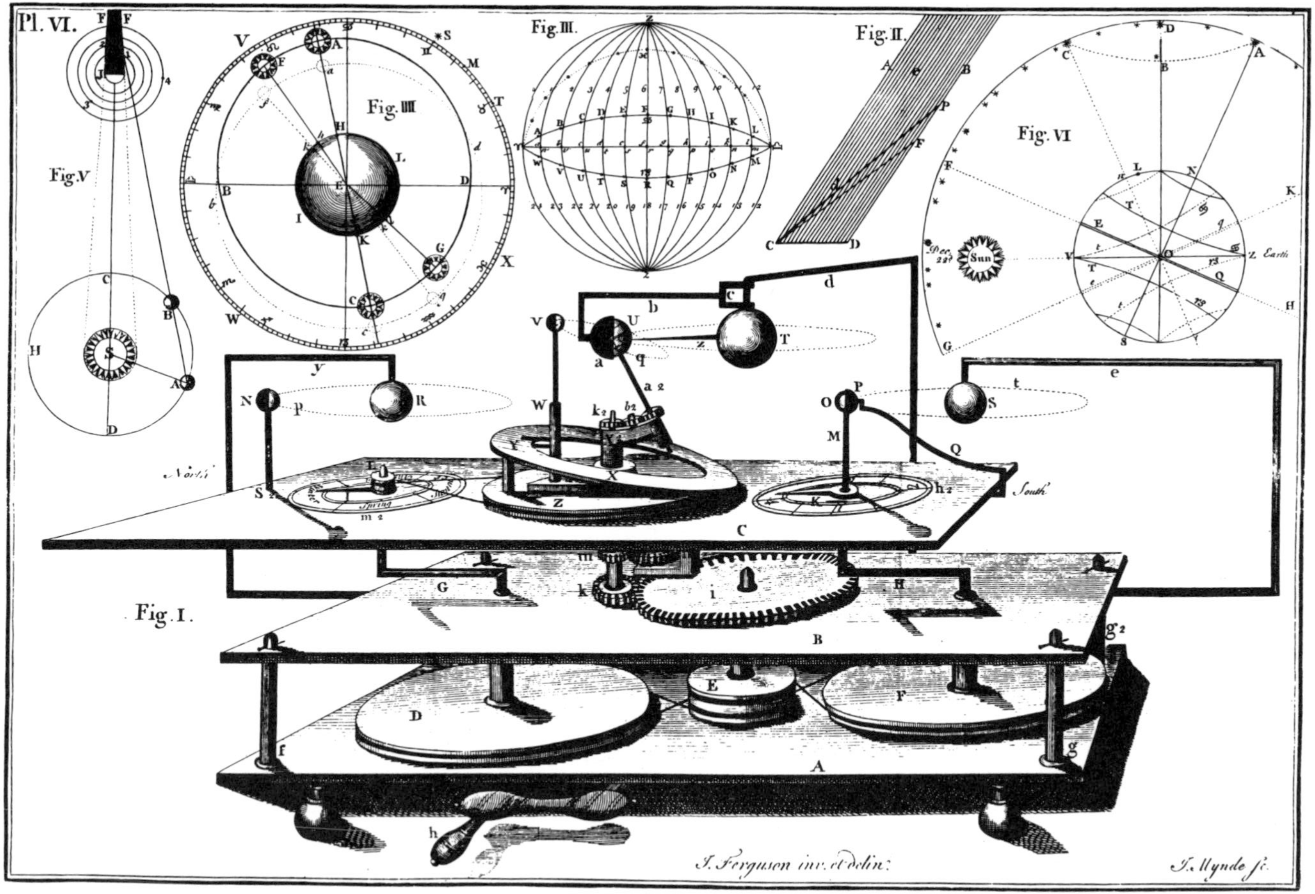

Pl. VI.

Fig. II

Fig.VI

Fig.III

Fig.III

Fig.V

Fig.I.

FIG. 1

J. Ferguson invt. et delin.

J. Mynde sc.

in their totality in some absolute sense, is hardly conducive to creative thinking. It is indeed, just as great a mistake to take the imperfections of our models too seriously as it is to ignore them altogether.

In order to make the present standpoint clear and to emphasize the vital role played by model-making in scientific research consider the physical representation of an astronomical model illustrated in Fig. 1. It yields answers by straight analogy. Once the model has been designed, the role of the mathematician is clear. In this case he has only to turn the handle gently, and all that the scientist has to do meanwhile is to watch on in the hope that he will recapture something from the past or predict some aspect of the future.

2. Examples of Animal Diffusion

Empirical science starts with observation. Let us therefore glance briefly at a few examples of natural phenomena to which diffusion theory might be applied.

Figure 2 illustrates just one of many known examples of an animal outbreak. It refers to the geographical expansion of the muskrat after its escape

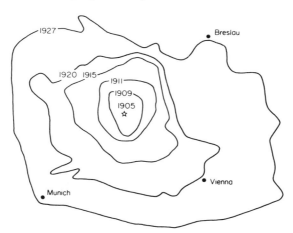

FIG. 2. After Ulbrich

from captivity in Central Europe. Despite the irregularities of the terrain, the rate of expansion of the contour enclosing its range is remarkably uniform. Such a result is to be expected theoretically when diffusion and population growth occur simultaneously.

Figure 3 illustrates just one of many similar examples of animal behaviour in the laboratory. *Paramoecium* is a minute and highly mobile aquatic organism, and at a uniform temperature the population appears to be

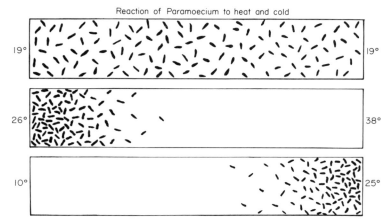

FIG. 3. Adapted from Jennings, after Mendelssohn

evenly distributed throughout the water in the tank. In temperature gradients, movement tends to be strongly biassed towards conditions at about 25 or 26°C. Notice how the tails of the distributions thin out.

Figure 4 shows three examples of vertical migration in marine organisms under changing diurnal conditions. Note that in darkness the Mysid is fairly evenly distributed over all depths, but that as light from above increases in intensity, so does the downward drift. The Copepod and Medusa exhibit somewhat similar mass behaviour, but in these cases there is apparently an additional and opposing source of drift associated with conditions which develop with depth. Note that in all cases the distributions have tails and that all individuals do not consistently occupy the preferred location.

In the examples shown, there is no great difficulty in describing mass behaviour in the language of diffusion theory. To be genuinely fruitful, however, such exercises need to be accompanied by studies on the loco-motory responses of the animals themselves or linked more closely to biological theory, particularly that which is concerned with the ways in which a living organism may be adapted to its environment.

3. The Texture of the Medium

To a biologist, a simple random walk model on an extended chess-board is not obviously relevant. It may be quite clear that asymptotic normality results and that the mean-square-deviation increases linearly with time, but he knows of no habitats which resemble rectangular lattices nor animals that move in unison in steps of fixed size confined to four directions.

Only after the model is reconstructed, say, on a triangular grid with six permissible directions and when substantially the same broad end-result

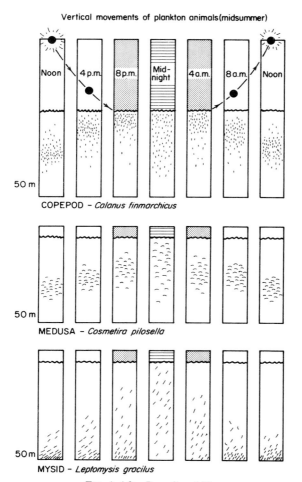

Vertical movements of plankton animals (midsummer)

COPEPOD – *Calanus finmarchicus*

MEDUSA – *Cosmetira pilosella*

MYSID – *Leptomysis gracilus*

FIG. 4. After Russell and Yonge

ensues, does it appear plausible to him that some of the detailed features of the model may not matter after all. But even when all directions are allowed, as in Karl Pearson's model (1906), it still seems unnatural to the biologist for all steps to be given the same size. This criticism, of course, is easily met by resort to the Central Limit Theorem, but even so several valid objections remain.

Every biologist knows that earthworms cannot burrow through boulders of granite, and that the habitats of animals are often more akin to a reticulum than to a continuous expanse. In order to demonstrate once again that the large-scale results of diffusionary processes are virtually independent of the fine texture of the medium, Monte-Carlo methods were employed to generate a large number of realizations of a random walk in a plane from which access to more than half the area was excluded by enclosing numerous small blocks within reflecting barriers. The outcome is shown in Fig. 5. The normal

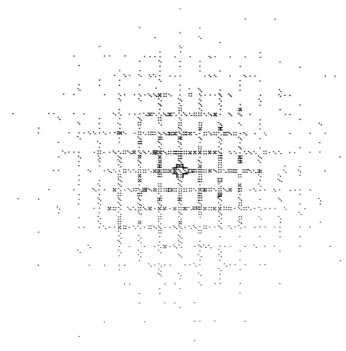

Fig. 5

function provides a most satisfactory graduation not only of the broad pattern in the plane as a whole but also of the population densities in the living-space between the enclosures.

4. The Pattern of Movement

Another serious objection against most current dynamic modelling in this area is the unrestricted use being made of the assumption of statistical independence. This is indeed a powerful and far reaching assumption with an unsavoury reputation for seeping into scientific theory in the most insidious ways. Dependence and independence, like order and chaos, are polarized notions, and in practice we are obliged to represent intermediate ideas by a judicious combination of the two. These notions are essential ingredients in the language now in use and cannot be avoided altogether.

The manner, however, in which they are introduced, does matter and must be acceptable in the scientific context in which they are employed.

Models of animal diffusion are often conceived in the same way that mathematicians first conceived of Brownian motion. The elementary theory in its most extreme form with infinitely wriggly paths and unbounded velocities is not totally acceptable even to a physicist. How much less so must this kind of diffusion model appear to a biologist who knows for certain that animals move with quite modest velocities in irregularly twisting paths and that displacements in successive seconds are strongly correlated.

The model illustrated in Fig. 6 may help to allay his fears. There is an analogy here to a cluster of animals which are induced to disperse in all

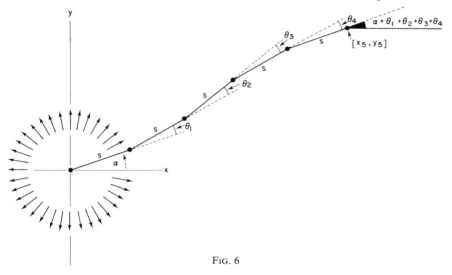

FIG. 6

directions by an explosive incident in their midst. Each individual track then suffers from the cumulative effects of successive navigational errors.

Eventually all directions of motion become manifest in each individual track, and the mean-square-dispersion increases linearly with time. An outline of the mathematical argument follows. Incidentally the vacuum created around the origin soon fills up and bivariate normality ultimately ensues.

The rth vector contributes $s \cos(\alpha + \theta_1 + \theta_2 \cdots + \theta_{r-1})$ and $s \sin(\alpha + \theta_1 \cdots + \theta_{r-1})$ to the x and y coordinates of the end-point of the track. It follows that

$$x_n + iy_n = s \sum_{r=1}^{n} \exp\{i(\alpha + \theta_1 + \theta_2 \cdots + \theta_{r-1})\}$$

and

$$x_n - iy_n = s \sum_{r=1}^{n} \exp \left\{ i(-\alpha - \theta_1 - \theta_2 \cdots - \theta_{r-1}) \right\}.$$

$R_n^2 = x_n^2 + y_n^2$ is given by forming the product. The n^2 individual terms simplify immediately.

If the random angular displacements, θ_j, are distributed independently and in a like manner with $\varepsilon\{\sin \theta_j\} = 0$ and $\varepsilon\{\cos \theta_j\} = \rho$, then

$$E\{\exp i\Sigma\theta_j\} = E\{\Pi \exp i\theta_j\} = \Pi E\{\exp i\theta_j\} + \Pi\rho.$$

It then follows that

$$E\{R_n^2\} = s^2[n + 2(n - 1)\rho + 2(n - 2)\rho^2 \cdots + 2\rho^{n-1}].$$

Excluding the trivial cases when $|\rho| = 1$, this result condenses to

$$E\{R_n^2\} = s^2 \left(\frac{1 + \rho}{1 - \rho} \right) \left[n - \frac{2\rho(1 - \rho^n)}{1 - \rho^2} \right],$$

whence,

$$E\{R_n^2\} - E\{R_{n-1}^2\} = s^2 \left(\frac{1 + \rho}{1 - \rho} \right) \left(1 - \frac{2\rho^n}{1 + \rho} \right).$$

As $n \to \infty$, $\rho^n \to 0$. Indeed, when n is only moderately large, ρ^n is often negligible.

5. The Induction of Displacement

There are two extreme limiting forms of the previous model according to whether directional disturbance is appropriately measured as so-much per second (β), or so-much per metre (γ). If $s \to 0$ and $\rho \to 1$ together, we find in these two extreme cases respectively that the rates of increase of the expected mean-square-deviation are

$$v^2 \cdot \frac{2}{\beta}(1 - e^{-\beta T}) \quad \text{and} \quad v \cdot \frac{2}{\gamma}(1 - e^{-\gamma S}),$$

where v denotes velocity, T denotes time from the beginning, and S denotes the length of the track.

It is not often that this distinction is emphasized. Indeed, the popular mathematical approach to diffusionary processes tends to disregard the very aspects which might throw light on the underlying mechanism. Animal displacements, whether conceived in stochastic terms or not, are induced or evoked by events in space-time (Fig. 7). Some agents are of short duration and wide extent, whilst others are localized and persistent. The level of activity and the velocity of locomotion are influenced by environmental

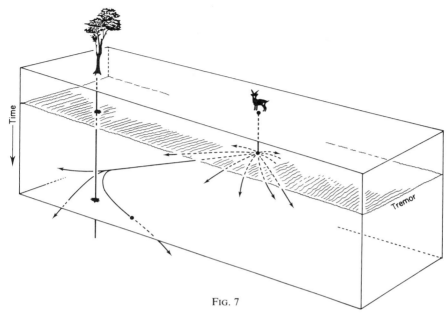

FIG. 7

conditions, both physical and biological. The manner in which the parameters of the diffusion process respond to changes in the operative factors
may furnish clues to the nature of the underlying mechanism. This thought,
touched on already, is perhaps better illustrated by the simple random-walk
scheme shown diagrammatically in Fig. 8. More is involved than just a
change of scale.

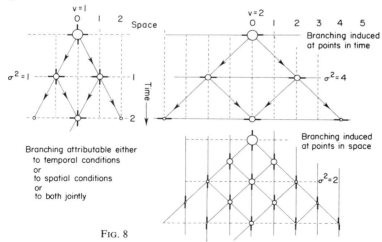

FIG. 8

Whether or not an animal moves or is moved from a position x_1 to an adjacent position x_2 depends on many things. If $\lambda(x_1, x_2)$ denotes the probability that in a small fixed interval of time such a displacement occurs, then in common sense terms λ is high when conditions at x_1 are "repulsive" or when stimuli emanating from x_2 are "attractive". Furthermore λ is high when the conditions between x_1 and x_2 are conducive to locomotion. These considerations are reflected in the partial derivatives of $\lambda(x_1, x_2)$. If λ_1 and λ_2 denote the partial derivatives with respect to the first and second variables respectively, it is immediately apparent that $\lambda_2 = 0$ if displacement is induced by "repulsion" only and that $\lambda_1 = 0$ if displacement is induced by "attraction" only. Furthermore, if the probability of displacement depends only on the conditions which facilitate movement in the interval regardless of direction, $\lambda(x_1, x_2) = \lambda(x_2, x_1)$, whence $\lambda_1 = \lambda_2$.

Consider now a random walk at time points h units apart on a row of stepping stones k units apart. If $m_t(x)$ denotes the probability mass at (x, t), the change during the time interval t to $t + h$ in the total mass to the left of a point x (midway between two stones) is

$$m_t(x + \tfrac{1}{2}k)\lambda_t(x + \tfrac{1}{2}k, x - \tfrac{1}{2}k) - m_t(x - \tfrac{1}{2}k)\lambda_t(x - \tfrac{1}{2}k, x + \tfrac{1}{2}k).$$

This scheme [with $f_t(x) = m_t(x)/k$, the probability density, and $\gamma = k^2/h$] leads to the following approximation when h and k are sufficiently small and the usual regularity assumptions are made:

$$\frac{\partial}{\partial t} \int^x f_t(u)\, du = \gamma \left[\lambda \frac{\partial f}{\partial x} + (\lambda_1 - \lambda_2)f \right].$$

The behaviour of the system is highly dependent on the nature of $\lambda(x_1, x_2)$, that is on the detailed considerations which induce or affect displacement.

When there is no directional preference [$\lambda_1 = \lambda_2$], the equation takes the form,

$$\frac{\partial f}{\partial t} = \gamma \frac{\partial}{\partial x}\left(\lambda \frac{\partial f}{\partial x} \right),$$

and in a closed system $f \to$ a constant.

When movement is induced by repulsion [$\lambda_2 = 0$], the equation takes the form,

$$\frac{\partial f}{\partial t} = \gamma \frac{\partial^2}{\partial x^2}(\lambda f),$$

and in a closed system $f \to$ a constant$/\lambda$, so that population density tends to be low in places which are "repulsive".

When movement is induced by attraction $[\lambda_1 = 0]$, the equation takes the form,

$$\frac{\partial f}{\partial t} = \gamma \frac{\partial}{\partial x}\left(\lambda^2 \frac{\partial}{\partial x}\left(\frac{f}{\lambda}\right)\right),$$

and in a closed system $f \to$ a constant $\times \lambda$, so that the population tends to accumulate in those places which are "attractive".

The results of this exercise agree with everyday experience and common-sense. If we are primarily concerned with animal behaviour studies or the fine-scale pattern of density, it is clearly much better to construct the diffusion model in relation to the realities of the grass-roots situation than to borrow some simple model, such as $\partial f/\partial t \propto \partial^2 f/\partial x^2$, from physical science or mathematical text-books and trust in its applicability. Admittedly, in many large-scale studies, local fluctuations in the value of $\lambda_1 - \lambda_2$ (which may sometimes be positive and sometimes negative) may nevertheless have only a minor effect on the broad picture, and the simple equation might prove adequate.

6. The Role of Velocity

The model considered above, like most random walk schemes, throws no light on the part played by those conditions which affect the velocity of locomotion, for example, the firmness of the substratum or the penetrability of the medium. Nevertheless everyday experience and simple common-sense arguments lead us to expect that densities will be low where velocities are high, other things, of course, being equal.

In the straight-forward case where displacement is time-induced at evenly spaced time-points, the role of velocity is most easily seen by trans-forming the space variable. Imagine a simple random-walk on a chain of staging posts, which owing to the irregularities of the terrain are not evenly spaced in a physical sense but are nevertheless located to ensure that adjacent posts are exactly one "day's journey" apart. Let their locations be denoted by x_n ($n = 0, \pm 1, \pm 2, \ldots$). The process is conceived as being uniform and unbiassed with respect to the variable n (whereby position is expressed in "journey-days"). This scheme yields the well-known approximation,

$$\frac{\partial m}{\partial t} \propto \frac{\partial^2 m}{\partial n^2}.$$

By writing $dx/dn = v$ (velocity) and $f = m/v$ (so that probability density is expressed per unit of physical space, x) the differential equation is readily transformed to

$$\frac{\partial f}{\partial t} \propto \frac{\partial}{\partial x}\left(v\frac{\partial}{\partial x}(vf)\right).$$

In a closed system $f \to$ a constant$/v$ as expected.

In the more difficult case where displacements are purely space-induced and velocity varies according to the location, it can be shown that if stochastic branching occurs only at fixed points evenly arranged in a line and is un-biassed and dichotomous, the appropriate differential equation is

$$\frac{\partial f}{\partial t} \propto \frac{\partial^2}{\partial x^2}(vf).$$

In a closed system $f \to$ a constant$/v$ as before, and as expected.

It may seem unrealistic to a biologist to regard a population of diffusing animals as homogeneous. Nevertheless it is interesting to note that if the population were a mixture of classes $[f(x, t) = \sum_i f_i(x, t)]$ each with its own characteristic velocity $[v_i(x, t)]$, the differential equation given above would still hold for the whole population provided that $v(x, t)$ were defined to be the mean velocity over all individuals at (x, t).

7. Seasonal Periodicity

When reproduction and dispersal are confined to only a fraction of the annual seasonal cycle, there are two contrasting ways of simplifying the picture for the sake of mathematical convenience, and both can be justified when they both lead to substantially the same conclusions. In both cases the time-scale is distorted, in one case by condensing the active phase into an instant and in the other by extending it to the whole year. There is, of course, nothing sacrosanct about physical time, and as far as biological activity is concerned, one day in summer may be worth far more than a week in winter.

The second procedure brings dispersal in a particular season into line with diffusion in continuous time. From a purely mathematical point of view the two alternatives are related in a well-known way, comparable to the relation between the factorial function and the gamma function. One can be defined by a whole number of convolutions occurring in discrete time, and the other is an extension of the discrete system to non-integer values.

8. Classical Diffusion

Figure 9 is designed to describe the progress and outcome (but not the mechanism) of a diffusion process which proceeds in accordance with classical conceptions. When conditions are uniform and progress unrestricted, the cumulants of the resulting distribution increase linearly with time.

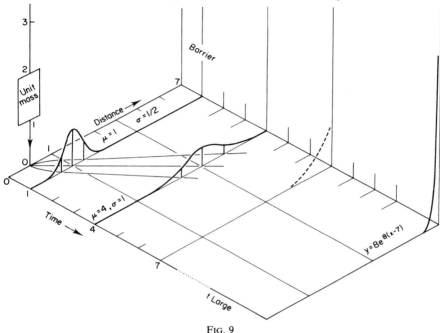

FIG. 9

When displacements are brought about by eddy currents in the atmosphere or in the oceans, the classical conception of diffusion is often useful but it is not entirely satisfactory as meteorologists well know.

One of the most helpful ways of conceiving and representing classical diffusion processes is by means of a succession of convolutions which are closely spaced in time (see Fig. 10). The equivalent differential equation can be derived from the integral expression by limit processes as follows.

Let $p(u, X, t, \Delta)\, du$ be the probability that a particle at position X at time t will be displaced by an amount $u \pm \frac{1}{2}\, du$ by time $t + \Delta$. Then

$$f(x, t + \Delta) = \int_{-\infty}^{\infty} f(x - u, t)p(u, x - u, t, \Delta)\, du.$$

When f and p as functions of x can be described to an acceptable degree of approximation by entire functions (e.g., exponential of a polynomial of

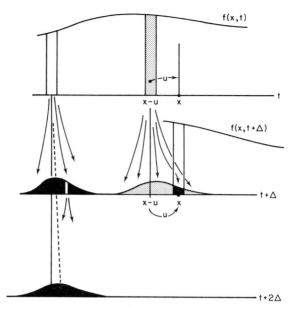

FIG. 10

very high degree) we may employ Taylor's theorem. Symbolically,

$$f(x, t + \Delta) = \int_{-\infty}^{\infty} e^{-u(D_1 + D_2)} p(u, x_2, t, \Delta)\, du \cdot f(x_1, t) \Big|_{\substack{x_1 = x \\ x_2 = x}}$$

$$= \sum_{r=0}^{\infty} \frac{(-)^r}{r!} (D_1 + D_2)^r \mu'_r(x_2, t, \Delta) f(x_1, t) \Big|_{\substack{x_1 = x \\ x_2 = x}}$$

If now $\kappa_r(x, t, \Delta)/\Delta \to \kappa_r(x, t)$ as $\Delta \to 0$, then $\mu'_r(x, t, \Delta)/\Delta$ has the same limit. Hence

$$\frac{\partial f}{\partial t} = \sum_{r=1}^{\infty} \frac{(-)^r}{r!} \frac{\partial^r}{\partial x^r} \{\kappa_r(x, t) f(x, t)\}.$$

Here the $\kappa_r(x, t)$ may be interpreted as the initial instantaneous rates of growth of the cumulants of the distribution generated by the diffusion process from a point-mass at (x, t).

Incidentally, owing to the realities of density dependence, the differential equation is only linear in appearance and is perhaps better regarded as a non-linear equation in disguise. The coefficients $\kappa_r(x, t)$ are then conceived as $\kappa_r(x, t, f)$, with the understanding that in determining the κ_r repeated use is made of auxilliary equations and that integration is effected in small successive steps.

All diffusion is not normal. Indeed the evidence that dispersal commonly leads to leptokurtic distributions is overwhelming. The higher even-order cumulants may not matter much in most applications, but their role in accelerating animal invasions or the spread of genes might be far from negligible.

9. The Concept of Dynamic Level

A statistician living in an etherial world of probability distributions might consider that the normal diffusion equation,

$$\frac{\partial f}{\partial t} = -\frac{\partial}{\partial x}(\kappa_1 f) + \frac{1}{2}\frac{\partial^2}{\partial x^2}(\kappa_2 f),$$

is already full of meaning, and that the over-riding objective is to obtain the mathematical solution. A biologist, however, in his search for systematizing concepts in his field of thought and experience might find an alternative form far more rewarding. By writing $\kappa_1/\kappa_2 = \frac{1}{2}(\partial/\partial x) \log \phi$, we obtain,

$$\frac{\partial f}{\partial t} = \frac{1}{2}\frac{\partial}{\partial x}\left\{\phi\frac{\partial}{\partial x}\left(\frac{\kappa_2 f}{\phi}\right)\right\}.$$

Compare this with the equations of heat conduction in a non-uniform bar.

$$\frac{\partial \theta}{\partial t} = \frac{1}{\mathscr{C}}\frac{\partial}{\partial x}\left\{\mathscr{K}\frac{\partial \theta}{\partial x}\right\},$$

$$\frac{\partial H}{\partial t} = \frac{\partial}{\partial x}\left\{\mathscr{K}\frac{\partial}{\partial x}\left(\frac{H}{\mathscr{C}}\right)\right\}.$$

In these equations θ denotes temperature and H denotes the amount of heat per unit length. $\mathscr{K}$ = Conductivity of material × Cross sectional area; $\mathscr{C}$ = Specific heat × Density of material × Cross sectional area. We bear in mind that it is the heat and not the temperature which diffuses.

If now we are prepared to make a straight analogy with temperature, and call the whole quantity, $\kappa_2 f/\phi$, the "dynamic level", we can then assert that animals diffuse from high to low dynamic levels, and that in an ecological system insulated against migration and closed to birth and death, the same dynamic level tends to be established everywhere. The level is raised by birth and lowered by death. Increased activity or increased velocity enhance diffusivity and thereby raise the level. When in addition there is a mass tendency to drift outwards by biassed responses which indicate the unattractiveness of a locality and are reflected in the relatively low value of ϕ, the dynamic level is again increased.

It may be noted that the difference between the values of ϕ at two points depend on the properties of the whole interval lying between them. It relates behaviour locally to the whole field. In this respect it shows some analogy to the notion of potential in a gravitational or electrical field.

The concept of "dynamic level" helps to explain commonplace diffusionary phenomena in terms of familiar ecological ideas. Furthermore, by making use of the so-called Weber–Fechner "Law",

$$\delta\Psi \propto \delta\Phi/\Phi = \delta \log \Phi$$

(where Ψ is a psychological effect or physiological or behavioural response and Φ is the intensity or concentration of the stimulus), we are led immediately to an interpretation of the field function ϕ which converts ϕ in effect into a resultant of all stimuli which evoke directional response. These are reflected in the value of κ_1. Both a gradient and a base line are needed for the detection of the difference between stimuli impinging from opposing sides. The base-line tends to be widened automatically when κ_2 is raised, as for example when a blow-fly is activated by the "smell" of carrion regardless of the location of the source.

10. Invasions

We now consider a class of models in which diffusion and growth occur simultaneously.

Figure 11 illustrates R. A. Fisher's conception (1937) of the increase in gene frequency of an advantageous mutation and its spread in a linear

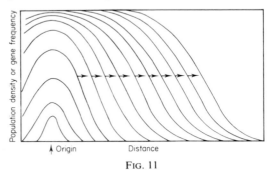

FIG. 11

habitat throughout which population density is uniform. Since there are two "wave-fronts", all future remarks will refer to that which slopes down-wards to the right. Fisher's equation may be written in the form

$$\frac{\partial y}{\partial t} = k\frac{\partial^2 y}{\partial x^2} + sy(1 - y),$$

where y denotes gene frequency $(0 < y < 1)$, k diffusivity, and $s > 0$ selective advantage. There are many possible variants depending on dominance, inbreeding and the dimensionality of the space.

The analytical approach is difficult, due in part to an unpleasant singularity associated with the leading edge of the "wave" and in part due to the indeterminacy of the velocity. Fisher conjectured that asymptotically the wave has constancy of form $[y = f(x - vt)]$ with velocity $v = 2\sqrt{(sk)}$, the minimum logically possible.

The problem can be approached numerically with the aid of a computer by replacing the differential equation by an appropriate finite-difference equation, for example,

$$y_{t+1}(x) = \tfrac{1}{6}[y_t(x - 1) + 4y_t(x) + y_t(x + 1)] + \tfrac{1}{10}y_t(x)[1 - y_t(x)].$$

When arguments and conjectures analogous to those of Fisher are used in this case, the supposed asymptotic velocity is given by the minimum value of

$$\frac{1}{\lambda}\log\left[\frac{2 + \cosh \lambda}{3} + \frac{1}{10}\right], \qquad (\lambda > 0).$$

The minimum value (0·24081 when $\lambda = 0·789136$) is somewhat less than the corresponding value (0·25820) for the differential equation with $k = 1/6$ and $s = 1/10$.

Computer runs carried out by M. D. Mountford more than 15 years ago revealed the interesting fact that the speed of the wave is greatly affected by rounding-off errors. The position reached in a given time using 2, 4 and 8 decimal place accuracy are compared in Fig. 12. Even between 12 and 27 decimal place accuracy there is a noticeable difference (Fig. 13).

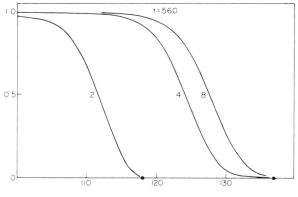

FIG. 12

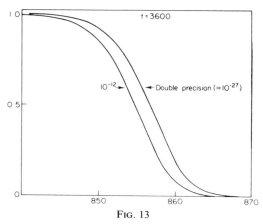

FIG. 13

The approach to the asymptotic state is slow (Fig. 14). Extrapolation by procedures which seem appropriate and acceptable yield results which convince me personally that Fisher's mathematical conjectures were correct.

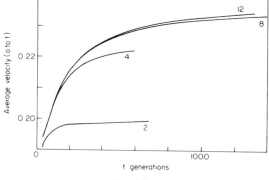

FIG. 14

Rounding-off errors blunt the leading edge of the wave and slow it down (Fig. 15). This observation has genuine biological relevance because in real populations stochastic errors would blunt the edge almost beyond recognition. Perhaps there is a rough analogy here with friction.

The use of continuous functions to represent population densities distorts the facts slightly because populations are not infinitely divisible. The issue is only side-stepped when we speak of theoretical values and perhaps think of them as being akin to expected values. The point at first seems trivial, but when we are dealing with the fringes of waves or the tails of distributions in realistic terms we find that the effective ranges are often small. For example, if the thousands of millions of people in the world

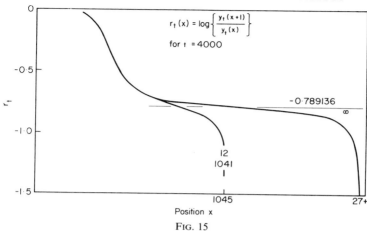

FIG. 15

today were transported to Antarctica and scattered independently around the South Pole in radial normal fashion with a root-mean-square deviation of 1 kilometre, so as to form a huge mound some hundreds of metres high, it could nevertheless be asserted beforehand with a probability that falls short of absolute certainty by less than one part in a hundred thousand million that no single individual would be located outside a radius of only 7 km.

In realistic terms invasions are started and maintained by the pioneers that actually materialize, not the pioneers that are expected. There is evidently a need for a full stochastic treatment of the subject. Excessive faith in the applicability of the mathematical concepts that we invent or choose to employ leads not to enlightenment but only to deception.

It is, of course, just as easy to be deceived by intuitive thinking. When we come to consider a more realistic version of Fisher's model, it seems intuitively plausible that a mutant gene arising near the edge of the geographical range of the species could easily be swamped, despite its selective advantage, by the influx of wild-type genes diffusing away from centres of high population density, and that where population density varies greatly from place to place (without being absolutely zero in between) the gene might not readily spread up steep gradients. How steep the gradients would have to be or how great the diffusivity or how small the selective advantage is not immediately apparent.

In this type of problem, where a purely empirical investigation would be highly laborious and extremely slow, there is no better way of clarifying our ideas than by formulating the model mathematically.

By bearing in mind that the elements which actually diffuse are living organisms and not ratios, equations which represent the relative sizes of the components of a competitive system of diffusing elements when no artificial restriction is imposed on the total population density may be derived as follows.

Two Competing Species
Let $A(x, t)$ and $B(x, t)$ denote the population densities of two species with the same diffusivity and let $r_A(x, t)$ and $r_B(x, t)$ denote their compound interest growth rates locally at those densities. The total population $N = A + B$, $A = yN$ and $B = (1 - y)N$. If

$$\frac{\partial A}{\partial t} = \frac{\sigma^2}{2} \frac{\partial^2 A}{\partial x^2} + Ar_A$$

and

$$\frac{\partial B}{\partial t} = \frac{\sigma^2}{2} \frac{\partial^2 B}{\partial x^2} + Br_B,$$

then

$$\frac{\partial y}{\partial t} = \frac{\sigma^2}{2} \left\{ \frac{\partial^2 y}{\partial x^2} + 2\eta \frac{\partial y}{\partial x} \right\} + sy(1 - y),$$

where

$$\eta = \frac{\partial}{\partial x} \log N(x, t) \quad \text{and} \quad s = r_A - r_B.$$

Two Competing Autosomal Alleles
A, B, C stand respectively for the population densities of the genotypes $[a, a]$, $[a, a']$, $[a', a']$ in the breeding population, all classes having the same mobility and mortality when adult. Selection takes place prior to recruitment into the breeding population, and Hardy–Weinberg proportions apply at fertilization. $N = A + B + C$ and $y = (2A + B)/(2N)$. If

$$\frac{\partial A}{\partial t} = \frac{\sigma^2}{2} \frac{\partial^2 A}{\partial x^2} - \mu A + \gamma_A y^2 N,$$

$$\frac{\partial B}{\partial t} = \frac{\sigma^2}{2} \frac{\partial^2 B}{\partial x^2} - \mu B + \gamma_B 2y(1 - y)N,$$

$$\frac{\partial C}{\partial t} = \frac{\sigma^2}{2} \frac{\partial^2 C}{\partial x^2} - \mu C + \gamma_C(1 - y)^2 N,$$

then

$$\frac{\partial y}{\partial t} = \frac{\sigma^2}{2} \left\{ \frac{\partial^2 y}{\partial x^2} + 2\eta \frac{\partial y}{\partial x} \right\} + y(1 - y)[s_1 y + s_2(1 - y)],$$

where

$$s_1 = \gamma_A - \gamma_B \quad \text{and} \quad s_2 = \gamma_B - \gamma_C.$$

As before

$$\eta = \frac{\partial}{\partial x} \log N.$$

When the diffusion component is written in the form

$$\frac{\partial y}{\partial t} = \frac{\sigma^2}{2} \frac{1}{N^2} \frac{\partial}{\partial x}\left(N^2 \frac{\partial y}{\partial x}\right),$$

an analogy can be drawn with heat conduction. At higher population densities both the "conductivity" and the "capacity" are greater. After all, it is the population which both carries and absorbs the genes.

In order to make comparisons with Fisher's model [for which $\eta = 0$] we take $s_1 = s_2 = s$, a constant. In Figs. 16 and 17, which show changes in

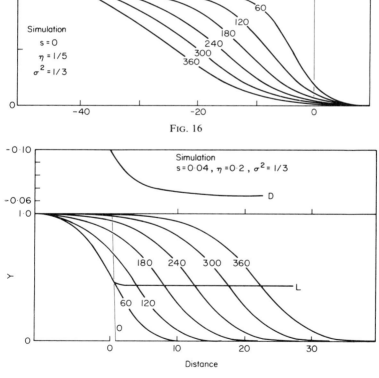

FIG. 16

FIG. 18

gene frequency, it is to be understood that population density is high on the right and falls off quickly and exponentially to the left. For the specific purpose of assessing the roles of opposing factors, the initial form given to the wave was steep and straight.

In the case illustrated in Fig. 16 the gene enjoys no selective advantage or disadvantage and is swept down the gradient as expected. Figure 17 illustrates a case where a gene with a moderate advantage successfully advances up a steep gradient. In actual cases everything depends on the precise values of the coefficients which apply. In this system (which tacitly implies extremely high densities even though they may vary greatly), the asymptotic velocity of advance is given by $v/\sigma = \sqrt{(2s)} - \sigma\eta$, at least in the case $(\sigma\eta)^2 < 2s$, for at the moment the situation when $(\sigma\eta)^2 > 2s$ is somewhat obscure.

Figure 18 refers to a linear habitat in which population density fluctuates spatially and is 22 thousand times denser at the peaks than at the troughs.

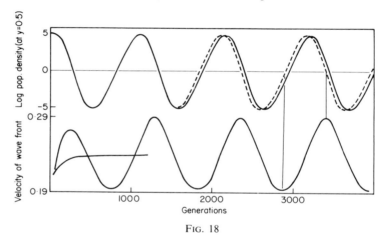

FIG. 18

The velocity of the wave is about half as fast again on the down-slopes as on the up-slopes. Even so, the form assumed by the wave-front changes surprisingly little. Furthermore the wave progresses through the spatially fluctuating population almost as quickly as it does when population density is constant (Fig. 19).

11. Concluding Remarks

Mathematical model-making in Population Dynamics looks easy, and so it is if we do not care about the uses and abuses to which our models may be put.

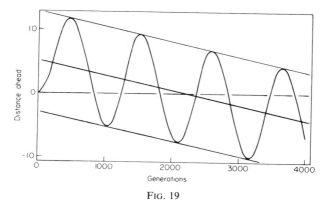

FIG. 19

In theory all that we have to do is to incorporate those things that matter and to ignore those things that do not matter, if we can, or if we can't, treat them in any arbitrary way that suits our mathematical convenience. By following the reassuring principle that conclusions hold more often than do the detailed premises, we can always entertain the prospect that results of wider applicability will emerge.

The snag is that it is not given to anyone to know beforehand which features matter and which do not, and in a real life context all sorts of things matter in varying degrees. The best we can do is to begin with schemes that may not be entirely realistic but which are at least meaningful to others as well as to ourselves.

It is important to scrutinize the models carefully and to explore numerous variants in order to allay fears and doubts created by the many artefacts we invariably introduce and glorify in the name of abstraction.

Finally we must seek empirical verification wherever and whenever possible, and in particular determine by empirical means the true range of applicability of the theoretical schemes and speculations that we have been bold enough to propose.

REFERENCES

FISHER, R. A. (1937). The Wave of Advance of Advantageous Genes. *Annals of Eugenics* **7**, 335–369.
PEARSON, K. (1906). Mathematical Contributions to the Theory of Evolution, XV Mathematical Theory of Random Migration. *Drap. Co. Res. Mem. biom. Ser.*, no. 3, 1–54.
SKELLAM, J. G. (1951). Random Dispersal in Theoretical Populations. *Biometrika* **38**, 196–218.

5. Food Distribution, Searching Success and Predator–Prey Models

G. Murdie and M. P. Hassell

Department of Zoology and Applied Entomology,
Imperial College, London, England

This paper is primarily concerned with insect predators or parasites (= parasitoids) and their prey or hosts, but our discussion may also have applications to other predatory animals. We shall refer to parasites and predators collectively as "natural enemies" and only refer specifically to each where it is necessary to distinguish between them. Although predator–prey and parasite–host interactions are similar because they involve common components, for example handling time (Holling, 1959) and searching efficiency, there are important differences in their biology which must be considered when developing population models. Firstly, the number of hosts parasitized defines parasite reproduction, whereas, the number of prey eaten only loosely determines the number of eggs laid by a predator. When few prey are eaten there may be a close relationship between the number eaten and predator fecundity, but the relationship will become less well defined when the intake of food approaches the amount needed for complete egg maturation. A second major difference is that the adult insect parasite is normally the only stage which searches for hosts, whereas all the larval stages, as well as the adults, of predators are often predatory, each having different searching characteristics. These features are important since all so-called predator–prey models to date deal with a single searching stage and assume that the number of progeny produced is dependent on the number of prey found. They are therefore, more descriptive of host–parasitoid interactions, even though the attack equations may be applicable to both in specific cases.

Our main concern in this study is to consider the importance of the spatial distributions of interacting populations of natural enemies and their prey. A major criticism of most "predator–prey" models is that they do not consider the relative distributions of the interacting species (Lotka, 1925; Volterra, 1928; Thompson, 1924; Nicholson and Bailey, 1935; Watt, 1959; Hassell and Varley, 1969). They assume that the natural enemies search at random with respect to their food species, implying equal risk of attack to each individual prey. The concept of random attack has probably been retained for its mathematical simplicity rather than for its biological realism. This is in spite of the fact that it is now widely recognized that most animal species are not randomly dispersed throughout their habitat (the negative binomial and other contagious distributions have successfully described the dispersion of many animal species (Bliss and Fisher, 1953; Neyman, 1939; Anscombe, 1949; Bliss, 1971)). We may expect, therefore, to find many natural enemies whose distribution is also non-random and numerically correlated with that of their prey due to differential aggregation where the prey are more numerous. This would have an obvious selective advantage; the predators will obtain more food or obtain their food more easily while the parasites would leave more progeny (Murdie, 1971, Hassell, 1971a). The evidence for such responses to prey distributions is steadily increasing in the literature although it is usually indirect and based on behavioural mechanisms which would logically lead to non-random distributions (see below).

There are few cases where the natural enemy distributions have been quantified (Hassell, 1971b). This is unfortunate since some recent work has demonstrated that spatial responses are important components of population models (Hassell and Rogers, 1972; Hassell and May, 1973). It was shown that the stability of two interacting populations can be affected as a result of the aggregative behaviour of natural enemies.

We illustrate this here using the random attack model described by Nicholson and Bailey (1935):

$$N_{t+1} = FN_t \exp{(-aP_t)} \tag{1}$$

where N = the host density in generations t and $t + 1$; P_t = the density of the searching parasites; F = the host rate of increase per generation; and a = the "area of discovery", a constant representing the searching efficiency of the parasite species. The subsequent parasite progeny are obtained directly from the number of hosts attacked. Typically this model produces unstable populations leading to the extinction of one or other of the species (Fig. 1a).

Now, rather than assume two homogeneous populations as in this model, let us distribute the hosts contagiously within n units of habitat and allow the searching parasites to aggregate differentially in response to the host density in each unit. Equation (1) now becomes:

$$N_{t+1} = FN_t \sum_{i=1}^{n} [\alpha_i \exp{(-\alpha\beta_i P_t)}] \tag{2}$$

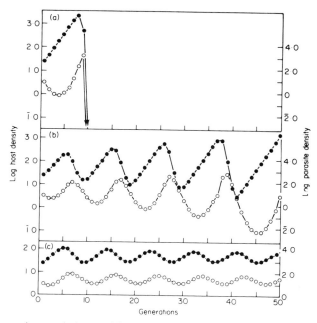

Fig. 1. Host-parasite population models. (a) Typical unstable Nicholson–Bailey model (area of discovery = 0·108; host rate of increase = 2·0). (b) As model (a) but hosts distributed contagiously and searching parasites distributed in proportion to host density per unit area in each generation. (c) As model (b) but parasite distribution supraproportional to host density. (*Note*—parasite numbers plotted on half scale.)

where α_i and β_i are the proportions of the total hosts and parasites respectively in the n unit areas (Hassell and May, 1973). This simple modification has been used to produce the population models in Figs. 1b and c, which differ only in the strength of the parasites' response to host density. The approach towards stability is marked.

There are two behavioural mechanisms which probably account for most examples of differential aggregation by insect natural enemies in areas of prey populations. In certain cases there is long range attraction in response to some volatile host product, for example as shown by various parasites and predators of bark beetles (Wood, 1972). The second mechanism causes the insect to remain in favourable areas; that is a restriction on re-dispersal. This is illustrated by Figs. 2a and b, which are examples of the paths taken by two insect parasites, a Braconid wasp, *Diaeretiella rapae*, (Curt.) and a Tachinid Fly, *Cyzenis albicans*, (Fall) searching in arenas. Both show increased turning rates after finding an aphid host or a sugar droplet respectively. (*Cyzenis* usually oviposits near sap fluxes on leaf surfaces,

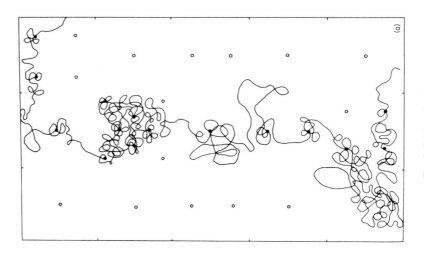

FIG. 2. Paths taken by (a) *Cyzenis albicans* in an arena with sugar droplets, and (b) *Diaeretiella rapae* in an arena with its aphid host (*Brevicoryne brassicae*). (Redrawn from Hafez, 1961.)

parasitism occurring when an egg is eaten by a winter moth caterpillar.) Any such aggregation by a natural enemy population is likely to lead to density dependent prey mortality (i.e. proportionate mortality increasing with density). For example, it is known that the highest incidence of winter moth parasitism by *Cyzenis* tends to occur on trees with the highest host larval densities (Hassell, 1968) (Fig. 3).

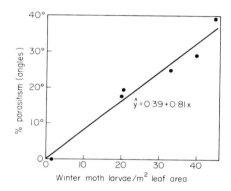

FIG. 3. The relationship between parasitism by *Cyzenis albicans* and winter moth density per unit leaf area for six trees at Wytham Wood (After Hassell, 1968).

Our aim in this study has been to analyse this kind of searching behaviour which involves increased turning rates associated with feeding. We have chosen the common housefly (*Musca domestica* (L.)) searching for sugar droplets in an arena as our experimental system since the housefly shows similar characteristic turning movements. Figures 4a and b show typical housefly tracks within the arena. We hoped to identify from such tracks easily measured behaviour patterns that are fundamental to this kind of searching. These characters could then be looked for in other species and perhaps used to characterize their searching movements in the vicinity of their hosts. We also intend to develop a model from the experimental results that will be used to simulate the fly's movements and which can be used to investigate further the effect of non random distributions on exploitation rates and feeding successes.

Methods

The experimental arena was perspex, 47 cm square and 2·5 cm deep with a fitted lid. Small droplets of sugar solution (10 % sucrose plus a little soluble starch), approximately 1 mm diameter, were distributed either evenly or in clumps of eight on the floor of the arena (Fig. 4). The densities chosen were 16, 32, 64 and 128 clumped droplets and 16, 36, 64 and 121 evenly spaced

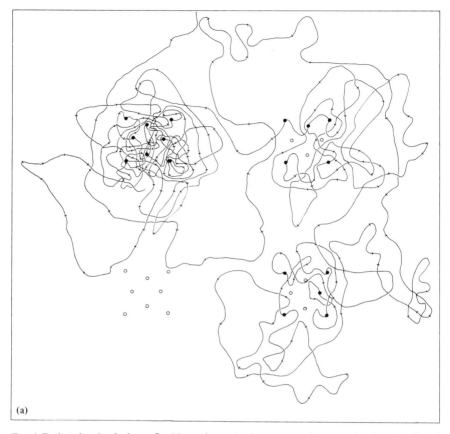

(a)

FIG. 4. Paths taken by the housefly, *Musca domestica*, in an arena with sugar droplets distributed (a) clumped or (b) evenly.

droplets with five replicates of each. The arena was evenly illuminated and held at 20°C.

Prior to each trial freshly emerged flies were fed for 24 hours on honey solution and were then starved for 24 hours. In each trial the fly was released into the arena and all walking movements of the fly on the floor of the arena were traced on the lid. Three-second time intervals were marked on the trace as well as the time spent resting, cleaning and feeding. Any flight interrupted the trace and a new one was started each time the fly resumed walking. The trials were terminated when the fly rested for prolonged periods or made frequent attempts at flight.

The details recorded for computer analysis were:

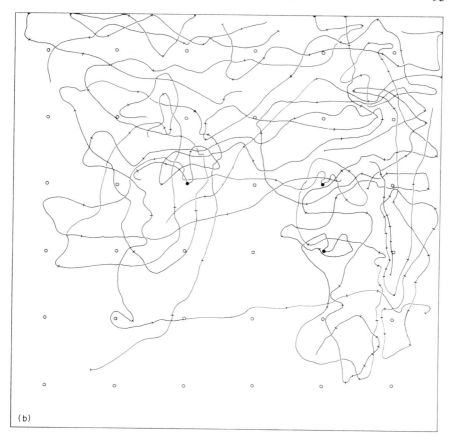

(b)

(i) co-ordinates of the fly's position at each 3-second time mark and at
     two intermediate positions considered as 1-second marks (assuming
     a constant speed in each 3-second period);
(ii) the track distance between 3-second marks;
(iii) time spent (a) feeding, (b) resting and (c) cleaning.
All these observations were made sequentially from start to finish of each
trial.

Analysis

The first stage in the analysis was to consider frequency of angles turned
and distance moved per unit time. The angles turned between 1-second
marks were computed and found to be distributed symmetrically about
zero mean (Fig. 5a). On this evidence a series of normal distributions of

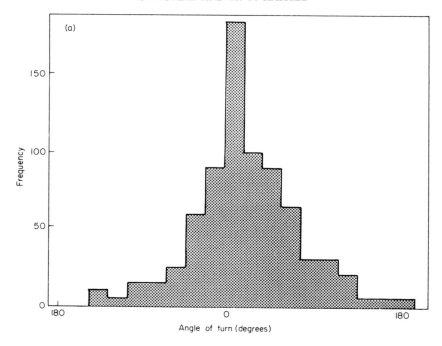

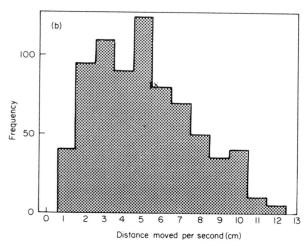

FIG. 5. Examples of frequency distribution of (a) angle turned and (b) distances moved per second by a housefly in a single arena experiment.

angles turned was used in the development of the simulation model. On the other hand the zero censored distribution of distances moved (Fig. 5b) is clearly asymmetrical, and has not been used in the simulation model.

These preliminary analyses of distributions do not discriminate between pre-and post-feeding patterns which are so obviously different (Fig. 4). Figure 6 shows the clear division between these pre- and post-feeding

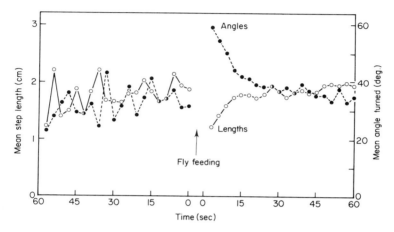

FIG. 6. Mean step lengths per second and angles turned in the 60-second periods before and after feeding. (Data pooled from 29 housefly experiments.)

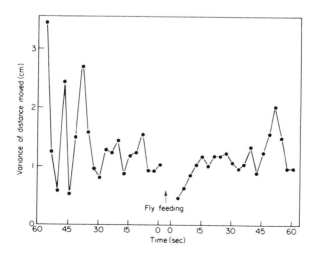

FIG. 7. Variance of step length per second in the 60-second periods before and after feeding. (Data pooled from 29 houseflies.)

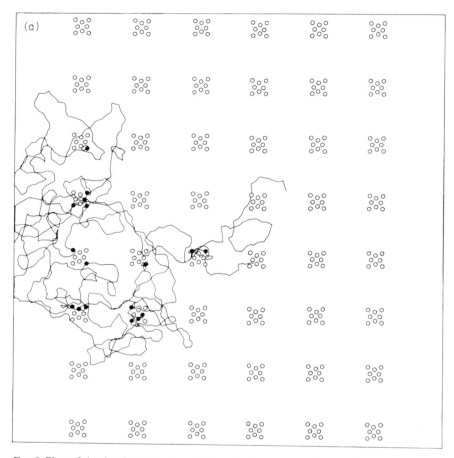

FIG. 8. Plots of simulated movements of a 'housefly' in an arena with sugar droplets distributed (a) clumped and (b) evenly. ● sugar droplets removed by fly.

phases for angles turned and distances moved. The plotted means for 20 three-second intervals after feeding are clearly time-dependent. Prior to feeding there is considerable variation in both the distances and turns which have no obvious pattern. Thus, immediately after feeding there is a sharp increase in the mean angle turned and a sharp decrease in the distance moved per second. Both responses act to keep the fly in the immediate area of feeding, increasing the probability of additional discoveries, when the food is clumped. The mean turning rate subsequently falls exponentially while the rate of movement increases exponentially until the pre-feeding patterns are resumed. Figure 7 shows another interesting feature of these

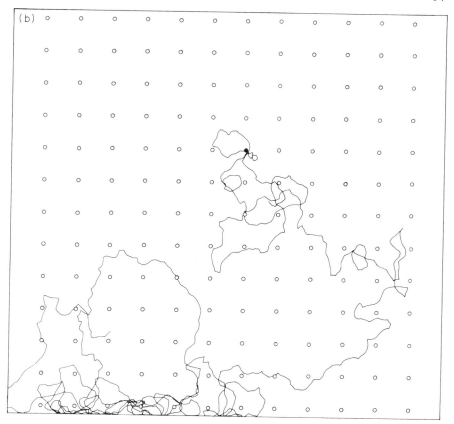

relationships. The variances of the distance moved are extremely variable prior to feeding. Immediately after feeding not only are the distances moved decreased but they are also restricted within smaller limits of variation and thus the distances moved are more rigidly controlled. These modified rates of turning and movement occupy only about 30 seconds. This short period strongly suggests that there is a central nervous programme which is initiated by feeding (Dr. John Brady, pers. comm.). The two functions of turning and movement after feeding are very important in determining the turning pattern. This was shown by a simple simulation model based only on these two relationships. Such a model, if made more realistic by the inclusion of satiation and handling time, could be used to investigate the different outcomes of search for random and non random prey.

The simulation model so far developed is a combination of deterministic and stochastic sub-models. Analytical functions to describe time-dependent responses will be developed when more information and analyses are available.

The "sugar droplets" were placed such as to mimic the actual spatial distributions in the experimental arenas. The mean angles and distances moved in any 1-second period were read from eye-fitted curves to the post-feeding data. The time elapsed from feeding determined the distance travelled. The asymptotic value was assumed for all steps before feeding and from 100 seconds after feeding (Fig. 6). The degree and direction of turn was obtained by random selection of an angle from a set of distributions of normal form (truncated at $\pm 2\pi$ radians) with zero means and variances which were time-dependent. Again asymptotic values were assumed for the pre-feeding and 100-second plus post-feeding values. If a "sugar droplet" was encountered in moving it was removed by the "predator" and the next movement originated from that point. The simulated traces and feeds were recorded graphically using a Calcomp plotter. Figure 8 shows the results of two such simulations for a clumped and even droplet distribution. Although the model used is very simple it produces fairly realistic tracks and clearly this justifies the use of two functions describing movement before and after feeding. Thus, although the outcome of the movements appears non-random the tracks can be produced by a series of random steps where the directions and distances have means and variances determined by feeding points in a time-dependent model.

It seems likely that this method of controlling movement is widespread among insect natural enemies. If this control is through a central nervous programme, then we might expect such functions to be species specific representing adaptations of particular predators to their prey distributions. It is possible that this programme will depend, not only on the average clump size, but also on the mean distance between individuals in the clump since this will determine the most effective turning pattern of the predator.

To the population biologist this behaviour is of interest because it shows that the outcome of predator search will be influenced by prey distribution as well as prey density. It is, therefore, surprising that most classical predator–prey models disregard the effect of both predator and prey distribution. That distribution is extremely important is illustrated by the population model described earlier in this paper, and by the results of the fly experiments. The form of the relationship between feeding rate and density (functional response) is markedly different when the flies search for clumped rather than evenly distributed sugar droplets (Fig. 9). In the even distribution we have a linear response, whereas the relationship is strongly curvilinear over the same density range for the clumped distribution. It is clear that flies are adapted to search for aggregated food particles since they are more efficient when food is clumped, except at the very highest density where the evenly

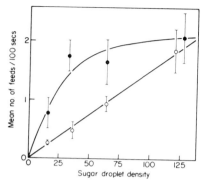

FIG. 9. Relationships between mean number of observed feeds by houseflies and sugar droplet density. ○ even distribution; ● clumped distribution of sugar droplets.

distributed sugar droplets are barely separated.

Functional responses for insect natural enemies have been frequently reported in the literature (usually for insect parasites) and several models have been proposed to describe these data; for example (Holling, 1959)

$$\frac{N_a}{P} = \frac{a'T_tN}{1 + a'T_hN} \tag{3}$$

and if exploitation is included (Rogers, 1972)

$$N_{ha} = N\left[1 - \exp\left(-\frac{a'T_hP}{1 + a'T_hN}\right)\right] \qquad \text{(for parasitism)} \tag{4}$$

$$N_{ha} = N[1 - \exp(-a'P(T_t - N_aT_h))] \qquad \text{(for predation)} \tag{5}$$

where N is the host density; N_a is the total encounters with hosts; N_{ha} is the number of hosts attacked; T_t is the total time available for search; T_h is handling time and a' is the coefficient of attack (an instantaneous rate).

These models depend upon search being random. They should not be used to describe the outcome of search where the natural enemies show aggregative behaviour in response to prey distribution. The importance of the searching patterns we have discussed here is that they provide a mechanism by which a natural enemy can spend more time searching in areas where prey density is relatively high. In such cases natural enemy–prey models require an extra level of complexity. Initially, the meaningful units of natural distributions need to be defined in terms of the natural enemy's behaviour. That is, the sampling units for prey should be subunits of the total area searched by individual natural enemies. The random attack models can now be used to describe exploitations, but only when the searching population is distributed between the different units of habitat considered, in response to the prey density in each. The model will now be based on a series of sub-units having a range of prey and predator densities. The total number of prey eaten will then be the sum of the numbers eaten in the separate sub-units (eqn. (2)). This outcome will differ from the situation where a given number of predators and prey interact through a single random attack model.

Most functional response experiments have used prey that have been evenly distributed. However, true values for the searching rate will be quite different if the prey are naturally clumped and the predators show an aggregative response. It is clearly important that experiments to estimate searching parameters should use prey distributions that are as natural as possible.

Summary

The stability of predator–prey population models is increased when the predator distribution is directly related to prey density per unit area. This is illustrated using simple population models.

The common housefly, *Musca domestica* (L.) shows increased turning rates after an encounter with food, very similar to those found in many insect predator and parasitoid species. The movements of the housefly in an area with various sugar droplet distributions were analysed. They showed that the turning behaviour could be explained by two exponential functions relating turning rate and distance moved per second to the time elapsed after feeding.

This behaviour increased the time spent by the housefly in areas with clumped sugar droplets. The resulting functional responses were very different depending on the sugar distribution. Flies were more efficient when searching for clumped droplets except at the highest droplet density.

The application of random attack equations where prey are distributed unevenly are discussed.

Acknowledgements
We are greatly indebted to Miss K. Pearce-Shorten for her help in carrying out the experiments and recording the results. The experiments were developed from a class exercise devised by Professor G. C. Varley to whom we are grateful. Computing was done on the University of London CDC 6600 via a link provided by the Office of Resource and Environment in the Ford Foundation.

REFERENCES

ANSCOMBE, F. J. (1949). The statistical analysis of insect counts based on the negative binomial distribution. *Biometriks* **5**, 165–173.

BLISS, C. I. (1971). The aggregation of species within spatial units. *In* "Statistical Ecology," Vol. 1. (G. P. Patil, E. C. Pielou, W. E. Waters, eds.), pp. 311–335. Penn. State Statistical Series.

BLISS, C. I. AND FISHER, R. A. (1953). Fitting the negative binomial distribution to biological data. *Biometrics* **9**, 176–200.

HAFEZ, M. (1961). Seasonal fluctuations of population density of the cabbage aphid *Brevicoryne brassicae* (L.) in the Netherlands, and the role of its parasite *Aphidius* (*Diaeretiella*) *rapae* (Curtis). *T. Pl.-Ziekten* **67**, 445–548.

HASSELL, M. P. (1968). The behavioural response of a tachinid fly (*Cyzenis albicans* (Fall.)). To its host, the winter moth (*Operophtera brumata* (L.)). *J. Anim. Ecol.* **37**, 627–639.

HASSELL, M. P. (1971a). Mutual interference between searching insect parasites. *J. Anim. Ecol.* **40**, 473–476.

HASSELL, M. P. (1971b). Parasite behaviour as a factor contributing to the stability of host-parasite interactions. *Proc. Adv. Study Inst. Dynamics Numbers Popul.* (Oosterbeek, 1970), 366–379.

HASSELL, M. P. AND MAY, R. M. (1973). Stability in insect host-parasite models. *J. Anim. Ecol.* **42**, (In press).

HASSELL, M. P. AND ROGERS, D. T. (1972). Insect parasite responses in the development of population models. *J. Anim. Ecol.* **41**, 369–383.

HASSELL, M. P. AND VARLEY G. C. (1969). A new inductive population model for insect parasites and its bearing on biological control. *Nature (Lond.)* **223**, 1133–1137.

HOLLING, C. S. (1959). Some characteristics of simple types of predation. *Can. Ent.* **91**, 385–398.

LOTKA, A. J. (1925). "Elements of Physical Biology." Dover Publications, New York.

MURDIE, G. (1971). Simulation of the effects of predator/parasite models on prey/host spatial distribution. *In* "Statistical Ecology," Vol. 1. (G. P. Patil, E. C. Pielou, W. E. Waters, eds.) pp. 215–233. Penn. State Statistical Series.

NEYMAN, J. (1939). On a new class of "contagious" distributions applicable in entomology and bacteriology. *Ann. Math. Stat.* **10**, 35–57.

NICHOLSON, A. J. AND BAILEY, V. A. (1935). The balance of animal populations. *Proc. Zool. Soc. Lond.* Part 1, 551–598.

ROGERS, D. (1972). Random search and Insect Population Models. *J. Anim. Ecol.* **41**, 369–383.

THOMPSON, W. R. (1924). La théorie mathématique de l'action des parasites entomophages et le facteur du hasard. *Annls. Fac. Sci. Marsailles* **2**, 68–89.

VOLTERRA, V. (1928). Variations and fluctuations of the number of individuals in animal species living together. *J. Cons. perm. int. Expl. Mer.* **3**, 3–51.

WATT, K. E. F. (1959). A mathematical model for the effect of densities of attacked and attacking species on the number attacked. *Can. Ent.* **91**, 129–144.

WOOD, D. L. (1972). Selection and colonization of Ponderosa pine bark beetles. *R. ent. Soc., Lond. Symp. No.* **6**, 101–117.

6. Geographic Variation in Host–Parasite Specificity

E. C. PIELOU

Biology Department, Dalhousie University,
Halifax, Nova Scotia, Canada

1. Introduction

It is probably true to say that no species of plant of extensive geographic range is genetically uniform. Each separate intrabreeding population of a sexually reproducing species is likely to differ, even if only slightly, from all others, leading to geographical variation in the properties of the species. Thus what a plant taxonomist would treat as a single species may consist of a number of distinct, internally homogeneous units of infraspecific rank, each occurring in a comparatively small geographic region; or else the variation in a species may be continuous. In either case, collections of specimens from widely separated localities are likely to differ from one another both morphologically and physiologically. Further, biochemical differences among the geographic races may affect their qualities as food for phytophagous insects especially those, such as aphids, with very specialized food requirements. Thus it would not be surprising to find that the strength of the associations between host plants and the aphid species that parasitize them exhibit geographic variations.

Nor are genetic variations among the host plants the only conceivable cause of geographic differences in host–parasite relations. Even if coevolution among plants and their associated aphid parasites tended to ensure that each aphid species continued to be matched to its host (so that for each geographical race of the host there were a corresponding race of the aphid) it seems unlikely that the correspondence would be perfect everywhere. The boundaries of the geographic ranges of both plants and animals are continually shifting. In response to climatic and other environmental changes any species at any time may be undergoing an expansion of range in some regions and a contraction in others. In short, range maps are not static and it seems inherently unlikely that fluctuations in a parasite's geographical pattern would always keep up with those of its favourite host.

These considerations led me to begin an investigation of the geographic ranges of the several species of a large plant genus; and, simultaneously, of the ranges of the aphid species that infest them. Initially the two chief objectives were: (i) to judge whether the ranges of the aphids were predominantly larger or smaller than those of their customary hosts; and (ii) to judge whether the strength of the association linking a plant–aphid pair was constant throughout the region of overlap of their ranges. Further, if it were found in answer to (i) that the range maps of the host plants and aphids were not perfectly congruent, one would also wish to discover what aphid attacked a plant outside the range of the plant's "usual" aphid; and, conversely, what plants served as food for a particular aphid species outside the range of that aphid's "usual" host plant.

The plant genus chosen for the work was *Solidago*, the goldenrods, and the observations to be described in this paper were made in the summers of 1969 and 1970. Work is still continuing and the present account is therefore an interim report.

2. Observational Materials and Methods

The genus *Solidago* (family Asteraceae) has nearly 100 species. All but a few are native to North America and the genus is richest in species in the eastern U.S.A. The definitive flora of northeastern North America (Gleason, 1952), which covers an area extending as far west as Minnesota and as far south as Virginia, describes 62 species. To a plant taxonomist *Solidago* is a "difficult" genus. There may be considerable intraspecific variation within species that do not differ markedly from one another; and many of the species are known to be capable of hybridizing. As a result, a botanist would have no difficulty in finding, if he wanted to, plenty of specimens of dubious affinities. Having said this, however, it must be emphasized that plants of debatable identity form only a small proportion of the millions of goldenrod plants to be found flowering any summer in northeastern North America. Anyone not deliberately searching for botanical oddities finds that the overwhelming majority of plants can be assigned with confidence to one species or another.

Most species of goldenrods are weeds of open, sunny, disturbed habitats such as roadsides, the borders of fields, old pastures, quarries, and clearings of all sorts. A few species are restricted to more distinctive habitats: for example, *Solidago sempervirens* is found only along the sea coast; *S. uliginosa* is a species of acid bogs; and *S. caesia* is found in shady woodlands. The 15 species of *Solidago* from which I collected aphids in 1969/70 are shown in Table 1. The names of the common species (in my collections) appear as column headings and those grouped in the last column as "others" (those with fewer than 15 occurrences) are named in the footnote to the table. All species of the genus are summer-flowering perennials.

The aphids found on these goldenrods belonged to 13 species in three genera; the names of the common species are the row headings in Table 1, and the less common ones (with fewer than five occurrences) are listed in the footnote. Three are new species and have been described by Richards (1972). All these species are monoecious (i.e. they do not infest two unrelated host plants alternately as do dioecious species), and they have life cycles typical of such aphids. They overwinter as eggs which hatch in spring. The resultant population is the first of a series of several generations of what are collectively known as "virginoparae", i.e. aphids that are parthenogenetic, viviparous, neotenous (reaching reproductive maturity while morphologically immature) and, for the most part, apterous (wingless) though alate (winged) ones capable of migrating to new plants are occasionally produced. In the virginoparae the duration of a generation may be extremely short. The daughter embryos in a gravid female contain within themselves the embryos of their own daughters, which are the gravid female's granddaughters. Thus three generations are nested in a single individual (Kennedy and Stroyan, 1959). At the end of summer a sexual generation is produced, of females incapable of parthenogenesis and males. Usually their mature stages are winged. They mate, and the females (which, unlike the virginoparae, are oviparous) lay the eggs that carry the population through the ensuing winter. A detailed account of the extraordinary complexities of aphid life cycles has been given by Hille Ris Lambers (1966).

As a rule aphids are narrowly specialized to a restricted number of plant hosts. Shaposhnikov (1961) states that most species are "found only on one genus or even one species of plant". According to Richards (1972 and pers. comm.) the aphids whose names appear in Table 1 have been found only on various species of the genus *Solidago* with the following exceptions: *Dactynotus gravicornis*; *D. erigeronensis* (found chiefly on species of *Erigeron*); *D. macgillivrayae* (found also on species of *Aster* and *Erigeron*); and *D. pieloui* (found also on species of *Aster*). It is probable that when more specimens are available for taxonomic study, *D. erigeronensis* and *D. macgillivrayae* will be found to consist of very similar, but distinguishable, species or subspecies each of which is confined to host plants of a single genus.

Solidago plants are not usually found to be heavily infested with aphids until July and infestations persist until late autumn. The aphids congregate in the inflorescences of the plants and on young succulent stems; (an exception is *Macrosiphum goldamaryae* which is commonly found on the backs of leaves). An aphid species was recorded as occurring on a *Solidago* species only if adult (i.e. reproductively mature) apterous individuals of the aphid were found on the plant. The presence of alate (winged) aphids was disregarded as the individuals observed might be pausing only temporarily while on migration to another host plant. Also disregarded were occurrences of nymphs (immature forms) in the absence of adults. An adult aphid will sometimes give birth on a plant species on which the young cannot grow to maturity and when this happens the plant species should obviously not be regarded as a possible host for the aphid.

All the observations to be described in what follows concern "occurrences" at different geographical "sites". To define precisely the meanings of these words as used here it is necessary to explain how the data were collected. The method was "biogeographical" rather than "ecological". When a region longer than 1400 miles in its longest dimension is to be inspected in a 10-week period it is obviously unreasonable to choose, ahead of time, the exact locations and boundaries of a set of sampling plots; and even if this were done, it would be wasteful to inspect these plots only, to the exclusion of all other places. One cannot tell in advance what species of *Solidago* will be found in a hitherto unseen plot, nor what their abundances will be, nor whether a piece of ground half a mile away would have been a better choice if the lie of the land had been known beforehand.

The data were collected during journeys (chiefly by car but also on foot and by canoe) in which stops were made at roughly equal intervals of distance; but the lengths of the intervals were adjusted to ensure that "inspection-stops" were made where the supply of *Solidago* plants was varied and abundant. The area of ground covered and the number of plants inspected during a stop depended on the density and diversity of *Solidago* plants present and on whether or not aphids were quickly found. Unsuccessful searches were not persevered with, since it was always more productive of data to move to a new locality than to intensify the search at an unproductive one.

In sum, the sampling procedure was wholly subjective. A "site" is a place of unspecified area which was searched (not necessarily exhaustively) for aphid-infested *Solidago* plants during an inspection-stop. No two sites were less than five miles apart and data were obtained from 123 sites. Their locations are shown on the map in Fig. 1. The density of the sites is uneven because sampling plans were modified in the light of experience. Results obtained in 1969, when the western end of the region was covered thoroughly, showed that in subsequent years it would be desirable to sample a much larger region much less thoroughly and this was done in the eastern part of the region in 1970. Also, unpremeditated sites were sampled whenever circumstances permitted, that is, whenever I was on a journey and saw a mass of goldenrod in flower.

An "occurrence" connotes the finding at a site of a particular plant–aphid combination. Thus an occurrence of one type was recorded only once for

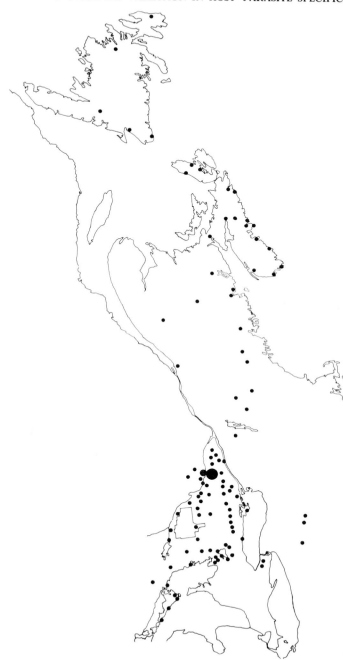

FIG. 1. Map of the 123 collecting sites. The two larger symbols represent 4 and 20 sites respectively.

any one site. No attempt was made to determine the abundances of the different aphids on the different host plants; abundances were probably governed as much by the amount of rainfall in the days preceding an inspection as by the suitability of the plant species for the aphid species; (aphid populations often crash, temporarily, after heavy rain). "Non-occurrences" were, of course, not recorded. In this respect the observations are typical of biogeographic, as opposed to ecological, studies. Thus if a particular site is not recorded as having, say, *Dactynotus caligatus* on *Solidago rugosa*, it does not necessarily follow that this aphid, and this goldenrod, were not found at the site; they may have been, but separately rather than together.

One final point to notice is that no precautions could be taken to avoid the (theoretically) possible confounding of geographical and seasonal effects. The large area covered made it impracticable to design a sampling scheme that ensured independence of dates and locations. No serious error is likely to result from this cause since there is no reason to believe there is any "succession" of aphid species on *Solidago* plants in the short aphid season. A consequence of seasonal change that can lead to difficulties, however, is the fact that morphological differences between the early-season and late-season aphids of a single lineage may be mistaken for taxonomic (i.e. genetic) differences. This problem will be mentioned below, in connexion with a particular example.

To conclude this account of the way in which data were collected, it is worth stressing that the sampling procedures used in biogeographical research are inevitably of a kind to horrify statistical perfectionists. It cannot be helped. The best solution, I believe, is to devise theoretical methods for the specific purpose of drawing firm conclusions from data of this kind. This is likely to be a far more rewarding approach than any attempt to reform data-gathering methods so that they satisfy the usual rules of sampling.

3. Results

(a) *The relative ranges of aphids and hosts*
A total of 482 occurrences were observed and they are summarized in Table 1. The table cannot be subjected to ordinary contingency table tests since the observations it records come from an extensive and presumably non-homogeneous region.

Now recall the first of the questions posed in the introduction: are the ranges of the aphids predominantly larger or smaller than those of their hosts? The area sampled was an east–west corridor and it seemed unlikely that this corridor was wide enough for north–south changes in the plants and aphids to be exhibited clearly. Attention was therefore focussed on the east-to-west extent of the ranges and the following approach was used.

TABLE 1. Observed frequencies of occurrence of aphid species on *Solidago* species

Aphid spp. \ *Solidago* spp.[a]	*canadensis* L.	*rugosa* Mill.	*nemoralis* Ait.	*juncea* Ait.	*gigantea* Ait.	*hispida* Muhl.	Other spp.[b]	Totals
Dactynotus nigrotuberculatus Olive	67	30	11	17	5	1	5	136
D. pieloui Richards	12	21	1	10	9	9	16	78
D. nigrotibius Olive	9	4	39	5	3	3	8	71
D. caligatus Richards	28	26	7	1	4	—	1	67
D. lanceolatus (Patch)	7	8	8	1	1	—	6	31
D. canadensis Richards	10	2	5	5	—	4	2	28
D. gravicornis (Patch)	9	2	4	5	4	—	1	25
Macrosiphum goldamaryae (Knowlton)	4	11	1	1	—	—	2	19
Cachryphora serotinae (Oestlund)	1	—	2	2	—	9	3	17
Other spp.[c]	2	—	—	1	—	—	7	10
Totals	149	104	78	48	26	26	51	482

[a] Nomenclature follows Gleason (1952).
[b] *Solidago bicolor* L., *S. caesia* L., *S. flexicaulis* L., *S. graminifolia* (L.) Salisb. *S. ohioensis* Riddell., *S. puberula* Nutt., *S. sempervirens* L., *S. squarrosa* Muhl. *S. uliginosa* Nutt.
[c] *Dactynotus erigeronensis* Thomas, *D. macgillivrayae* Olive, *D. pseudochrysanthemi* Olive, *D. solirostratus* Richards.

The sites were first ranked in order of decreasing west longitude. Each plant–aphid pair was then considered in turn, and two related questions asked concerning it. To take a concrete example, consider *Solidago rugosa* and *Dactynotus caligatus*. The questions are:

(a) is *S. rugosa* equally likely to be infested with *D. caligatus* everywhere in its (*S. rugosa*'s) range?

and

(b) is *D. caligatus* as likely to be found on *S. rugosa* (as on any other goldenrod) everywhere in its (*D. caligatus*'s) range?

First take question (a). Plants of *S. rugosa* infested with aphids (of any species) had been found at a total of $N = 67$ sites (as shown in the top row of Table 2). Consider these sites only. At $A = 26$ of them (see Table 1) *D. caligatus* had been found on the infested *S. rugosa* plants. From the original data it was known that the "span" of these 26 occurrences (relative to the $N = 67$ sites now being considered) was $S = 44$. The span of an observed set of occurrences connotes the number of sites (counting only the possible sites for this occurrence) between and including those of the easternmost and westernmost occurrences (see Fig. 2).

FIG. 2. To illustrate the definition of the *span* of aphid X relative to plant Y. Assume 13 sites (squares) were inspected and Y had been found at the $N = 8$ shown shaded. X was found on Y at the $A = 3$ shaded sites so marked. The span of these occurrences is $S = 7$.

We now wish to find the probability that if A sites are chosen at random and without replacement from N possibilities to be the sites of occurrences, the span of these occurrences will be at least S. Call this probability $P(S)$.

Denote by p_r the probability that the span of the A occurrences is exactly r; or, equivalently, that there are $(r - 2)$ sites between the two outermost occurrences.

There are $\binom{A}{2}$ ways of choosing two of the occurrences to be the outermost.

Of the $\binom{N}{2}$ possible pairs of sites for these two outermost occurrences, only $(N - r + 1)$ are such that exactly $(r - 2)$ sites lie between them; therefore the probability that this will happen is $2(N - r + 1)/[N(N - 1)]$.

To contain the remaining $(A - 2)$ occurrences there are $(N - 2)$ sites left. For the span of all A occurrences to be r, these remaining $(A - 2)$ occurrences must fall in the $(r - 2)$ sites between the two outermost occurrences. The probability that this will happen is

$$\frac{(r - 2)^{(A - 2)}}{(N - 2)^{(A - 2)}} = \frac{(r - 2)! \; (N - A)!}{(r - A)! \; (N - 2)!}.$$

Finally therefore,

$$p_r = \frac{A(A - 1)}{2} \frac{2(N - r + 1)}{N(N - 1)} \frac{(r - 2)! \; (N - A)!}{(r - A)! \; (N - 2)!}$$

$$= A(A - 1)(N - r + 1)\frac{(r - 2)!}{(r - A)!} \frac{(N - A)!}{N!};$$

and

$$P(S) = \sum_{r = A}^{S} p_r.$$

Clearly, $P(N) = 1$.

(This result has also been used by Moses (1952) in a different context.)

If $P(S)$ is found to be very small for a particular plant–aphid pair, we can conclude that the plant is attacked by the aphid over only part of the plant's range. (This seems to be true of *D. caligatus* on *S. rugosa* for which $P(S)$ is 5×10^{-6}.) Whereas if $P(S)$ is large there is no reason to suppose that the aphid does not attack the plant over the whole of the plant's range.

In Table 2 are shown the values of $P(S)$ for every plant–aphid pair (when neither was classed as "others") which occurred at two or more sites. The two columns on the right of the table show the observed span of each aphid's occurrences relative to all 123 sites at which *any* aphid-infested *Solidago*'s were found, with the corresponding value of $P(S)$; and a verbal description of the implication of this numerical result. It is worth noting that the only aphid found in Newfoundland (during this investigation) was *Macrosiphum goldamaryae*. However, species are described as "ubiquitous" in Table 2 if their spans are not significantly less than expectation at the 10% level.

Next consider question (b). It is obviously analogous to question (a) and can be answered by a formally identical test. We wish to determine whether the span of the $A = 26$ occurrences of *D. caligatus* on *S. rugosa* was unexpectedly short relative to the sites at which *D. caligatus* was found on any host plant. The number of these sites was $N' = 44$, as shown in the top row of Table 3. The observed span of the occurrences we are considering, relative to the 44 possible sites, was $S' = 42$ (see Table 3) and the probability under the null hypothesis of obtaining a span at least as great as this is $P(S') = 0.362$. Table 3, which is of the same form as Table 2, lists all the calculated $P(S')$ values; (the spans of occurrences relative to the aphid *Cachryphora serotinae* have been ignored since the aphid was found at only 11 sites).

TABLE 2. Spans of the occurrences relative to the *Solidago* sites, S, and values of $P(S)$

(The whole numbers are the observed spans. The $P(S)$ values are given in parentheses below, in italics if $0.01 < P(S) < 0.05$, and in boldface if $P(S) < 0.01$.)

Solidago sp.	canadensis	rugosa	nemoralis	juncea	gigantea	hispida	All species of *Solidago*	Range of aphid sp as judged from its occurrence on *Solidago*
Number of sites of the *Solidago* sp., N	92	67	57	34	19	16	123	
Dactynotus nigrotuberculatus	90 (0.18)	63 (0.25)	37 (0.03)	32 (0.50)	13 (0.37)	—	118 (**0.009**)	Absent only from Newfoundland
D. pieloui	77 (0.34)	62 (0.38)	—	31 (0.67)	18 (0.79)	15 (0.70)	118 (0.182)	Ubiquitous
D. nigrotibius	62 (0.13)	16 (0.03)	52 (0.01)	16 (0.10)	10 (0.13)	8 (0.09)	86 (4×10^{-9})	Western
D. caligatus	72 (**0.002**)	44 (5×10^{-6})	37 (0.19)	—	17 (0.90)	—	87 (5×10^{-8})	Central
D. lanceolatus	69 (0.41)	56 (0.56)	57 (1.00)	—	—	—	108 (0.133)	Ubiquitous
D. canadensis	42 (**0.003**)	8 (**0.001**)	26 (0.11)	11 (0.02)	—	6 (0.06)	70 (**0.0002**)	Central
D. gravicornis	37 (**0.002**)	17 (**0.007**)	27 (0.24)	22 (0.35)	6 (0.04)	—	73 (**0.001**)	Western
Macrosiphum goldamaryae	86 (0.97)	43 (0.03)	—	—	—	—	116 (0.695)	Ubiquitous
Cachryphora serotinae	—	—	12 (0.03)	15 (0.03)	—	13 (0.19)	92 (0.164)	Ubiquitous

TABLE 3. Spans of the occurrences relative to the aphid sites, S', and the values of P(S').

(Details as in Table 2)

Aphid sp.	D. nigro-tuberculatus	D. pieloui	D. nigrotibius	D. caligatus	D. lanceolatus	D. canadensis	D. gravicornis	Macrosiphum	All aphids	Range of Solidago sp. as judged from the present study
Number of sites of the aphid sp., N'	86	53	47	44	24	18	16	17	123	
Solidago canadensis	85 (0.40)	48 (0.59)	43 (0.76)	44 (1.00)	15 (0.10)	17 (0.71)	12 (0.08)	5 (0.02)	121 (0.175)	Ubiquitous
S. rugosa	81 (0.31)	50 (0.48)	23 (0.26)	42 (0.36)	22 (0.75)	12 (0.08)	10 (0.07)	16 (0.60)	118 (0.068)	Ubiquitous
S. nemoralis	40 (**0.001**)	—	47 (1.00)	30 (0.23)	21 (0.59)	10 (0.15)	14 (0.79)	—	98 (2×10^{-7})	Western
S. juncea	79 (0.50)	44 (0.39)	23 (0.14)	—	—	12 (0.32)	12 (0.48)	—	115 (0.231)	Ubiquitous
S. gigantea	41 (0.14)	34 (0.07)	17 (0.05)	36 (0.82)	—	—	9 (0.29)	—	86 (**0.005**)	Western
S. hispida	—	26 (**0.007**)	32 (0.29)	—	—	9 (0.21)	—	—	73 (**0.001**)	Western

As in Table 2, the two columns on the right in Table 3 refer to the spans, relative to all 123 sites, of occurrences in which a *Solidago* species was host to an aphid of any species. Since *Solidago* plants are subject to aphid infestations everywhere, these columns relate to the geographical range, within the area surveyed, of each plant species as a whole irrespective of the aphids that attack it. However, it would be incorrect to make a precisely analogous comment regarding the two right hand columns of Table 2. Table 2 deals with the ranges of aphids occurring on *Solidago* and some, at least, of the aphid species are known to attack plants of different though related genera (*Aster* and *Erigeron*). The lack of complete symmetry between the two tables arises from the fact that on field collecting trips the objects sought were *Solidago* plants, which were then inspected for aphids; I did not search for aphids as such and subsequently identify the plants they were feeding on. It should also be remarked that the ranges of the *Solidago* species given in Table 3 are those implied by my own studies and are, of course, much smaller than those given in a standard flora such as Gleason's (1952) in which the stated range of a species embraces all recorded occurrences from the beginning of botanical exploration. According to Gleason, all the *Solidago* species listed in Table 3 occur throughout the area of the present study except that *S. nemoralis*, *S. juncea* and *S. gigantea* are absent from Newfoundland.

In Tables 2 and 3 values of $P(S)$ and $P(S')$ between 0·05 and 0·01 are in italics; and values less than 0·01 in boldface. Considering the two tables together, it is clear that the ranges of the aphids and the hosts they attacked were frequently mismatched. For a particular plant–aphid pair, a low value of $P(S)$ implies that the pair occurred in only part of the plant's range; and a low value of $P(S')$, that it occurred in only part of the aphid's range. It is quite possible for a pair to have very low values of both $P(S)$ and $P(S')$. Two different sets of circumstances could lead to this. In the first place, it would happen if the geographic ranges of the aphid and the plant were nearly but not quite disjunct and the aphid attacked the plant throughout the small region of overlap. In the second place, it would happen if, even though the geographic ranges of the plant and the aphid were co-extensive over a large area, the aphid attacked the plant only within a subregion of this area. This may explain the simultaneously low values of $P(S)$ and $P(S')$ (0·03 and 0·001 respectively) of the pair *D. nigrotuberculatus* on *S. nemoralis*, though the evidence is not very strong.

Clearly, if a *Solidago* species and an aphid species have coextensive geographical ranges, and the aphid attacks the plant throughout their joint range, we should expect $P(S)$, and also $P(S')$, to be rectangularly distributed in (0, 1). It is therefore easy to test whether the pattern of the aphid ranges is congruent with that of the plant ranges. There are two null hypotheses:

H_0, that every aphid species that attacks a given *Solidago* species does so throughout the range of the *Solidago* species (in which case the $P(S)$ values in Table 2 will be rectangularly distributed); and H'_0, that every *Solidago* species eaten by a given aphid species is available throughout the range of the aphid (in which case the $P(S')$ values in Table 3 will be rectangularly distributed). Table 4 shows the observed distributions of the 38 values of $P(S)$ in Table 2 and the 35 values of $P(S')$ in Table 3. The expected frequencies in each of the five classes are obviously 7·6 for the $P(S)$ distribution and 7·0 for the $P(S')$ distribution. Testing the goodness of fit of the rectangular distribution to the observed distributions gives

(i) for $P(S)$, $\chi^2 = 39·9$ with 4 DF; $P < 10^{-5}$

(ii) for $P(S')$, $\chi^2 = 8·57$ with 4 DF; $0·05 < P < 0.10$.

It thus appears that H_0 should be rejected whereas H'_0 can, perhaps, be accepted. The greater frequency of low values in the $P(S)$ than in the $P(S')$ distributions shows that the spans of plant–aphid occurrences tended to be shorter with respect to the range of the plant member of the pair than with respect to the aphid member.

TABLE 4. The observed distributions of $P(S)$ (from Table 2) and $P(S')$ (from Table 3)

	Values of the cumulated probabilities					Totals
	$\leq 0·2$	0·21 to 0·40	0·41 to 0·60	0·61 to 0·80	0·81 to 1·00	
$P(S)$	23	6	3	3	3	38
$P(S')$	12	10	6	4	3	35

An alternative way of testing this conclusion is to compare the means of the observed $P(S)$ and $P(S')$ values, which is most easily done using the Mann-Whitney U test. It is found that $U = 475$, $E(U) = 665$ and $\sigma_U = 90·653$. Thus the mean of the observed $P(S)$ is significantly smaller ($P < 0·02$) than the mean of the observed $P(S')$ whence, again, we may infer that the geographic ranges of the aphid species are less extensive than those of the *Solidago* species on average.

There are three (at least) possible explanations for this. They are: (i) that the geographical ranges of the *Solidago* species are expanding at a faster rate than those of the aphids; (ii) that each plant species is made up of a number of genetically distinct geographical races and no one aphid is equally well adapted to all the races; and (iii) that the plants are more tolerant than the aphids of environmental variation so that at the periphery of a plant's range conditions are unsuitable for the aphids that infest it near the centre of its range. Work to determine the relative importance of these mechanisms is now in progress.

(b) *Geographical changes in the strengths of plant–aphid associations*
We come now to the second of the two questions posed in the introduction, namely, for a given plant–aphid pair, say *Solidago* X and aphid Y, does the strength of their association vary geographically? First, suppose it were known *not* to vary. It would then be permissible, if a measure of the association between X and Y were desired, to condense the observations in Table 1 into a 2 × 2 table of the form:

	Infested *Solidago*	
	X	not X
Feeding { Y	a	b
aphid { not Y	c	d

Some appropriate function of these frequencies may be used as a measure of the association between X and Y. The measure chosen must be one that is not influenced by the relative sizes of the table's marginal totals. These are entirely arbitrary since they were determined as much by the relative durations of data-gathering forays in different geographical regions as by the relative abundances of the different plant and aphid species throughout the area. For a measure of association to be independent of the marginal totals it must, as Edwards (1963) has shown, be a function of the cross-product ratio ad/bc. Of the various measures that meet this requirement, Yule's $Q = (ad - bc)/(ad + bc)$ is convenient and will be used in what follows. Its variance is

$$\text{var}\,(Q) = \tfrac{1}{4}(1 - Q^2)^2\left(\frac{1}{a} + \frac{1}{b} + \frac{1}{c} + \frac{1}{d}\right).$$

We now wish to judge whether the value of Q for any chosen plant–aphid pair remains constant throughout the area surveyed. As before, we shall treat the area as a narrow east-to-west corridor and shall make no attempt to look for north–south trends.

In principle, an obvious way to look for east–west trends would be to divide the whole area, regarded as a strip, into a number of short segments of equal length. One could then pool the observations from all the sites within any one segment, use the pooled data to calculate Q for the segment, and thus obtain a set of values of Q forming an ordered sequence from west to east. This approach does not work in practice, however. If the segments are to contain enough sites, and hence sufficiently many occurrences, for

Q to have an acceptably low variance, it is necessary to make them long; but a long segment is unlikely to be sufficiently homogeneous for its sites to be pooled legitimately. Conversely, if the segments are made short enough to be homogeneous the number of sites in each will be low and var (Q) will be high. It therefore seemed best to calculate a series of "moving Q's" arrived at in a manner analogous to that used in calculating moving averages.

The raw data (for each plant–aphid pair) consisted of 123 2×2 tables, one for each of the 123 sites. This body of data can be written as a 123 $\times$ 4 matrix in which each row lists the cell frequencies of the 2×2 table for a single site; i.e. each row consists of a vector of the form $[a \ b \ c \ d]$ with the cell frequencies of a 2×2 table as its elements. For example the first five rows of the data matrix for the pair *S. nemoralis–D. nigrotibius* are:

$$0 \quad 0 \quad 0 \quad 4$$
$$0 \quad 0 \quad 0 \quad 1$$
$$1 \quad 1 \quad 1 \quad 4$$
$$0 \quad 0 \quad 0 \quad 1$$
$$0 \quad 1 \quad 0 \quad 2$$
$$\cdots\cdots\cdots$$

These rows relate to sites $1, \ldots, 5$; site 1 is the westernmost. The moving Q's, each of which was based on the data from 15 sites, were calculated as follows.

Let the jth row of the data matrix be $[a_j \ b_j \ c_j \ d_j]$.

Put $A_m = \sum_{j=m}^{m+14} a_j$ and define B_m, C_m and D_m analogously. Then Q_m, the mth in the sequence of moving Q's, is given by

$$Q_m = (A_m D_m - B_m C_m)/(A_m D_m + B_m C_m).$$

Sequences of Q_m values with $m = 1, 2, \ldots, 109$ were calculated for five of the plant–aphid pairs. The pairs chosen are those which seemed (from Table 1) to show some evidence of being positively associated. They are named in Fig. 3 in which are shown the five trend lines of Q_m versus m. The m's are site numbers. It should be recalled that the sites were ranked by longitude and were not separated by equal intervals. The ranking was from west to east and thus the left end of each graph is west and the right east. Curves (i), (ii) and (iii) cannot be extended as far east as curves (iv) and (v) owing to the restricted geographic ranges of the aphids *D. nigrotibius* and *D. caligatus* which were absent from the most easterly sites. As a result, 2×2 tables involving one of these aphid species, and compiled from data on occurrences in the east of the area, had $a = 0$ and $b = 0$ and hence Q indeterminate.

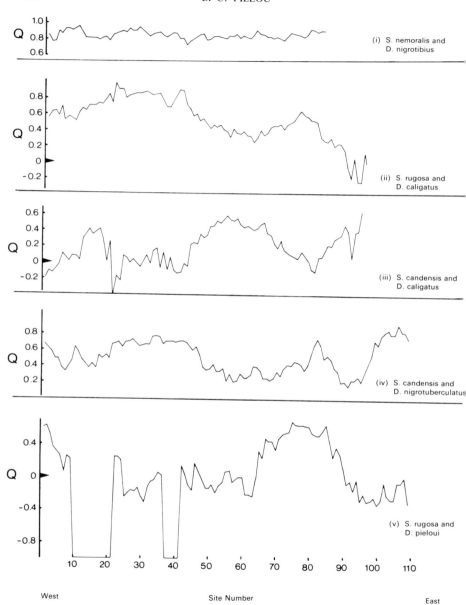

FIG. 3. Q_m versus m for five plant–aphid pairs.

The choice of 15 as the number of sites to combine for each calculation of Q was arbitrary; it was a compromise between taking too small a number (which would cause var (Q) to be unduly large) and too large a number (which would cause excessive smoothing of the trend line). It should also be noticed that in calculating a moving Q each of the 15 sites contributing to it was weighted equally. To have done otherwise, for example to have used anything analogous to Simpson's 15-point formula, would have been inappropriate for these data because the numbers of occurrences varied from site to site.

Judging the implications of the trend lines is unavoidably somewhat subjective. Much of the apparent oscillation is no doubt a manifestation of the Slutzky–Yule effect (Kendall and Stuart, 1966). Oscillations of this kind are inevitable since the successive Q's are serially correlated, but I have been unable to derive their expected period and amplitude. It is obvious, however, that the larger the value of Q the smaller will be the amplitude of Slutzky–Yule oscillations since var (Q) decreases with increasing Q. This effect seems to be well illustrated by the contrast between curves (i) and (v) in Fig. 3. Thus curve (i), which shows the moving Q values for the strongly associated pair *S. nemoralis–D. nigrotibius*, exhibits only minor fluctuations. As the curve gives no evidence of any trend in the association of this pair, it is reasonable to pool the data from all the sites. One can then use the cell frequencies of the resultant 2×2 table, for which $[a\ b\ c\ d] = [39\ 32\ 39\ 372]$, to calculate an "overall" Q. This presumably gives an estimate of the strength of the association between these species wherever they occur. The result is $\hat{Q} = 0.8416$, var $(\hat{Q}) = 0.0018$, whence the 95 % confidence interval for Q is 0.758 to 0.925. The same numerical result would have been arrived at, of course, if the curve of Q_m versus m had never been plotted. But until the curve had been inspected the possibility existed that Q varied from region to region in which case an overall value obtained from the combined data would have had no biological meaning.

Curve (v), which relates to the pair *S. rugosa–D. pieloui*, contrasts with curve (i) both in the much greater amplitude of its oscillations and in its smaller mean value. The curve gives no evidence of any trend, so again we may combine all the observations into a single 2×2 table. Its vector of cell frequencies is $[21\ 57\ 83\ 321]$ whence, with 95 % probability, $-0.94 < Q < 0.444$. Thus in spite of the fact (see Table 1) that *D. pieloui* was found more often on *S. rugosa* than on any other goldenrod, there may well be no true association between them.

Now consider curve (ii), that for the pair *S. rugosa–D. caligatus*. Of all the five curves in Fig. 3, this one gives the clearest manifestation of a trend. It appears that the strength of the association between the aphid and the goldenrod declines from west to east. It is interesting to divide the data into two sets: those from sites west of and including site 49 and those from sites east of and including site 50. (The division comes at 76° West, approximately the longitude of Ottawa). Pooling the data from each set gives two 2×2 tables with the following vectors of cell frequencies:

$$\text{West, } [a_W \quad b_W \quad c_W \quad d_W] = [8 \quad 10 \quad 17 \quad 145];$$

$$\text{East, } \quad [a_E \quad b_E \quad c_E \quad d_E] = [18 \quad 31 \quad 61 \quad 192].$$

Following Plackett (1962), one may test these tables for second order interaction. For each table, compute

$$z = \ln \frac{ad}{bc}; \quad \text{and } u \text{ where} \quad \frac{1}{u} = \frac{1}{a} + \frac{1}{b} + \frac{1}{c} + \frac{1}{d},$$

Next, obtain the test statistic

$$X^2 = \Sigma u z^2 - (\Sigma u z)^2 / \Sigma u$$

where the summation is over the two subregions, east and west. If there is no second order interaction, X^2 is asymptotically distributed as χ^2 with 1 DF. In the present case

$$z_W = 1.9204; \quad z_E = 0.6030;$$

$$u_W = 3.4397; \quad u_E = 9.1395.$$

Then

$$X^2 = 4.337 \quad \text{and} \quad 0.02 < P < 0.05.$$

The numerical result thus lends support to the hypothesis that the association between S. rugosa and D. caligatus is stronger in the western part of the region surveyed than in the eastern. Admittedly, the observations used in testing the hypothesis were the ones that originally suggested it and therefore an objective test is impossible. This scarcely matters, however. Even if the test were statistically respectable one would certainly need additional evidence, obtained by sampling the region more thoroughly, before accepting the hypothesis unreservedly.

The whole object of the present analysis is, indeed, to suggest hypotheses that may be worth focussing attention on in future. Thus curves (iii) and (iv) of Fig. 3 lead to interesting speculations. They suggest that where S. canadensis is closely associated with D. nigrotuberculatus it is not associated with D. caligatus and vice versa. Further, a comparison of curves (iii) and (ii) suggests that perhaps the region where there is no association between S. canadensis and D. caligatus coincides with the region where S. rugosa and D. caligatus are most strongly associated. Returning to curves (iii) and (iv), obviously the relationship between them may be only apparent; the oscillations in both may be due to the Slutzky–Yule effect and the apparent negative cross-correlation may be artificial and a result of the serial correlation within each series of Q_m values. All the same, speculations worth further study are the desired outcome of this initial work. There were no earlier observations to serve as a guide before this study began and collecting data such as these is troublesome and time-consuming. It is therefore desirable to formulate tentative hypotheses as early as possible. Future observations can then be planned with the express purpose of testing them. For instance it will be interesting to collect aphid-infested *S. rugosa* from two big groups of closely-spaced sites, one near the eastern and the other near the western end of the study area. It will then be possible to test whether the proportions of *D. caligatus* among the infesting aphids differ in the two areas, and hence settle conclusively the speculation prompted by inspection of curve (ii).

(c) *Are Dactynotus nigrotibius and D. lanceolatus different species?*
Although in all the foregoing analysis *D. nigrotibius* and *D. lanceolatus* have been treated as different species it is possible that they are not, in fact, distinct. Thus Richards (1972) writes that there appears to be a complete inter-gradation of the two species with respect to diagnostic characters and that the differences among specimens may be due to seasonal changes in the aphids. The data collected cast some light on this taxonomic problem.

First, to test the possibility of a seasonal effect, the data (for 1969 only) were divided into those collected before August 1 and those collected on or after that date. The result is:

	D. lanceolatus	D. nigrotibius
Early	5	30
Late	7	48

The null hypothesis that the difference between aphid specimens are not attributable to seasonal changes is supported by this table. Also in support is the fact that the earliest and latest dates (July 29 and August 28) on which *D. caligatus* was found fall between the corresponding dates (June 21 and September 4) for *D. nigrotibius*.

Next we consider whether the two species (if they are two) differ in respect of their chosen host plants. In particular, is *D. lanceolatus* as closely associated as *D. nigrotibius* with *S. nemoralis*? The relevant 2 × 2 table is:

	D. lanceolatus	D. nigrotibius
S. nemoralis	8	39
Other Solidago's	32	23

A χ^2 test gives $X^2 = 6.24$, $0.01 < P < 0.02$.

Fig. 4. Sites of occurrences of *D. nigrotibius* only (•); of *D. lanceolatus* only (○); and of both species (■).

Thus the evidence from this source supports rejection of the null hypothesis that the two species are the same.

Additional evidence tending to the same conclusion is supplied by the map in Fig. 4 from which it appears that the geographic range of *D. nigrotibius* is smaller than, and possibly contained within, that of *D. lanceolatus*. It would be premature to theorize on the biological implications of this result since so many possibilities suggest themselves. It should be noticed that the map by itself could be misleading if no test had been done to rule out seasonal change as an explanation of the differences among aphid specimens. As remarked in Section 2, practical considerations made it impossible to randomize the temporal sequence in which the various collecting sites were visited, so that in looking at maps like that in Fig. 4 one must always guard against confounding spatial and temporal effects.

Acknowledgements

This work was supported by a grant from the National Research Council of Canada. I thank Mr. Albert Dugal for help with collecting the data, and I am also indebted to Dr. W. R. Richards for identifying the aphids.

REFERENCES

EDWARDS, A. W. F. (1963). The measure of association in a 2 × 2 table. *J.R. statist. Soc. A*, **126**, 109–114.

GLEASON, H. A. (1952). "Illustrated Flora of the Northeastern United States and adjacent Canada." Hafner, New York.

HILLE RIS LAMBERS, D. (1966). Polymorphism in Aphididae. *Ann. Rev. Ent.* **11**, 47–78.

KENDALL, M. G. and STUART, A. (1966). "The Advanced Theory of Statistics," Vol. 3. Griffin, London.

KENNEDY, J. S. and STROYAN, H. L. G. (1959). Biology of aphids. *Ann. Rev. Ent.*, **4**, 139–160.

MOSES, L. E. (1952). A two-sample test. *Psychometrika*, **17**, 239–247.

PLACKETT, R. L. (1962). A note on interactions in contingency tables. *J.R. statist. Soc. B*, **24**, 162–166.

RICHARDS, W. R. (1972). Review of the *Solidago*-inhabiting aphids in Canada with descriptions of three new species. *Can. Ent.*, **104**, 1–34.

SHAPOSHNIKOV, G. Kh. (1961). Host-specificity and adaptation to new hosts in aphids in the process of natural selection. *Entomol. Obozr.*, **40**, 739–762.

7. Some Examples of the Mechanisms that Control the Population Dynamics of Salmonid Fish

E. D. Le Cren

*Freshwater Biological Association, The River Laboratory,
East Stoke, Wareham, Dorset, England* *

1. Introduction

Many of the models now being developed to aid the understanding of biological populations, ecosystems and parts of ecosystems, are built up out of a series of relatively simple components each of which describes a single functional relationship. Although many of these functional relationships can be expressed in simple mathematical terms, if the model is to be successful they must be realistic and reasonably accurate approximations to reality. Further, if the model is to be used for prediction the relationships must hold over the whole range of conditions that may be involved; and it is in the nature of predictions that these conditions may extend beyond those normally experienced. It is also often desirable that the functional relationships be described in terms that provide insight as well as realism.

This paper concerns the problems of obtaining the biological information necessary to describe functional relationships in population ecology, and uses as examples researches into the population dynamics of some salmonid fishes.

2. General Features of Salmonid Population Dynamics

Most salmonid fishes have a life-history which involves the laying of a moderate number of eggs in clean gravelly situations in streams or on lake littorals. The larvae have large yolk sacs which provide them with sustenance after hatching. Usually there is a migration at some stage from the cool, relatively unproductive environment in which the eggs are laid to richer feeding grounds in river, lake or ocean where growth occurs and whence

* Present address: Freshwater Biological Association, Windermere Laboratory, The Ferry House, Ambleside, England.

the adults migrate back to spawn. Evolution into a range of relatively similar genera, species and sub-specific taxa has occurred since the last ice-age, within which a small number of trends to occupy similar niches in different regions can be recognized (Benke, 1972).

The spawning migrations of the adults often support fisheries and it is thus of practical importance in conservation to understand the population mechanisms that operate in the life history stages between the adult spawning migration and migration back to the feeding grounds by the resultant progeny. Although the feeding stage of the life history may include functional relationships of significance to population control, it would seem that for most species those operating in the early stages are more powerful. It is thus of considerable practical as well as of theoretical interest to determine the relationships between the numbers of spawners and the numbers of their progeny. For several of the major salmon fisheries, this determines the size of the catch that can safely be taken on a long-term basis. The continued existence, over a great many years, of fisheries for relatively discrete stocks is, of itself, an indication that density-dependent mechanisms must exist. If progeny abundance were directly and proportionately related to parental abundance, such fisheries would long ago have declined to virtual extinction. Ricker (1954) has discussed the relationships between stock and recruitment and various theoretical and empirical "reproduction curves". The general problem is rather differently discussed by Beverton and Holt (1957) and the topic is further developed in Parrish (1973).

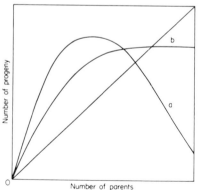

FIG. 1. Reproduction curves. Two main extreme types are shown as (a) and (b).

The relationships between stock and recruitment can be simply expressed graphically by plotting progeny against parents (Fig. 1). Such graphs can include a 45° line which, with suitable scales, can represent a situation where there is exact replacement of one generation by the next. There is evidence for at least two types of curve: (a) a dome shaped curve with a falling right hand limb indicating that high densities of parents or spawning can lead to lower numbers of progeny than moderate densities of parents, and (b) a curve flattening out to a maximum with a relatively constant production of progeny over a wide range of higher parental numbers. Both these curves are shown in Fig. 1. This type of graph is also useful for the study of successive periods of the life history when numbers of survivors present at the end of a period are plotted against the numbers at the beginning.

An aspect of population dynamics which becomes important when salmonids (and indeed most fish) are being studied is the size of the individuals and the great plasticity of growth-rates. For example brown trout (*Salmo trutta*) in Windermere grow up to 5 kg in weight whereas in some of the small streams in the same drainage area adults weighing over 100 g may never be found. Growth-rate is influenced by a great many environmental factors, but it is well known that there is often an inverse relationship between growth and population density. This can also affect the population in terms of numbers, both through fecundity, which is proportional to size, and where mortality may be size linked and the speed of growth through such a vulnerable stage determines the total mortality experienced.

3. Methods of Approach to Functional Relationship Studies in Population Ecology

The population dynamics of salmonid fish can be used to illustrate the four main ways in which the problems of functional relationships can be approached. These four, intergrading approaches are:
   (i) by theory based on general ecological principles and aided by mathematics;
   (ii) by observation of natural populations;
   (iii) by laboratory experiment;
   (iv) by field experiment.
These approaches will be discussed below.

The theoretical approach
Modern ecological theory demands some form of density relationships to maintain population abundance under some degree of long-term control, and mechanisms resulting in curves of the type shown in Fig. 1 are necessary for such control. It is relevant to note, however, that theory alone may not be of much help in deciding what, out of a range of possibilities, the shape of the curve actually is. Theoretical considerations, and mathematical models will often provide evidence on the implications of a particular relationship which may aid the planning of further observational and experimental work. Ricker (1954) has shown how some dome shaped curves (Fig. 1) can lead to population oscillations, and such oscillations sometimes

seem to occur in salmonid populations. (Other papers in this symposium illustrate well the use of such theoretical approaches).

The observation of natural populations
Some observation of natural populations is obviously essential, if only to avoid natural history "howlers". A great many data on salmon populations are available from the catches of fisheries, and also from the counts of migrating adults and sometimes fry or smolts returning to the sea. A number of examples of such data for the sockeye salmon (*Oncorhynchus nerka*) are given by Johnson (1965). Such data may, in effect, be analogous to experimental data, in that observations are collected over a period of time when the independent variable, e.g. population density, or environmental temperature, varies naturally. This approach is, however, usually chronophagous as it may take a good many years before an adequate range of population densities or temperatures are experienced. To quote a hypothetical but nevertheless relevant practical example; it is proposed to site a thermal power station on a lake and it is therefore necessary to predict the effects of warm water on the growth rate of the fish. It may be only once in thirty years that summer temperatures will reach naturally the predicted average after the power station has been built. Further, over the range naturally experienced it is observed that growth increases with temperature, but it is possible that at temperatures above the normal range an optimum might be exceeded. Extrapolations based on either the empirical data naturally available or on a theoretical Q_{10} law may both be dangerous.

Another draw-back to empirical natural data is that they tend to be very variable and include a great deal of "noise". The data presented by Ricker (1954) show that some "faith" or knowledge from other sources is often needed to draw curves through the points available (see Ricker, 1954, Figs. 18, 20, 24).

Laboratory experiments
The laboratory approach is valuable because it allows control of most of the environmental factors not under consideration, and thus reduces "noise" in the data, and also because it allows a whole range of the variables to be studied fairly rapidly. For example, it would not be difficult to grow fish under a range of different temperatures, including those rarely experienced naturally, in order to determine the effects of temperature upon growth in the laboratory.

On the other hand, unless care is taken to provide the experimental animals with conditions which resemble their natural environment in terms of "creature comforts" and "social life" they may behave abnormally. Hoar (1969) describes two examples of this. The young of Pacific salmon (*Oncorhynchus* spp) are normally schooling fish and their reactions to light intensity and water current depend upon this. Hoar did not obtain the correct results for sockeye smolts (*O. nerka*) unless he provided adequate space for them to form their normal vertically thin but laterally wide schools. Nor would the fry of pink salmon (*O. gorbuscha*) react normally until they had had experience of schooling with their own brethren. The laboratory approach can therefore have its pitfalls unless the experimenter is skilful enough to make his fish "happy" and to think of all the factors that might be important.

Field experiments
The advantages of field experiments are that the animals can be living under natural or nearly natural conditions while one or two of the important factors in their environment are varied, often over a wide range. They thus occupy a half-way position between the laboratory experiment and field observation. The value of field experiments has long been recognized in some branches of applied biology, e.g. agriculture and fish culture, and elaborate experimental and statistical techniques developed. Though the experimental conditions cannot be controlled as closely as in the laboratory, some quite precise results can be obtained. For example, Wolny and Grygierek (1972) describe experiments on the food chains in carp ponds which yielded correlation coefficients (r) between phytoplankton photosynthesis and fish production as high as 0·98. The main disadvantage of field experiments is that they can be expensive, especially if a large number of replicates is required.

Some experiments on the effects of population density on the survival and growth of the fry of trout (*Salmo trutta*) will illustrate the kind of results that can be obtained from this approach.

Field experiments with trout fry
Observations and population estimates on a number of becks in the north of England suggested that the population density of trout fry (i.e. fish 3–5 months after hatching) and the production of trout in these streams was much less variable than would be suggested by the variation in types and numbers of parental spawners and the probable numbers of eggs laid. Observations on the behaviour of the fry by myself and others (e.g. Kalleberg 1958) suggested that aggressive territorial behaviour might be an important factor. I therefore divided a natural beck into a series of small stretches by fry-proof screens and stocked these with varying numbers of newly hatched trout or eggs about to hatch; and observed the survival and growth over the next few months.

The results showed (Backiel and Le Cren 1967, Le Cren 1965, Le Cren 1973):

(a) that above a certain minimum density when mortality was low, mortality was strongly density dependent with the coefficient of mortality (Z) varying directly with the logarithm of initial population density (Fig. 2);

(b) that specific growth rate (G) was inversely proportional to the logarithm of population density, though at high population densities there was a relatively constant low growth rate (Fig. 2);

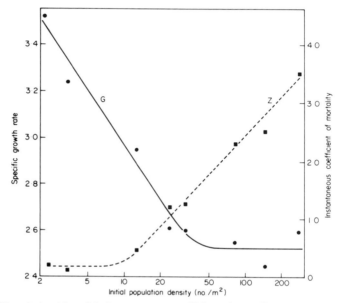

FIG. 2. The relationships of the instantaneous coefficient of mortality (Z) ■ – – – ■ and specific growth rate (G) ● — ● to logarithm of initial population density for experiments with trout fry over their first five months. Lines drawn by eye. (Based on Le Cren 1965, 1972 and Backiel and Le Cren 1967).

(c) the resultant final numbers, and biomass and production were relatively constant above a fairly low initial stocking density (Fig. 3).

The results could be expressed in another way by plotting the production per fish stocked against the area available per individual fry stocked (Fig. 4). This showed that a direct relationship resulted and area was the important factor determining survival (and growth and production). The fry fight each other for territories as soon as they start to feed, and fry in excess of those for whom territories are available disperse downstream or die of starvation. The numbers and production of young trout is thus determined by the area of suitable rearing ground available. (Later on in life it would appear that defence of a "station" or "lie" (usually under cover) is more important than a spatial territory.)

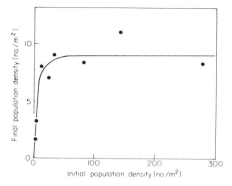

FIG. 3. The final population density of experimental populations of trout fry after five months plotted against the initial density. Line drawn by eye. (Based on the same experiments as Fig. 2.)

Kalleberg (1958) showed that flowing water was essential for the normal territorial behaviour of trout fry and so any experiments carried out in the usual still-water aquaria would be likely to have produced erroneous results. Some artificialities were inevitably introduced into these experiments; for example the screens prevented full down-stream dispersion and in most of the experiments yearling or older fish were excluded; but other experiments have explored the effects of these conditions. The range of population densities used in the experiments exceeded those normally occurring in nature.

It can be concluded that experiments of this kind can effectively explore some of the basic functional relationships in fish population dynamics.

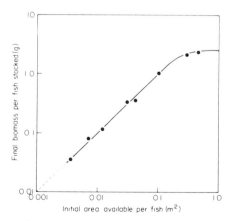

FIG. 4. Experimental trout fry populations. The biomass recovered at the end of five months per fish stocked plotted against the area available to each fish when stocked. Logarithmic scales, line drawn by eye. (Based on the same data as Fig. 2.)

The relationships can be accurately described in very simple mathematical terms and the mechanisms that control them understood.

4. Interspecific Variations and Relationships

Although the basic mechanism of population density functions in trout fry appears to be simple other experiments with other species of stream-living salmonids have indicated the variations that can be found.

The fry of Atlantic salmon (*Salmo salar*) show a similar territorial behaviour, but are rather less aggressive than those of the trout. In one experiment I found that the mortality rate of salmon in mixed populations of salmon and trout fry depended upon the total population of trout and salmon, whereas mortality of the trout depended upon the density of trout alone. There is also observational evidence that in small streams and the shallow areas of larger streams trout dominate salmon, whereas in larger and deeper streams salmon may predominate (Lindroth, 1955).

Hartman (1965) has experimented with the behaviour of young coho salmon (*Oncorhynchus kisutch*) and steelhead trout (*Salmo gairdneri*) in stream aquaria. In summer the steelheads tend to occupy riffles whereas the coho tend to live in pools, but in winter both inhabit pools and are less aggressive. Here there is clearly a habitat selection intensified by inter-specific aggressive behaviour. Hartman has also shown that age and temperature affect aggressive behaviour in young steelhead trout. In field experiments with young coho and steelheads Fraser (1969) obtained relatively little inter-specific effect except at high population densities, though both growth and mortality were related to intra-specific densities.

Various workers (e.g. Stuart 1953) have shown that visual stimuli are important in the ethology of aggressive behaviour and territorial defence, so the effects of this behaviour can be modified by physical objects. More trout fry can be crowded into a habitat where it is more difficult for them to see one another.

Food supply is a factor which would be expected to influence territorial behaviour and density-dependent effects on growth and survival. Chapman (1966), Mason (1966) and others have shown that food can have some influence on territory size in young chinook salmon (*Oncorhynchus tshawytscha*) at least. In young brown trout my impression is that territory is not directly related to feeding. Certainly density-dependent effects on both growth and mortality occur at widely different levels of food production (and thus average growth rate). Johnson (1965) has shown that with young sockeye salmon (*O. nerka*) there are similar inverse relationships between growth and population density at four quite different growth rates at similar ranges of population density in four lakes with very different basic productivities.

Stream living species feed mostly on drifting invertebrates brought to the trout hovering over its station by the water current. I suspect that the influence of population density on growth is an indirect one; fish that spend time on fighting have less for feeding. Food spent on providing energy for fighting is not available for growth either, but further experiments are needed to explore these relationships. This discussion has dealt only with the young fish after hatching; in general the number and size of the eggs laid, and parental factors would seem to have less influence but Bagenal (1969) has shown how the size of eggs can influence subsequent fry survival.

Any good model purporting to predict the number of young salmonids reared in a particular stream would have, therefore, to take note of a number of factors. These would include population density of eggs, their size, the area of stream, its division into riffles and pools and other physical factors, the food supply available, the species present, etc. Some of the factors would have to be arranged in sequential order, and some would involve thresholds and upper limits. However, it would seem that many of the actual functional relationships could be quite simple.

5. Discussion and Conclusions

The example of population dynamics and regulation in the young stages of salmonid fish has highlighted, I believe, some of the problems in building realistic and informative population models.

Observations alone are unlikely to discover the basic functional relationships, at least over the full range of variables that may be necessary. Enough observational data are also likely to take a long time to collect. The aberrant behaviour of young salmonids when in laboratory situations shows that care has to be taken in drawing conclusions from purely laboratory experiments. The ethology of fish can have a profound influence on their reactions to changes in even physical environmental factors, and also be the basis of population mechanisms. Field experiments can be revealing and provide many of the data necessary to describe functional relationships but must be designed with this in view. The design of such experiments may be based on fundamental theory; (for example classical Nicholsonian ideas about density-dependent factors were the basis for my trout fry experiments).

The observations and experiments with different species of salmonids show that while the basic relationships can be quite simple there may be important modifications with changes in environmental conditions. Further, the differences between population relationships and ethology of closely related species may be just those factors that control their habitat selection and the effects of competition between them. Thus there may be difficulties in trying to combine species into "functional groups" in an effort to simplify ecosystem models and reduce the number of their components.

The experiments with trout fry also showed that the relationships over the middle range of population densities did not necessarily hold at extreme ranges and thus extrapolation beyond the limits usually encountered could be risky.

In spite of the complications and variations in the population functional relationships of young salmonids it would seem that detailed study could eliminate some of the less important factors and lead to simplification into quite briefly stated functions which could be components of relatively simple yet realistic and informative models.

Summary

Building realistic and informative population or ecosystem models requires good understanding of component functional relationships. There are problems in obtaining such knowledge and these have been discussed and illustrated by reference to the population dynamics of young salmonid fish. The theoretical and laboratory approaches may suffer from ignorance of basic natural history and ethology; field observations take time and may not encompass an adequate range of variables. Field experiments have many merits and have revealed relatively simple population mechanisms based on territorial behaviour in young stream salmonids. These are complicated by physical conditions and inter-specific behaviour and may be different at extreme ranges of population density.

REFERENCES

BACKIEL, T. AND LE CREN, E. D. (1967). Some density relationships for fish population parameters. *In* "The Biological Basis of Freshwater Fish Production." (Gerking, S. D. ed.) 261–293. Blackwell, Oxford.

BAGENAL, T. (1969). Relationship between egg size and fry survival in brown trout *Salmo trutta* L. *J. Fish Biol.* **1**, 349–353.

BENKE, R. J. (1972). The systematics of salmonid fishes of recently glaciated lakes. *J. Fish. Res. Bd Canada* **29** (6), 639–671.

BEVERTON, R. J. H. AND HOLT, S. J. (1957). On the dynamics of exploited fish populations. *Fish. Invest. Lond.* **19** (2), 1–533.

CHAPMAN, D. W. (1966). Food and space as regulators of salmonid populations in streams. *Amer. Nat.* **100**, 345–357.

FRAZER, F. J. (1969). Population density effects on survival and growth of juvenile coho salmon and steelhead trout in experimental stream-channels. *In* "Symposium on Salmon and Trout in Streams." (Northcote, T. G. ed.) 253–266. *Macmillan Lect. Br. Columb. Univ.*

HARTMAN, G. F. (1965). The role of behaviour in the ecology and interaction of under yearling coho salmon (*Oncorhynchus kisutch*) and steelhead trout (*Salmo gairdneri*). *J. Fish. Res. Bd Canada* **22** (4), 1035–1081.

HOAR, W. S. (1969). Comments on the contribution of physiological and behavioural studies to the ecology of salmon in streams. *In* "Symposium on Salmon and Trout in Streams." (Northcote, T. G. ed.), 177–180. *Macmillan Lect. Br. Columb. Univ.*

JOHNSON, W. E. (1965). On mechanisms of self-regulation of population abundance in *Oncorhynchus nerka*. *Mitt. int. Verein. theor. angew. Limnol.* **13**, 66–87.

KALLEBERG, H. (1958). Observations in a stream tank of territoriality and competition in juvenile salmon and trout (*Salmo salar* L. and *S. trutta* L.) *Rep. Inst. Freshw. Res.*, *Drottningholm* **39**, 55–98.

LE CREN, E. D. (1965). Some factors regulating the size of populations of freshwater fish. *Mitt. int. Verein theor. angew. Limnol.* **13**, 88–105.

LE CREN, E. D. (1973). The population dynamics of young trout (*Salmo trutta*) in relation to density and territorial behaviour. *In* "Symposium on Stock and Recruitment." Aarhus 1970. (Parrish, B. B., ed.), *Rapp. P-v. Reun. Cons. perm. int. Explor. Mer* **164**.

LINDROTH, A. (1955). Distribution, territorial behaviour and movements of sea trout fry in the river Indalsälven. *Rep. Inst. Freshw. Res. Drottningholm* **36**, 104–119.

MASON, J. C. (1966). Behavioral ecology of juvenile coho salmon (*O. kisutch*) in stream aquaria with special reference to competition and aggressive behavior. Ph.D. thesis Oregon-State Univ. 195 pp.

PARRISH, B. B. (Ed.) (1973). "Symposium on Stock and Recruitment." Aarhus 1970. *Rapp. P.-v. Reun. Cons. perm. int. Explor. Mer* **164**.

RICKER, W. E. (1954). Stock and recruitment. *J. Fish. Res. Bd Canada* **11** (5), 559–623.

STUART, T. A. (1953). Spawning migration, reproduction and young stages of loch trout. *Freshwat. Salm. Fish. Res.* **5**, 39 pp.

WOLNY, P. AND GRYGIEREK, E. (1972). Intensification of fish ponds production. *In* "Productivity Problems of Freshwaters." (Kajak, Z. and Hillbricht-Ilkowska, A., eds.), 563–571. PWP Polish Scient. Pub., Warszawa-Krakow.

Part III

Population Genetics

8. Deterministic Models in Population Genetics

W. F. Bodmer

Genetics Laboratory, Department of Biochemistry,
University of Oxford, Oxford, England

A major goal of population genetics is to predict the frequencies of phenotypes in future generations knowing their frequencies in the present and perhaps in past generations. Since, however, evolution is necessarily the change in the genetic constitution of a population, a primary requirement for setting up a population genetic model is knowledge of the relationship between phenotype and genotype, as one proceeds from one generation to the next, or in other words, the *pattern of inheritance*. Mendel's laws, supplemented by information on genetic linkage and mutation, provide the basis for the description of inheritance patterns. Natural selection is, of course, considered to be the major agent of evolution. Knowledge concerning the *selective forces* acting on the phenotypes that are being studied is therefore a second essential ingredient of a population genetic model. The causes of differential natural selection may involve complex interactions between fertility and survival, but for the purposes of model building only the overall magnitudes of the effects are required and these can often be specified in a relatively straightforward way. The third essential ingredient of a population genetic model is knowledge concerning the population structure. By this is meant the probabilities with which matings occur between the various phenotypes, which includes specification of sub-divisions of a population and of the patterns of migration between these sub-divisions. Also included in population structure are such factors as the effects of random sampling due to finite population size, which involves the setting up of stochastic rather than deterministic models.

The complete description of evolutionary changes given these three categories of input information is clearly a formidable, if not impossible task. For a model to be, even to some extent, mathematically tractable many simplifications often have to be made. But experience shows that the models can neverthless be useful and that evolution is, to this extent anyway, amenable to quantitative study.

One major assumption that is common to many models is that of *discrete generations*. This means that we assume that all individuals mate at the same given time and are then completely replaced by their offspring. The frequencies of phenotypes and genotypes can then be thought of as referring to some particular time in the life cycle: often for example, the time of onset of reproduction. If we add to the assumption of discrete generations, that of infinite population size, thus eliminating random sampling effects and so stochastic considerations, then the model can be written as a general vector recurrence equation of the following form.

$$\mathbf{u}_n = S(\mathbf{u}_{n-1}). \tag{1}$$

Here $\mathbf{u}_n$ is the vector of frequencies of genotypes in generation n and $\mathbf{u}_{n-1}$ that in generation $n-1$, the relation between the two being specified by a vector function S which depends on the three ingredients of a population genetic model discussed above. Equilibria are obtained by equating $\mathbf{u}_n$ to $\mathbf{u}_{n-1}$. Perturbations about an equilibrium are studied by setting

$$\mathbf{u}_n = \mathbf{u}_e + \mathbf{X}_n \tag{2}$$

where $\mathbf{u}_e$ is the equilibrium vector and $\mathbf{X}_n$ represents a small displacement from the equilibrium. Ignoring squared terms in the vector $\mathbf{X}$ gives rise to a system of linear equations

$$\mathbf{X}_n = \mathbf{A}\mathbf{X}_{n-1} \tag{3}$$

whose matrix $\mathbf{A}$ determines the stability of the equilibrium $\mathbf{u}_e$. Generally, if the dominant latent root of this matrix $\mathbf{A}$ is less than 1, then the equilibrium is stable and vice versa. In most models the function S describing the changes in frequency from one generation to the next is non-linear, often a ratio of polynomials. This means that usually even the simplest of such models cannot be completely solved analytically. Information is then usually sought by studying special cases and by examining behaviour in the neighbourhood of equilibria, especially those on the boundaries of the space within which $\mathbf{u}_n$ is constrained to lie. In the remainder of this paper we shall illustrate some of the approaches to studying discrete deterministic population genetic models by reference to two examples. The first is the problem of the evolution of two alleles at each of two genetically linked loci subject to natural selection in a random mating population. The second involves the introduction of fertility effects and reproductive compensation into the standard one locus two allele random mating model.

Linkage and Selection: the Two Locus, Two Allele Random Mating Diploid Model

Let us first review briefly the situation for one gene locus with two alleles (alternative forms of the gene) A and a. There are three different genotypes

AA, Aa, aa, since we are assuming a diploid organism in which chromosomes, and so genes, occur in pairs. If we assume respective frequencies u, v, w ($u + v + w = 1$) for these three genotypes then it is a classic result, called the Hardy–Weinberg Law after its discoverers, that assuming random mating, no selection and discrete generations in an infinite population, the population reaches in one generation an equilibrium specified by

$$u = p^2 \qquad v = 2pq \qquad w = q^2 \tag{4}$$

where $p = 1 - q = u + \frac{1}{2}v$ is the gene frequency of A in the population. This equilibrium is equivalent to the result obtained by choosing pairs of genes at random from a pool containing pA and qA genes. This process is known as "random union of gametes" since A and a represent the alternative genetic contents of a gamete (egg or sperm) at the locus in question. The Hardy–Weinberg law is thus equivalent to saying that random mating gives the same result as random union of gametes. This result applies quite generally to much more complicated genetic situations and greatly simplifies population genetic models because it reduces them from considering genotypes to considering gametes, or in the case of one locus, just 2 alleles A, a instead of three genotypes (AA, Aa, aa).

Selection is most simply introduced in terms of relative weightings S_1, S_2, S_3 on the three genotypes which are proportional to the probability of survival from being a fertilized zygote (newly fused egg + sperm) to being a reproductively mature individual. This weighting changes zygote genotype frequencies (the immediate products of mating) from

$$p^2, 2pq, q^2 \quad \text{to} \quad \frac{S_1 p^2}{\bar{w}}, \frac{2S_2 pq}{\bar{w}}, \frac{S_3 q^2}{\bar{w}}$$

at maturity where $\bar{w} = S_1 p^2 + 2S_2 pq + S_3 q^2$ is a normalizing factor to make the new genotype frequencies add up to one. This now specifies the model for changes over one generation, a model which was first solved by Fisher in 1922. He proved the now well known result that a stable equilibrium with non-zero p and q exists only if $S_2 > S_1, S_3$; that is if the heterozygote Aa is fitter than both homozygotes AA and aa.

In the case of two loci each with two alleles, say A, a and B, b there are four gametes that can be formed, taking one allele from the first locus and one from the second

$$Ab, \quad Ab, \quad aB \quad \text{and} \quad ab.$$

These four gametes can form ten genotypes which is the number of pairwise combinations of gametes allowing for the possibility of picking the same gamete twice. Two of these, the so called *double heterozygotes*, AB/ab and Ab/aB, are of special interest in connection with the problem of genetic linkage which arises if both loci are on the same chromosome. In this case the gametic outputs of these genotypes take the complementary forms

$$AB/ab \rightarrow \tfrac{1}{2}(1 - r)AB + \tfrac{1}{2}(1 - r)ab + \tfrac{1}{2}rAb + \tfrac{1}{2}raB$$

$$Ab/aB \rightarrow \tfrac{1}{2}(1 - r)Ab + \tfrac{1}{2}(1 - r)aB + \tfrac{1}{2}rAB + \tfrac{1}{2}rab \tag{5}$$

| Parental | Recombinant |
| Combinations | Combinations |

where r, the recombination fraction, is the overall proportion of recombinant (non-parental) combinations produced by either of the two double hetero-zygotes. This quantity, r, is a measure of the "genetic distance" between the loci, which ranges from 0 for complete linkage to $\tfrac{1}{2}$ for independence, corresponding to the situation when the genes are in fact on different chromosomes.

The major questions raised by the two locus problem concern the inter-action between r, the recombination fraction, and the effects of selection on the various genotypes. This problem was first discussed verbally by Fisher in 1930, who suggested that a particular form of interaction between selection and linkage might be of great importance. The quantitative aspects of the problem were first considered to a limited extent by Wright (1952) and later expanded by Kimura (1956), Lewontin and Kojima (1960) and Bodmer and Parsons (1962). Following these first four papers a considerable literature has developed around this subject and references to this can be found in the papers by Bodmer and Felsenstein (1967) and by Karlin and Feldman (1970) which form the basis for the discussion presented in this paper.

Let the frequencies of the four gametes AB, Ab, aB, ab at any given genera-tion be x_1, x_2, x_3 and x_4 respectively ($\sum_i x_i = 1$) and let w_{ij} be the fitness of the genotype consisting of gametic types i and j. Thus, for example, w_{14} is the fitness of genotype AB/ab, by which is meant that the probability an individual of this genotype survives birth to reproduce as a parent is w_{14}. The extension of the Hardy–Weinberg law to this situation means that the genotype frequency array can be expressed formally as

$$(ABx_1 + Abx_2 + aBx_3 + abx_4)^2.$$

Thus the relative frequencies of genotypes AB/ab and AB/AB before selec-tion, for example, are $2x_1x_4$ and x_1^2 respectively and after selection $2w_{14}x_1x_4$ and $w_{11}x_1^2$. The gametic outputs of the various genotypes follow from Mendel's laws, and for the double heterozygotes from the rules for re-combination between a pair of linked loci as given by expression (5). Thus, for example, the contribution made by genotype AB/ab to AB gametes in the next generation is

$$2w_{14}x_1x_4 \times \tfrac{1}{2}(1 - r)$$

or

$$w_{14}(1 - r)x_1x_4.$$

Summing the contributions made by each genotype to each gamete type gives rise to the equations which describe the expected changes in the frequencies in the gametes from one generation (x_i) to the next (x_i'). These equations take the form

$$\bar{w}x_1' = x_1w_1 - r(w_{14}x_1x_4 - w_{23}x_2x_3)$$

$$\bar{w}x_2' = x_2w_2 + r(w_{14}x_1x_4 - w_{23}x_2x_3)$$

$$\bar{w}x_3' = x_3w_3 + r(w_{14}x_1x_4 - w_{23}x_2x_3) \tag{6}$$

$$\bar{w}x_4' = x_4w_4 - r(w_{14}x_1x_4 - w_{23}x_2x_3)$$

where

$$w_i = \sum_{i=1}^{4} x_jw_{ij} \quad \text{for } i = 1, 2, 3, 4 \quad \text{and} \quad \bar{w} = \sum_{i=1}^{4} x_iw_i = \sum_{i=1}^{4}\sum_{j=1}^{4} x_ix_jw_{ij}$$

is a normalizing factor such that $\sum_{i=1}^{4} x_i' = 1$.

This deceptively simple looking system of equations (6) provides the basis for most of the analysis of the interaction between selection and linkage which has so far been done. The equations with general values of w_{ij} are rather intractable. As a result a great deal of work has been done on the so-called symmetrical model which takes the form

$$w_{14} = w_{23} = 1; \quad w_{11} = w_{44} = 1 - \delta; \quad w_{22} = w_{23} = 1 - \alpha$$

$$w_{12} = w_{34} = 1 - \beta \quad \text{and} \quad w_{13} = w_{24} = 1 - \gamma.$$

Equation (6) can then be written in the form

$$\bar{w}x_1 = x_1 - \delta x_1^2 - \beta x_1x_2 - \gamma x_1x_3 - rD$$

$$\bar{w}x_2 = x_2 - \beta x_1x_2 - \alpha x_2^2 - \gamma x_2x_4 + rD$$

$$\bar{w}x_3 = x_3 - \gamma x_1x_3 - \alpha x_3^2 - \beta x_3x_4 + rD \tag{7}$$

$$\bar{w}x_4 = x_4 - \gamma x_2x_4 - \beta x_3x_4 - \delta x_4^2 - rD$$

where $D = x_1x_4 - x_2x_3$. The quantity D turns out to be central to the analysis of this system of equations, and also to equation (6) when $w_{14} = w_{23}$.

D has been called the linkage disequilibrium parameter or, more appropriately, the gametic association, since it is a measure of the association between the two loci on the gametes. Thus, when $D = 0$ the loci behave, effectively, as if they were independent, but not otherwise. This follows from the fact that the gamete frequencies can be expressed in the form

$$x_1 = Pp + D$$
$$x_2 = Pq - D$$
$$x_3 = Qp - D$$
$$x_4 = Qq + D$$

where $P = x_1 + x_2$ and $p = x_1 + x_3$ are the gene frequencies of alleles A and B respectively so that $Q = 1 - P$ and $q = 1 - p$ are the frequencies of alleles a and b. Thus when $D = 0$ the gamete frequencies are just products of the frequencies of their constituent alleles (e.g. Pp for AB). It is a classical result, due to Jennings (1917), that in the absence of selection, when $w_{ij} = 1$ for all i and j, D tends to zero at a rate $(1 - r)$, so that ultimately at equilibrium the two loci will be effectively independent however closely they may be linked.

There are at least three major questions raised by this model, namely:

(1) Are there conditions under which the value of r will initially influence the evolutionary dynamics of the system, especially at its boundaries? For example, are there conditions on r which will determine whether or not a newly formed gamete may increase in frequency?

(2) What are the conditions for the existence and stability of equilibria for which D is appreciably different from zero?

(3) Can one make out a case, under at least some circumstances, for tighter linkage between two loci being advantageous? In other words, can selection influence the relative position of genes on the chromosome? This is a fundamental evolutionary problem but its consideration is beyond the scope of the present paper. Here we shall simply give some examples of answers to the first two questions using the symmetric viability model (equation 7), and illustrate the complexities which can arise even from these much simplified equations.

Suppose that ab is the predominant gamete while Ab, aB and AB are all rare, in other words x_4 is near 1 and x_1, x_2 and x_3 are all small. Then it can be shown, for example, that if $\delta < \beta, \gamma$ then x_1, x_2 and x_3 will only increase in frequency if

$$r < \frac{\delta}{1 - \delta}. \tag{9}$$

This provides a clear example of a situation where new types, in this case alleles A, and B, will only increase in frequency if the two loci are sufficiently closely linked.

It is easily shown that symmetrical equilibrium solutions for equation (7) must exist which can be expressed in the form $x_1 = x_4 = \frac{1}{4} + D$, $x_2 = x_3 = \frac{1}{4} - D$.

It can also be shown that the equilibrium values of D are given by the solutions of a cubic equation whose coefficients are simple functions of α, β, γ and δ. When r is small these solutions tend to be near zero, near $+\frac{1}{4}$ and near $-\frac{1}{4}$. Only the solution giving a value of D close to zero gives rise to an equilibrium situation of near independence between the loci. Let us, for the sake of simplicity, consider the case when $\alpha = \delta$. The three equilibrium solutions are then

$$\text{(a)} \ D = 0 \quad \text{and} \quad \text{(b)} \ D = \pm \frac{1}{4} \left(1 - \frac{8r}{l} \right)^{\frac{1}{2}} \tag{10}$$

where $l = 2(\beta + \gamma) - (\alpha + \delta)$.

When $r > l/8$ equilibria (b) do not exist, and the equilibrium with $D = 0$ is stable only if $\alpha = \delta > |\beta - \gamma|$, while if $\delta < |\beta - \gamma|$ there will be no equilibrium with all four gametes present at non-zero frequencies. When $r < l/8$ then the equilibria (b) with non-zero D exist and can be shown always to be stable provided r is sufficiently small. The conditions on r for stability away from $r = 0$ can however be quite complex depending on the relation between δ and $|\beta - \gamma|$. Thus it can be shown that there exist conditions under which the equilibria (b) are stable in the ranges $0 < r < r^*$ and $r^{**} < r < l/8$ (where $0 < r^* < r^{**} < l/8$), but unstable in the intervening range $r^* < r < r^{**}$.

In spite of the obvious symmetry of equations (7), Karlin and Feldman (1970) have been able to show that there can exist asymmetrical equilibria which can also be stable. Even when these asymmetric equilibria are unstable, they will be important in helping to define the domains of attraction to the stable equilibria.

The four gametic frequencies can be conveniently represented as a point inside a regular tetrahedron whose vertices represent the four possible fixed states including only a single gamete thus

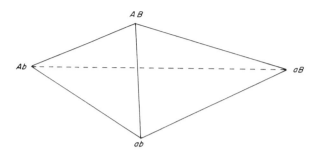

The behaviour of the complete system can then be considered from the point of view of the stability of the boundaries of the tetrahedron, for if no boundary point is stable then there must exist a stable interior point. The tetrahedron has three types of boundaries—the vertices, the edges and the faces. The stability of the vertices is readily investigated and depends to a large extent on conditions such as (9). No face can ever have a stable point if $r \neq 0$ since recombination will always produce the fourth gamete. It is possible however that equilibria exist near a face, corresponding to the situations where one gamete is much rarer than the other three. There can, similarly, be no stable point on the two edges $AB - ab$ and $Ab - aB$. Stable points near these edges correspond to situations for which D is decidedly non-zero. Stable points can exist on the remaining four edges corresponding to a reduction of the system to one locus. Explicit conditions for the stability of such points can be derived thus completing our picture of the existence and stability of boundary points. It can incidentally be shown that the stability of points on the edges may depend on the existence of sufficiently small values of r. There can never exist more than two internal stable points. However, conditions can be found under which, in addition to two stable internal points, there are four stable points on the boundary. This raises the important question of defining the domains of attraction to these equilibria, namely predicting which equilibrium will be approached from any given starting point in the tetrahedron. This problem has so far been quite intractable analytically. All that can be said is that unstable equilibria must lie on boundaries between domains of attraction. Some mapping of the domains can, of course, be done by computer simulation.

Apart from the evolutionary interest in trying to understand why it might be advantageous for certain genes to be close to each other on the chromosomes, there is the practical question of explaining population associations which may be due to genetic linkage. Statistical associations between phenotypes determined by two linked loci will, in general, depend entirely on the existence of non-zero values of D. Recently, important examples of such associations have been described in human populations. These involve association between the HL–A white cell blood group system (or tissue types) and certain diseases, which are explained by the existence of one or more genes affecting susceptibility to the diseases being closely linked to the genes of the HL–A system. An understanding of the reasons for these associations depends on an understanding of the possible basis of the existence of non-zero D values involving alleles from the two loci.

Fertility Models in Population Genetics

In the two locus model discussed above, natural selection was represented by the weighting of genotype frequencies by w_{ij}, corresponding to an effect only on viability. Fertility, however, is a major component of natural selection. It is important therefore to consider what differences from the simple viability models might arise in a model which allows specifically for fertility differences. The complication here is that fertility is, in general, a property of a mating (i.e. a genotype pair) and so, as we shall see, the Hardy–Weinberg law is no longer applicable and genetic changes must be described in terms of genotype rather than gene frequencies. Comparatively little work has so far been done with fertility models. The results to be described here are based on Bodmer (1965) and Feldman *et al.* (1969).

Let us again consider the simple case of one locus with two alleles A and a. If we take into account which genotype is male and which female, there are nine mating types that can be formed from the three genotypes AA, Aa and aa. A general fertility model assigns arbitrary fertilities F_1, F_2 etc. to each mating where these quantities represent an overall relative weighting on the output of the mating. Such a general fertility model is illustrated in Table 1.

TABLE 1. Mating types and fertilities

Mating type		Fertilities		
Female	Male	General	Multiplicative model	Frequency assuming random mating
AA	$\times$ AA	F_1	$f_1 m_1$	u^2
AA	$\times$ Aa	F_2	$f_1 m_2$	uv
Aa	$\times$ AA	F_2^1	$f_2 m_1$	uv
AA	$\times$ aa	F_3	$f_1 m_3$	uw
aa	$\times$ AA	F_3^1	$f_3 m_1$	uw
Aa	$\times$ Aa	F_4	$f_2 m_2$	v^2
Aa	$\times$ aa	F_5	$f_2 m_3$	vw
aa	$\times$ Aa	F_5^1	$f_3 m_2$	vw
aa	$\times$ aa	F_6	$f_3 m_3$	$\dfrac{w^2}{1}$

u, v and w are the respective frequencies of the genotypes AA, Aa and aa.

Thus, assuming random mating as indicated in the table, the total output of the mating female $AA \times$ male Aa is, for example, $F_2 uv$ and so its contributions to the AA and Aa genotypes in the next generation are each $\frac{1}{2}F_2 uv$. In addition to fertility differences we can assign relative viability weightings a, h and b to the three genotypes AA, Aa and aa respectively. Proceeding in the usual way to calculate the combined contributions to the genotypes of the offspring generation, we obtain the following equations relating the new frequencies u', v' and w' to those in the previous generation.

$$\frac{Tu^1}{a} = F_1 u^2 + \tfrac{1}{2}(F_2 + F_2^1)uv + \tfrac{1}{4}F_4 v^2$$

$$\frac{Tv^1}{h} = \tfrac{1}{2}(F_2 + F_2^1)uv + (F_3 + F_3^1)uw + \tfrac{1}{2}F_4 v^2 + \tfrac{1}{2}(F_5 + F_5^1)vw \qquad (11)$$

$$\frac{Tw^1}{b} = \tfrac{1}{4}F_4 v^2 + \tfrac{1}{2}(F_5 + F_5^1)vw + F_6 w^2.$$

Here T is, as usual, a normalizing factor such that $u' + v' + w' = 1$. Only the sums of the fertilities of reciprocal matings (i.e. $F_2 + F_2'$ for $AA \times aa$ and $aa \times AA$) enter into the equation so that the number of independent parameters needed to describe fertility differences can be reduced from 8 to 5. Equations (11) are more or less intractable as they stand and certainly cannot be expressed in terms which correspond in any way to the Hardy–Weinberg Law. This requires that the quadratic forms on the right hand side of equation (11) can be expressed as suitable products of linear functions in u, v, w which correspond in some way to the gene frequencies. If the fertilities take the multiplicative form indicated in Table 1, then it is easily shown that equations (11) can be written in a form which corresponds to an extension of the Hardy–Weinberg law. The multiplicative model assumes that the fertility of a mating is the product of the fertilities of the genotypes taking part in that mating, allowing for differences between male and female fertilities. This reduces the number of independent fertility parameters from 5 to 4. Writing

$$p = \frac{m_1 u + \tfrac{1}{2}m_2 v}{m_1 u + m_2 v + m_3 w}, \qquad q = 1 - p$$

and $\qquad\qquad\qquad\qquad\qquad\qquad\qquad\qquad\qquad\qquad\qquad\qquad\qquad$ (12)

$$P = \frac{f_1 u + \tfrac{1}{2}f_2 v}{f_1 u + f_2 v + f_3 w}, \qquad Q = 1 - P$$

and replacing F_1 by $f_1 m_1$, $F_2 + F_2^1$ by $f_1 m_2 + f_2 m_1$ etc., equations (11) may be rewritten in the form

$$Tu^1 = apP$$
$$Tv^1 = h(pQ + Pq) \qquad (13)$$
$$Tw^1 = bqQ.$$

Here, again, T is such that $u^1 + v^1 + w^1 = 1$.

Here P, Q and p, q represent the gene frequencies of A and a in females and males respectively after weighting genotypes by their fertilities. Equations (13) thus represent an extension of the Hardy–Weinberg law, namely that random mating is equivalent to random union of gametes. It can be shown that this generalization applies to arbitrarily complex genetic situations provided always that the fertility of a mating can be represented by the product of the fertilities of the constituent genotypes. For the investigation of equations (13), it is convenient to work in terms of the male and female "gene ratios" $u_1 = p/q$ and $u_2 = P/Q$ respectively. The equations in u_1 and u_2 involve only the relative values of the products af_1, hf_2, bf_3 and am_1, hm_2, bf_3 respectively. Thus, it can be shown that without loss of generality equations (13) can be written in the form

$$u_1^1 = \frac{(1 - \mu_1)u_1 u_2 + \frac{1}{2}(u_1 + u_2)}{(1 - v_1) + \frac{1}{2}(u_1 + u_2)}$$

$$u_2^1 = \frac{(1 - \mu_2)u_1 u_2 + \frac{1}{2}(u_1 + u_2)}{(1 - v_2) + \frac{1}{2}(u_1 + u_2)} \qquad (14)$$

involving only four independent parameters, where

$$af_1 = 1 - \mu_2 \qquad hf_2 = 1 \qquad bf_3 = 1 - v_2$$

and

$$am_1 = 1 - \mu_1 \qquad hm_2 = 1 \qquad bm_3 = 1 - v_1.$$

This in fact corresponds to a viability model in which viabilities are different in the two sexes, namely

	AA	Aa	aa
Males	$1 - \mu_1$	1	$1 - v_1$
Females	$1 - \mu_2$	1	$1 - v_2$

The equilibria of this system are given by a cubic equation in the ratio u_1/u_2. So long as the male and female parameters are not too different there is no major difference from the standard viability model which ignores sex differences. There are however two novel features of this model. First when $\mu_1, v_1 > 0$ (male heterozygote advantage) and $\mu_2, v_2 > 0$ (female heterozygote disadvantage) conditions can be found which give rise to *two* internal (i.e. non-trivial) equilibria which are stable. Second, a stable internal equilibrium can exist when $\mu_1, v_2 < 0$, $\mu_2, v_1 > 0$ corresponding to a situation in which allele A is favoured in males but a is favoured in females. In terms of fertility effects this could, for example, arise if fertility differences existed only in the female and viability acted in the opposite direction to fertility in both sexes. This result is important because it is an example of a stable equilibrium with a comparatively simple model which is not the result of an advantage of the heterozygote over both homozygotes.

Reproductive Compensation in the Rhesus Blood Group System

As an extension of the previous discussion on fertility models we shall now consider briefly a situation in which the viabilities are, to some extent, a function of the mating type (see Feldman *et al.*, 1969). This arises from an attempt to construct a model for the well-known rhesus blood group system and its associated disease of the new-born which is caused by an immunological reaction of the mother against her foetus. The $Rh(+)$, $Rh(-)$ difference is, in simple terms, controlled by a dominant gene R such that the genotypes RR and Rr are $Rh(+)$ while rr is $Rh(-)$. The selective effect of rhesus incompatibility is when mothers who are $Rh(-)$ react against offspring who are $Rh(+)$. This gives rise to a selective disadvantage for the Rr genotype only in the matings female $rr \times$ male RR or Rr. There are good biological reasons for believing the selection against Rr may not be the same in these two matings. It has, furthermore, been suggested that $Rh(-)$ mothers might compensate for the loss of $Rh(+)$ offspring by an increased fertility. This would, for example, correspond to a desire for a given number of offspring irrespective of their Rh genotype, so that lost offspring are always replaced. A model allowing for these two special types of selective effects, as well as overall relative viabilities of $1 - s, 1, 1 - t$ for genotypes RR, Rr and rr respectively, is illustrated in Table 2. The parameter k in the mating $rr \times RR$ represents the combined effects of incompatibility selection and reproductive compensation. Only one parameter is needed since only one genotype is produced by the mating.

The parameter l represents incompatibility selection against Rr in the mating $rr \times Rr$, while f represents reproductive compensation in this mating. Once again, the Hardy–Weinberg law does not apply. It is readily

TABLE 2. Mating types and selection scheme for Rh blood groups

| Mating | | Offspring distribution | | | |
Female	Male	Frequency at nth generation	RR	Rr	rr
$RR \times RR$		u_n^2	1	0	0
$RR \times Rr$		$u_n v_n$	$\frac{1}{2}$	$\frac{1}{2}$	0
$Rr \times RR$		$u_n v_n$	$\frac{1}{2}$	$\frac{1}{2}$	0
$RR \times rr$		$u_n w_n$	0	1	0
$rr \times RR$		$u_n w_n$	0	$1 - k$	0
$Rr \times Rr$		v_n^2	$\frac{1}{4}$	$\frac{1}{2}$	$\frac{1}{4}$
$Rr \times rr$		$v_n w_n$	0	$\frac{1}{2}$	$\frac{1}{2}$
$rr \times Rr$		$v_n w_n$	0	$\frac{1}{2}(1 - l)f$	$\frac{1}{2}f$
$rr \times rr$		w_n^2	0	0	1
Over-all relative viabilities			$1 - s$	1	$1 - t$

Note—RR and Rr are Rh-positive, and rr is Rh negative.

shown that the equations relating the frequencies of the three genotypes in successive generations take the form

$$\frac{T_n u_{n+1}}{1 - s} = u_n^2 + u_n v_n + \frac{v_n^2}{4},$$

$$T_n v_{n+1} = u_n v_n + u_n w_n (2 - k) + \frac{v_n^2}{2} + \frac{v_n w_n}{2}[1 + f(1 - l)], \quad (15)$$

$$\frac{T_n w_{n+1}}{1 - t} = w_n^2 + \frac{v_n w_n (1 + f)}{2} + \frac{v_n^2}{4},$$

where T_n is a normalizing factor such that $u_{n+1} + v_{n+1} + w_{n+1} = 1$.

Numerical calculations actually show that the system of equations (15) is often very close to a Hardy–Weinberg representation. This is perhaps not surprising considering the fact that only half of the mating $rr \times Rr$ really represents a deviation from the assumption of the Hardy–Weinberg law. As is nearly always the case with such non-linear systems of equations, general solutions cannot be found. In the absence of overall viability differences ($s = t = 0$) the signs of k and $f(1 - l) - 1$ are the decisive factors for the behaviour of the system. When $k < 0, f(1 - l) > 1$ there is "overcompensation" for the incompatibility selection and a stable internal equilibrium exists. Conversely when $k > 0, f(1 - l) < 1$ only the marginal states ($u = 1$ or $w = 1$) are stable. The other cases, namely when k and

$f(1 - l) - 1$ have the same signs, are complicated and the existence of stable non-trivial equilibria depends on the relative magnitudes of k and $f(1 - l) - 1$.

The stability of the marginal equilibria can be readily determined in the general case. When u is near 1, and v and w are small, so that the population is almost all RR and $Rh(+)$, the allele r increases in frequency if $s > 0$. Conversely when w is near 1 and u, v are small, so that the population is nearly all rr and $Rh(-)$, the allele R increases if

$$f(1 - l) > 1 - 2t$$
or
$$t > \tfrac{1}{2}(1 - f(1 - l)). \tag{16}$$

In the first case, when R is the prevalent allele, incompatibility selection is negligible. However, when r is the prevalent allele, R is subject to incompatibility selection in one half the matings in which it occurs, namely those in which the female parent is rr. Thus, as implied by expression (16), for R to increase in this case, unless there is overcompensation ($f(1 - l) > 1$), which is biologically unlikely, the excess viability of Rr over rr ($t > 0$) must first of all exceed the incompatibility selection against Rr. These conditions alone suggest that R may be the original allele and r the newcomer, a hypothesis which fits the world distribution of R and r quite well. $Rh(-)$ individuals are relatively localized, being found predominantly in Caucasoid populations.

Conclusions

The examples discussed in this paper illustrate just one aspect of the analysis of population genetic models. Even though greatly over simplified in comparison with reality, as they must be, they nevertheless provide valuable insight into interesting biological situations. Increasing reality, which is needed to provide models that can be used for more quantitative rather than qualitative fits to observed data leads to much more complex models. Taking account of the stochastic element is, of course, one of the major complicating factors and also one of the greatest mathematical challenges to population genetic theory. Some aspects of stochastic analysis are considered by the authors of genetical papers in this volume.

REFERENCES

(*For general references to population genetics—see the paper in this volume by Bartlett.*)

BODMER, W. F. (1965). Differential fertility in population genetics models. *Genetics* **51** (3), 411–424.

BODMER, W. F. AND FELSENSTEIN, J. (1967). Linkage and selection: Theoretical analysis of the deterministic two locus random mating model. *Genetics* **57**, 237–265.

BODMER, W. F. AND PARSONS, P. A. (1962). Linkage and recombination in evolution. *Advan. Genet.* **11**, 1–100.

FELDMAN, M. W., NABHOLZ, M. AND BODMER, W. F. (1969). Evolution of the Rh polymorphism: A model for the interaction of incompatibility, reproductive compensation and heterozygote advantage. *Amer. J. of Hum. Genet.* **21** (2), 171–193.

FISHER, R. A. (1922). On the dominance ratio. *Proc. Roy. Soc. Edinburgh* **42**, 321–341.

FISHER, R. A. (1930). "The Genetical Theory of Natural Selection." Oxford University Press (2nd Edition 1958, Dover).

JENNINGS, H. S. (1917). The numerical results of diverse systems of breeding, with respect to two pairs of characters, linked or independent, with special relation to the effects of linkage. *Genetics* **2**, 97–154.

KARLIN, S. AND FELDMAN, M. W. (1970). Linkage and Selection: Two locus symmetric viability model. *Theoretical Population Biol.* **1**, 39–71.

KIMURA, M. (1956). A model of a genetic system which leads to closer linkage by natural selection. *Evolution* **10**, 278–287.

LEWONTIN, R. C. AND KOJIMA, K. (1960). The evolutionary dynamics of complex polymorphisms. *Evolution* **14**, 458–472.

WRIGHT, S. (1952). The genetics of quantitative variability. *In* "Quantitative Inheritance," (K. Mather, ed.), pp. 5–41. Her Majesty's Stationery Office, London.

9. Sex and Infinity: a Mathematical Analysis of the Advantages and Disadvantages of Genetic Recombination

SAMUEL KARLIN*

The Weizmann Institute of Science, Rehovot, Israel
and
Stanford University, Stanford, California, U.S.A.

1. Introduction

Recent literature has witnessed much interest and controversy engaged with the problem of discerning the evolutionary advantages of sexual recombination. This subject has intrigued and perplexed a number of different authors; the classical perspective stems from some cryptic comments of Fisher (1930) elaborated upon by Muller (1932). We can summarize their thesis as follows: the major evolutionary feature of genetic recombination is that it facilitates the accumulation in a single individual of advantageous mutations which arose separately in different individuals.

The following is a simple model used to investigate this contention. Consider a series of favourable mutations occurring at a set of loci, say two for definiteness. Let A, a and B, b denote the possible alleles at locus 1 and 2 respectively. Assume mutation is unidirectional transforming

$$a \to A \quad \text{and} \quad b \to B$$

at rates μ and λ respectively, where for ease of exposition we take $\mu = \lambda$. The double mutant chromosomal type AB is the most fit and the haplotype (gamete) ab represents the wild type. The restriction of the model encompassing unidirectional but recurrent mutation is widely adopted as a reasonable approximation to reality. Thus, reverse mutation events are assumed as negligible.

The Fisher–Muller line of reasoning on the advantage of recombination is approximately as follows. Two different advantageous mutations in an asexual population can only be incorporated into a single individual if

* Supported in part by N.I.H. Grant USPRS 10452-09.

one of the mutations occurs in a descendant of an individual in which the other mutation occurred. For a sexually reproducing population (i.e., one with a recombination mechanism) the two mutations which arise in different individuals can be combined in the same genome by recombination. Thus in a population with recombination the fixation of the different mutants will be more or less independent.

Fisher's statement, "Recombination would accelerate evolution in the incorporation of favourable mutations" is open to a variety of interpretations. It is in fact a very vague statement. Does it mean "Recombination hastens the appearance of a first double mutant when single mutants exist and are selectively advantageous?" Or alternatively, can it be that "recombination causes the earlier fixation of favoured double mutants?" If the single mutants are disadvantageous but are advantageous in combination, what then? Does incorporation of mutants mean that they reach 5%, 50% or 100% of the population? There has been a tendency to accept the "favourability" of recombination, whatever that means, and reject models whose conclusions do not, apparently, endorse it. Recombination disrupts associations between mutually favourable advantageous genes while at the same time helping the formation of new advantageous combinations. This balance between recombination producing new types and ultimately perhaps breaking up advantageous types has been widely discussed by Dobzhansky, Darlington, Mather and others.

Continuing the verbal argument it is commonly stated that:
(1) Asexual reproduction exploits the immediate advantages with which the most fit types are endowed;
(2) On the other hand, recombination provides means for the "quick" production of new types capable of adapting to changing environments.

The problem as to whether recombination speeds up or slows down the total and/or first incorporation of the desired type is not easily settled in view of the contrasting motivations of properties (1) and (2). Another familiar assertion is that perhaps the optimum strategy for evolution would alternate sexual and asexual phases. In fact, many organisms encompass in their life cycle both an asexual and a sexual stage of different relative durations.

In any qualitative or quantitative interpretation of Fisher and Muller it is also essential to take account of population size, the range of fitness parameters, order of mutation rates, niche effects, and other factors.

Muller (1958) and then Crow and Kimura (1965) attempted to refine the earlier ideas of Muller and set forth a quantitative framework for evaluating the evolutionary advantages of asexual and sexual populations. The underlying assumptions and accompanying analysis of their models are tenuous. They assert that the "advantage" conferred by recombination is greater when the population is large and mutation small. Their main conclusion in the deterministic version of the problem (see later for details) was brought into contention by Maynard Smith (1968) by means of a numerical counter example. Eshel and Feldman (1970) rigorously validated Maynard Smith's observations. In their model the relative fitness of AB, Ab, aB and ab were σ_1, σ_2, σ_3, 1 ($\sigma_2 = \sigma_3$). The surprising outcome of this model is that if $\sigma_1 > \sigma_2^2 > 1$, so that the double mutant is more favourable than the independent combination of single mutants and if the initial state consists entirely of "ab" individuals, then in subsequent generations the frequency of the double mutant is always larger *without* recombination than with it. A criticism of this model has been that it commences with all individuals "ab" before the mutation starts. In our treatment of the problem below, we find the evolution of the process quite sensitive to the specification of the initial frequency vector of the population's genetic structure. Owing to this fine dependence on initial conditions, the disagreement of Crow and Kimura with Maynard Smith can be amicably resolved (see Section 2).

Another case of considerable biological interest is that where the two single mutants are unfavourable but the double mutant has a selective advantage. Eshel and Feldman (1970) investigated this case imposing unnecessarily strong restrictions on the fitness coefficients. Karlin and McGregor (1971), stimulated by their work, have established that if $(1 - r)\sigma_1 < 1$, a stable mutation selection balance can be set up which prevents the subsequent increase of the double mutant. On the other hand without recombination the double mutant proceeds to fixation. Actually, the condition $(1 - r)\sigma_1 < 1$ was shown to be necessary and sufficient for the existence of a mutation selection balance (see Section 3).

Since haploid systems cannot, under the action of selection alone, maintain polymorphisms, this result could imply that recombination is also advantageous in the sense of maintaining potential variability. Deleterious mutations could be stored, until they are of benefit single rather than merely in combination (see Section 3).

Another source of controversy in the Fisher–Muller thesis concerns the interactive effect of finite population size and the recombination force on the rate of incorporation. Crow and Kimura inferred that increased population size and increased recombination rate operate concomitantly to increase the rate of incorporation of favourable mutant types while Bodmer (1970) argued to the contrary that the advantage of recombination is greater, the smaller the population size. Bodmer regards the problem of the time until the formation of the first double mutant type as the important criterion by which to gauge the advantage of recombination. Where a number of single mutants are already existing it appears that small population size in the presence of recombination will encourage the quicker formation of new combinations. Bodmer strongly contends that this is the most common situation. He states that most data indicate that the most obligate sexual systems exist in organisms with relatively small population size.

Intuitively, if a state exists where ab and the single mutants Ab and aB are present together in the population, the formation of the first double mutant will be expected earlier if recombination exists as against no recombination. In fact, double mutants can arise from double mutation and also with recombination, if possible.

But the problem does not end there. Does the double mutant remain at a higher frequency in the population with recombination? Evidently not. Our calculations (see Section 5) produce the conclusion that, in a finite population, fixation in the favoured double mutant is expected to take longer with recombination than without, whereas the formation of the first double mutant was expected earlier. Intermediate proportions of double mutants are difficult to track; nevertheless the problem arises—at what proportion of double mutants does one define "incorporation" to be fulfilled? Bodmer, unfortunately treats an infinite deterministic population where the notion of first passage time is not well defined. Crow and Kimura and also Maynard Smith (1971) appear to be concerned with both the first passage time and with the time for total fixation. Crow and Kimura and Maynard Smith essentially ignored the dependence of first passage time on the recombination fraction. These authors attempt to evaluate the combined effects of population size and recombination strength on the rate of evolution through the study of a set of hybrid deterministic and stochastic models. The assumptions underlying the models are obscure and the analyses are quite rough. A recent paper of Felsenstein (1973) on this subject suffers the same difficulties.

Maynard Smith (1971) has emphasized certain ecological factors. He envisages frequent situations where two genetically different populations migrate into a new environment such that the favoured genotype is now a mixture of the genes from the invading populations, and sexuality (rather than recombination) will promote the mixing.

This paper offers a careful formulation of the basic concepts and models taking due account of varying fitness parameters, population size effect and magnitudes of mutation rates. We are concerned with the role of sexuality expressed through the extent of recombination and its impact on the "rate of evolution". There are other *fascinating* aspects of the concept of sexuality and recombination with its attendant theory, observation and speculation. It may be worth mentioning some of these related problems not dealt with in the paper:

1. How to explain the varying manifestations of sex determining mechanisms, and the related problem of the evolution of general incompatibility systems.

2. How to account for the abundance of populations exhibiting an approximate 1:1 sex ratio.
3. To explain the evolutionary significance and representations of sexual dimorphism contrasted to sexual monomorphism. There is the associated problem of delineating the advantages and disadvantages of haploidy versus diploidy.
4. The relevance of the difference in sizes of male and female sex gametes.
5. Why not three or more sexes?

We will not attempt to cite from the voluminous references on these topics.

The concepts and problems examined in this work have relevance to the study of all the above problems; hopefully they will add insight to the prohibitive problem of explaining the evolution of sex.

Sections 2 and 3 set forth the basic deterministic model studied by Crow and Kimura (1965), Maynard Smith (1968) and Eshel and Feldman (1970). A more thorough discussion of its structure and properties is given. We will see that the results depend sensitively on the initial conditions.

In Section 4 we develop a continuous time version of the model; this is restricted in that no fitness differences among gamete types are introduced.

Section 5 formulates the stochastic finite population size version of this model. The expected time until the formation of the first double (and pth order) mutant are essentially evaluated. The expected total fixation time is calculated in asymptotic terms in Section 6. Recourse to diffusion approximations are exploited in order to work our some of these calculations. Computer runs and simulations are also relied upon to yield insights.

The final section contains a summary of the main points of the results. Throughout the body of the text discussion and interpretations are interspersed. The complete mathematical analyses validating the results reported will be published elsewhere.

In concluding this introduction, it might be interesting to present a fuller quotation from Fisher (1930) on the subject of the "advantages" of the sexual over the asexual form of reproduction:

"A consequence of sexual reproduction which seems to be of fundamental importance to evolution theory is that advantageous changes in different structural elements of the germ plasm can be taken advantage of independently; whereas with asexual organisms either the genetic uniformity of the whole group must be such that evolutionary progress is greatly retarded, or if there is considerable genetic diversity, many beneficial changes will be lost through occurring in individuals destined to leave no ultimate descendants in the species. In consequence an organism sexually reproduced can respond so much more rapidly to whatever selection is in action, that if placed in competition on equal terms with an asexual organism similar in all other respects. the latter will certainly be replaced by the former."

Crew (1965), in his book on Sex Determination (p. 2) summarizes Fisher's viewpoint as follows:

"The function of sexual reproduction is to quicken the pace of evolutionary progress by making it possible to obtain an accumulation of advantageous mutational steps without having the respective multiplication of these steps occur in series."

The lengthy introduction of this section and our later developments strongly point up the difficulties in interpreting the above "plausible" statements.

2. The Basic Deterministic Model Case (all Favorable Mutations)

Most of the authors cited in Section 1 take as the basic model a *two-locus* infinite haploid population consisting of four gamete types with possible alleles a, A and b, B at the first and second locus respectively. It is generally believed that the evolution of any sexual expression necessarily commended in a haploid stage, and this is undoubtedly the partial justification for concentrating on a haploid formulation. The parameters are listed in the table below.

Gamete	AB	Ab	aB	ab
Fitness coefficients	σ_1	σ_2	σ_3	1
Frequencies in a given generation of haploid individuals	x_1	x_2	x_3	x_4

The recombination fraction is denoted by r. The mutation rate of $a \to A$ or $b \to B$ is μ in both cases with mutations occurring independently at each locus.*

For definiteness we postulate, as is done in most of the literature cited (although this assumption is irrelevant to our qualitative conclusions), that the effects occur in the order

mutation → random union of gametes → segregation → selection.

Thus the population can be envisaged as consisting of mature haploids which produce gametes to be fertilized; at this stage mutations occur. After segregation, selection operates.

The recursion relations connecting the gamete frequencies in two successive generations $(x_1, x_2, x_3, x_4) \to (x'_1, x'_2, x'_3, x'_4)$ are derived in the standard

---

* All our analysis and results carry over allowing different mutation rates at each locus; specifically rate μ_1 for $a \to A$ and μ_2 for $b \to B$.

way yielding

$$Wx'_1 = \sigma_1\{x_1 + \mu(x_2 + x_3) + \mu^2 x_4 - r(1 - \mu)^2 D\},$$

$$Wx'_2 = \sigma_2\{(1 - \mu)x_2 + \mu(1 - \mu)x_4 + r(1 - \mu)^2 D\},$$

$$Wx'_3 = \sigma_3\{(1 - \mu)x_3 + \mu(1 - \mu)x_4 + r(1 - \mu)^2 D\},$$

$$Wx'_4 = (1 - \mu)^2 x_4 - r(1 - \mu)^2 D,$$

(2.1)

where $D = x_1 x_4 - x_2 x_3$ (the disequilibrium expression) and W as usual stands for the sum of the right hand members of the four equations. W is a quadratic function of the variables x_1, x_2, x_3, x_4.

We stipulate throughout that

$$\sigma_1 > \max(1, \sigma_2, \sigma_3)$$

(2.2)

so that the double mutant gamete AB is most fit. In considering the advantage of recombination on the rate of incorporation of the AB-gametes into the population, we will dwell on two principal situations:

(a) All single mutations are not unfavourable, $\sigma_2 \geq 1, \sigma_3 \geq 1$
(b) Single mutations are each deleterious, $\sigma_2 < 1, \sigma_3 < 1$

(2.3)

but the double mutant carries a fitness advantage in accordance with (2.2).

We focus in this section on (a) (situation (b) is treated in Section 3). There are three cases to be distinguished:

(i) $\sigma_1 > \sigma_2 \sigma_3 > 1$ supermultiplicative viabilities
(ii) $\sigma_1 = \sigma_2 \sigma_3 \geq 1$.multiplicative viabilities
(iii) $1 < \sigma_1 < \sigma_2 \sigma_3$ submultiplicative viabilities.

The interpretation of (ii) is manifest. Each mutation contributes an *independent* positive value to the fitness of its carrier. Thus, the fitness of a double mutant is inherited in a multiplicative manner from the fitness advantage of the separate mutation events at the two loci.

The supermultiplicative postulate (case (i)) expresses an interaction effect where the double mutation event confers a greater fitness value than the independent benefits derived from single mutations. This assumption is probably of most biologic significance and we will accordingly concentrate our attention primarily on it. Related results for the multiplicative case are also indicated since many of the writers (e.g., Crow and Kimura, Maynard Smith, Bodmer), because of mathematical convenience, restrict consideration primarily to the case (ii).

The submultiplicative case is not of special interest; a mathematical treatment of it is contained in Karlin (1972).

It is convenient to introduce the notation

AB	Ab	aB	ab
$x_1^{(n)}$	$x_2^{(n)}$	$x_3^{(n)}$	$x_4^{(n)}$

for the frequencies of the gamete types in the nth generation, corresponding to $r > 0$ while

$$y_1^{(n)}, \quad y_2^{(n)}, \quad y_3^{(n)}, \quad y_4^{(n)}$$

will denote the associated frequencies when $r = 0$ (no recombination). We sometimes exhibit the dependence on r by writing explicitly $x_i^{(n)}(r)$ rather than $x_i^{(n)}$.

We shall henceforth refer to the development of the frequency vector $\mathbf{x}^{(n)} = (x_1^{(n)}, x_2^{(n)}, x_3^{(n)}, x_4^{(n)})$ $n \geq 1$ as the *sexual process* (involving a recombination mechanism) and to that of $\mathbf{y}^{(n)} = (y_1^{(n)}, y_2^{(n)}, y_3^{(n)}, y_4^{(n)})$ as the *asexual process*.

Since the forces of selection and mutation concordantly favour the AB gamete subject to the stipulation of cases (i) and (ii), it is intuitively expected that the population tends to fixation of the AB gamete, independently of the extent of recombination, $r \geq 0$. This fact is established rigorously in Karlin (1972). Formally, we have

$$x_1^{(n)} \to 1 \qquad y_1^{(n)} \to 1 \qquad n \to \infty. \tag{2.4}$$

The concepts of "rate of evolution" or "rate of incorporation" can be formalized by comparing the "rates and manner of convergence" of $x_1^{(n)}$ and $y_1^{(n)}$ to 1. Suppose for a given initial frequency vector

$$x_i^{(0)} = y_i^{(0)}, \qquad i = 1, 2, 3, 4$$

there would result the inequalities

$$x_1^{(n)} < y_1^{(n)} \quad \text{for all } n \geq 1. \tag{2.5}$$

This inequality tells us that in each generation following the initial one, the frequency of the double mutant type in an asexually reproducing population exceeds that of the corresponding sexually evolving population. In the circumstance (2.5), we can state unequivocally that the rate of asexual evolution is greater than the rate of sexual evolution. If the reverse inequality to (2.5) is maintained, then sexual evolution can be regarded unambiguously as more efficient than asexual evolution in incorporating the AB gamete.

Eshel and Feldman (1970) demonstrated that if

$$x_1^{(0)} = x_2^{(0)} = x_3^{(0)} = 0, \qquad x_4^{(0)} = 1 \quad \text{and} \quad x_i^{(0)} = y_i^{(0)} \tag{2.6}$$

i.e., the initial population consists entirely of wild types, then subject to the supermultiplicative assumption (i), we have the relations

$$x_1^{(n)} < y_1^{(n)} \qquad n \geq 1 \tag{2.7}$$

and incorporation of the double mutant type proceeds faster for asexual reproduction than for sexual reproduction. (This fact appeared to us quite surprising at first, and contrary to the ubiquitous phenomenon of the sexual mechanism.) Actually the above authors proved more: namely, when

$$\mathbf{x}^{(0)} = (x_1^0, x_2^0, x_3^0, x_4^0) = \mathbf{y}^{(0)},$$

is such that

$$D^{(0)} = x_1^{(0)} x_4^{(0)} - x_2^{(0)} x_3^{(0)} \geq 0 \tag{2.8}$$

then already (2.7) prevails. The quantity D called the disequilibrium function is a measure of association (or "correlation") of the alleles A and B. Indeed, determine a random variable X to be 1 or 0 according as the first locus carries the allele A or a respectively and Y is defined analogously with respect to the second locus. For a given frequency array a direct calculation reveals the identity $\text{Cov}(X, Y) = D$. This interpretation is very familiar.

For the situation of multiplicative fitnesses (case (ii)) where $D^{(0)} = 0$, we obtain the identity

$$x_1^{(n)} = y_1^{(n)} \quad \text{for all } n \tag{2.9}$$

and recombination neither enhances nor impedes (it has no effect) the rate of fixation of the AB chromosomal type. The conclusion (2.9) and the results connected with (2.7) appear to contradict some of the claims of Crow and Kimura (see also Maynard Smith, 1968). We can secure the further inference from our analysis that if $D^{(0)} \geq 0$, and the super multiplicative assumption holds (case (i)), then

$$x_1^{(n)}(r) < x_1^{(n)}(r^*) \quad \text{for all } n \geq 1 \tag{2.10}$$

provided $0 \leq r^* < r$ (Karlin, 1972).

Thus, for an initial frequency vector $x^{(0)}$ with non-negative linkage disequilibrium $D^{(0)} \geq 0$ increasing recombination progressively slows the rate of incorporation of the desired AB haplotype.

The results associated with (2.7) and (2.10) were predicated on an initial state with $D^{(0)} \geq 0$. A pertinent inquiry as to the nature of the progress of the frequency vectors $x^{(n)}$ and $y^{(n)}$ for initial conditions with $D^{(0)} < 0$ is in order. It can be proved generally that for any initial starting point (excepting the fixation state $(1, 0, 0, 0)$ we ultimately obtain for an appropriate N

$$D^{(n)} > 0 \quad \text{for all } n \geq N(x^{(0)}) \tag{2.11}$$

i.e., the population achieves a positive linkage disequilibrium following the lapse of a sufficient number of generations. Comparing the fact of (2.11) and the result associated with (2.7) and (2.9), it is tempting to conjecture that for any initial $x^{(0)} = y^{(0)} \not\equiv (1, 0, 0, 0)$ ultimately we would have

$$x_1^{(n)} < y_1^{(n)} \quad \text{for } n \geq N^*(x^{(0)}). \tag{2.12}$$

With some surprise we have established the following opposite result.

THEOREM 1. *Let*

$$(x_1^0 = 0, x_2^{(0)} = x_3^{(0)} > 0, x_4^{(0)} \geq 0), \tag{2.13}$$

$\sigma_2 = \sigma_3$, $\sigma_1 > \sigma_2^2 > 1$ *and suppose* r *satisfies* $(\sigma_1/\sigma_2^2)(1 - r) > 1$. *Then for sufficiently small mutation rate, we have*

$$x_1^{(n)} > y_1^{(n)}, \qquad n \geq 1. \tag{2.14}$$

It should be emphasized that whereas for $D^{(0)} \geq 0$ always (2.7) prevails there are cases where $D^{(0)} < 0$ and (2.7) continues to hold. The above theorem tells us that there are also other cases with $D^{(0)} < 0$ and suitable r and μ for which the complete reverse to (2.7), that is, (2.14) is maintained.

Under the specifications of the theorem involving an initial state of negative linkage disequilibrium, we see that sexual recombination leads to quicker fixation of AB than would occur in a corresponding asexually evolving population. This conclusion is in sharp contrast to the result associated with (2.7) assuming (2.8). In view of these differences, it is therefore important to ascertain the relevance and naturalness of an initial state of the kind (2.13) entailing negative linkage disequilibrium versus an initial state satisfying (2.8).

Biological situations pointing to an initial state of type (2.13) arise as follows. Imagine two isolated populations exhibiting a polymorphic equilibrium at the first and second locus respectively, perhaps maintained by mutation-selection balance. In particular, the first population involves only the chromosomal types ab and Ab while in the second population only the chromosomal types ab and aB are represented. Changes in the environment may so have improved the fitnesses of Ab and aB that they came to be advantageous with respect to ab. Further events also cause the two populations to come together and mix. At this moment the population state is of the kind (2.13). Maynard Smith (1968) and (1971) conceived that such phenomena might occur with considerable likelihood and significance for evolutionary theory. Actually he used the above perspective to argue in favour of an initial conditions with $D^{(0)} = 0$ feeling this was possibly okay if the population was infinite. We feel the initial condition $D^{(0)} < 0$ is a more logical consequence of his heuristic reasoning. In a recent personal communication he seems to agree with this conclusion.

Another means of achieving an initial state prescription expressing negative linkage disequilibrium emanates from considerations of population size effect. In a finite population, mutation events would be likely to produce many single mutants of both types before any double mutant individuals were formed. At this point, the state (2.13) is the appropriate configuration.

Founder or fluctuation effects may also be an important factor producing an initial $D^{(0)} < 0$. It appears that Bodmer (1970) thought along these lines.

The relevance of the initial state specification (2.6) (that of a pure wild type population) can possibly be justified as follows. In determining the effects of some new influence imposed on an existing genetic system (e.g., mutation effects) it is reasonable to take the initial state in the extended system to be a stable equilibrium point of the original system. It is known that the only possible stable equilibria for a two-locus haploid selection model are those of fixation states, i.e., pure populations (see Feldman, 1971). The frequency vector $(0, 0, 0, 1)$ can be such an equilibrium state where only the wild type ab is represented.

Also, if the population size N is originally small, it is likely that the natural initial gamete numbers correspond to $(0, 0, 0, N)$ since it is unlikely that single mutant types have been formed in a moderate time period, especially if $\mu = 10^{-8}$ and $N \leq 10^4$. Now suppose a sudden increase in population size takes place before the onset of mutation events then the resulting initial frequency state $(0, 0, 0, 1)$ appears in order.

We sum up the salient conclusions of the preceding discussion.

Case (i) Supermultiplicative fitnesses
The relative advantage or disadvantage of recombination on the rate of evolution depends sensitively on the state of the initial frequency vector $\mathbf{x}^{(0)}$. Thus if $D^{(0)} \geq 0$ then the asexual process constantly maintains more of the favoured double mutant haplotype and brings quicker fixation of AB than the corresponding sexual process. For cases of $D^{(0)} < 0$ (for example where $x_1^0 = 0$, $x_2^0 = x_3^0 > 0$, $x_4^0 \geq 0$) increased recombination accelerates the incorporation of the haplotype AB provided μ is sufficiently small and r satisfies $\sigma_1(1 - r) > \sigma_2^2$. Thus a sufficiently small mutation rate is more likely to be associated with an "advantage" to sexual evolution.

The above analytic findings are even more pronounced. Thus, when $D^{(0)} \geq 0$, then in each generation the asexually evolving population has a higher frequency of the double mutant gamete than a population commencing from the identical initial state but subject to an additional recombination pressure. When $D^{(0)} < 0$ the complete reverse relationships can prevail giving an "advantage" to recombination.

The above paragraph emphasizes the fact that both sexual and asexual populations can persist because each is endowed with advantages depending on the nature of the initial frequency state of the population. It was explained earlier how certain biological conditions motivate an initial state configuration satisfying $D^{(0)} < 0$ while in other circumstances $D^{(0)} \geq 0$ is natural. The appropriate initial state may differ and be determined by changing environmental conditions, founder effects or sudden interruptions of certain existing stable configurations in small or large populations.

Case (ii) Multiplicative fitnesses (This includes the possibility of no selection effects.)

If $D^{(0)} = 0$ then $x_1^{(n)} = y_1^{(n)}$ for all $n \geq 1$ and recombination exercises no effect on the progress towards fixation.

For $D^{(0)} > 0$ we have $x_1^{(n)} < y_1^{(n)}$, $n \geq 1$ and the assumption $D^{(0)} < 0$ entails $x_1^{(n)} > y_1^{(n)}$ for $n \geq 1$.

The above result indicates that the effectiveness of the recombination mechanism in accelerating the incorporation of the double mutant decisively depends on the sign of the initial value of $D^{(0)}$. It is suggested that in any finite population version of the model where the population number N is sufficiently large (and therefore approximates the deterministic setting) the influence of N is only of secondary importance while the sign of $D^{(0)}$ is a primary factor for the estimation of the advantages or disadvantages of recombination. Section 5 discusses this aspect of the problem in greater depth.

3. The Deterministic Model (Deleterious Single Mutations but Favourable Double Mutants)

Fisher enunciated his dictum on the value of sexual reproduction in the context of all advantageous mutations. It is of interest, and possibly of greater importance, to investigate the nature of the evolution of the model of Section 2 for the case where

$$\sigma_2 < 1, \quad \sigma_3 < 1) \quad \text{(single mutation events are deleterious)} \quad (3.1a)$$

while

$$\sigma_1 > 1 \quad \text{(the double mutant type is advantageous).} \quad (3.1b)$$

The fitness array (3.1), manifestly satisfies the supermultiplicative assumption, case (i), Section 2.

A description of the development of the process is contained in Karlin and McGregor (1971). Some special situations were analysed earlier by Eshel and Feldman (1970). Under the assumptions of (3.1) it is possible for fixation of the AB haplotype not to occur, and for a two-locus mutation selection balance to be established with the wild type frequency persisting in abundance. The precise result is as follows:

THEOREM 2: *Let the conditions* (3.1) *hold. If*

$$(1 - r)\sigma_1 < 1 \tag{3.2}$$

then there exists a positive μ_0 such that for all μ ($0 < \mu < \mu_0$) there exists a stable polymorphism (rather mutation selection balance) $x^* = (x_1^*, x_2^*, x_3^*, x_4^*)$ where $x_1^* + x_2^* + x_3^* \leq \varepsilon_0$ and $\varepsilon_0(\mu_0)$ tends to zero as μ_0 tends to zero. When $(1 - r)\sigma_1 > 1$ holds, fixation of the AB gamete occurs independent of the rate of mutation.

The qualitative implications of this theorem are now concisely stated.

1. For mutation events which are singly deleterious but manifesting an advantageous (double) mutant, an increase in the recombination rate would require increasing fitness advantage of the double mutant type to insure fixation of the *AB* gamete. Equivalently, increased recombination has an increased deterrent effect in establishing the desired *AB* haplotype.

2. With unidirectional recurrent small mutations operating ultimately in favour of the most fit gamete and fitness coefficients as in (3.1) it is essential that recombination be sufficiently loose (to the extent that inequality (3.2) holds) in order that a stable balanced polymorphism be possible. Observe that the larger the value of r the greater the likelihood for the validity of (3.2), and concordantly more opportunities for mutation selection balance for even larger mutation rates.

Our result on mutation selection balance further underlines the importance of recombination as a device for maintaining potential and actual variability.

If the loci are tightly linked so that $(1 - r)\sigma_1 > 1$ in the haploid case, then the population independent of the mutation rate will always fix on a pure state consisting of only *AB* gametes. In fact, additional mutation pressure toward the *AB* gametes can only increase the rate of its fixation.

3. The result of the theorem also has relevance to the findings of Feldman (1971) whose work suggests that in a haploid two-locus population, selection pressures alone cannot maintain a stable polymorphism. Thus, to achieve polymorphism some other influences apart from selection pressures should be operating.

Some ways of producing stable polymorphisms for multi-locus haploid systems could involve the following mechanisms and structures:

(i) an incompatibility mating behaviour,
(ii) selection coupled with certain assortative (and not only disassortative or incompatibility) patterns of mating,
(iii) multi-niche sub populations subject only to selection forces with small gene flow.

4. It is noteworthy that for the haploid or diploid model, when at least one of the single type mutants is of neutral character (in the sense that the

fitness coefficient is the same as that of the wild type) but where the double mutant is selectively advantageous, then it can be shown that the existence of mutation selection balance is precluded. Thus, the deleterious character of each single mutation with the beneficial effect of the double mutant such that (3.1) holds, is an essential requirement for the feasibility of a stable polymorphic equilibrium.

For other implications on Theorem 2, its formal validation and development of results for the diploid version of the model, see Karlin and McGregor (1971).

4. A Continuous Time Model without Selection

The studies of Sections 2 and 3 concerned frequency changes occurring in discrete non-overlapping generations. We now treat the model of Section 2 in continuous time but without selection. It is expected that the qualitative results to be described remain correct in the presence of small selection pressures.

Let $X_1(t)$, $X_2(t)$, $X_3(t)$, $X_4(t)$ be the frequencies of the gamete types AB, Ab, aB, ab respectively at time t. The pressures acting on the population include unidirectional mutations at a rate μ and recombination with rate r. The transformation relations for $X_i(t)$ are now expressed through the differential equations

$$
\begin{aligned}
X'_1(t) &= \mu X_2(t) + \mu X_3(t) - r(1 - 2\mu)D(t) \\
X'_2(t) &= -\mu X_2(t) + \mu X_4(t) + r(1 - 2\mu)D(t) \\
X'_3(t) &= -\mu X_3(t) + \mu X_4(t) + r(1 - 2\mu)D(t) \\
X'_4(t) &= -2\mu X_4(t) - r(1 - 2\mu)D(t)
\end{aligned}
\tag{4.1}
$$

where

$$
D(t) = X_1(t)X_4(t) - X_2(t)X_3(t).
$$

It is readily checked from (4.1) that if $X_i(0) \geq 0$, and $\sum_{i=1}^{4} X_i(0) = 1$, these relations are preserved for $t \geq 0$.

A direct calculation leads to

$$
D'(t) = -[2\mu + r(1 - 2\mu)]D(t).
$$

Let $\gamma = 2\mu + r(1 - 2\mu)$; then we readily derive

$$
D(t) = D(0)\,e^{-\gamma t}.
\tag{4.2}
$$

Substituting into the equations of (4.2) and solving successively gives

$$
X_1(t) = 1 - X_4(t) - Y(t) \qquad Y(t) = X_2(t) + X_3(t)
\tag{4.3}
$$

and

$$X_4(t) = D(0)\,e^{-\gamma t} + [X_4(0) - D(0)]\,e^{-2\mu t}$$
$$Y(t) = -2D(0)\,e^{-\gamma t} + 2[D(0) - X_4(0)]\,e^{-2\mu t} + [Y(0) + 2X_4(0)]\,e^{-\mu t} \quad (4.4)$$

Examining the explicit formula for $X_1(t)$, and noting that r enters only through γ, we may infer the following facts:

For $D(0) > 0$ and $0 \leq r < r^*$,

$$X_1(t, r) > X_1(t, r^*) \quad \text{for all } t > 0 \tag{4.5}$$

$(X_1(t, r) = X_2(t)$ where the dependence on r is exhibited).

For $D(0) < 0$,

$$X_1(t, r) < X_1(t, r^*) \tag{4.6}$$

and when $D(0) = 0$,

$$X_1(t, r) = X_2(t, r^*) \quad \text{independent of } r. \tag{4.7}$$

The conclusions of (4.5) and (4.7) conform with the corresponding results associated with (2.7) and (2.9) of the discrete generation model.

The quantity $X_1'(t)$ can be interpreted loosely as the flux at time t of the production of the AB gamete. Roughly, we have that $X_1'(t)\,dt$ is the proportion of AB gametes produced during the time epoch t to $t + dt$. Accordingly, we define

$$T = \int_0^\infty t X_1'(t)\,dt$$

as the elapsed time for the population to fix on the AB gamete. This is an underestimate of the actual time to fixation; T is evaluated by substituting from (4.3) and (4.4) yielding

$$T = \frac{D(0)}{2\mu\{1 + [2\mu/r(1 - 2\mu)]\}} + \frac{Y(0)}{\mu} + \frac{3X_4(0)}{2\mu}.$$

Therefore, if

$$D(0) > 0, \quad T \text{ increases with } r$$

$$D(0) = 0, \quad T \text{ is independent of } r$$

$$D(0) < 0, \quad T \text{ decreases with } r.$$

A version of this continuous time model, taking account of selection differences among gametes, can also be formulated. We will not write out the associated differential equation system as they appear intractable. But we are convinced that the qualitative properties of the process will parallel those described in Section 2.

5. The Effect of Finite Population Size on the Relative Advantage of Sexual over Asexual Reproduction

Our present objective will be to evaluate the impact of population size on the rate of formation and incorporation of the advantageous double mutant type. We start by setting down a proper stochastic version of the model of Section 2 involving no selection differences among the gamete types. It is quite recognized that it would be most desirable to incorporate selection into the models. This appears to be intractable at present. However, it is emphasized that probably all our qualitative findings for the no-selection case will continue to be valid in the presence of small selection parameters.

Discrete time stochastic model
Let N be the constant population number and let (i_1, i_2, i_3, i_4) denote the numbers of AB, Ab, aB and ab gametes respectively in the present generation. Manifestly, the constraints $i_v \geq 0$, $\sum_{v=1}^4 i_v = N$ apply.

For a given current state, define the quantities

$$p_1 = \frac{i_1}{N} + \mu \frac{i_2}{N} + \mu \frac{i_3}{N} + \mu^2 \frac{i_4}{N} - r(1 - \mu)^2 D$$

$$p_2 = (1 - \mu)\frac{i_2}{N} + \mu(1 - \mu)\frac{i_4}{N} + r(1 - \mu)^2 D$$

$$\text{(5.1)}$$

$$p_3 = (1 - \mu)\frac{i_3}{N} + \mu(1 - \mu)\frac{i_4}{N} + r(1 - \mu)^2 D$$

$$p_4 = (1 - \mu)^2 \frac{i_4}{N} - r(1 - \mu)^2 D$$

where $D = (i_1 i_4 - i_2 i_3)/N^2$.

Comparing (5.1) to (2.1) these equations express the deterministic transformation law for a set of frequencies

$$(x_1, x_2, x_3, x_4) = \left(\frac{i_1}{N}, \frac{i_2}{N}, \frac{i_3}{N}, \frac{i_4}{N}\right)$$

involving no selection differences, where the left hand side computes the resulting frequencies of the next generation. The stochastic model is structured by superimposing a multinomial sampling scheme where the average changes are determined consonant to (5.1). Specifically we construct a time homogeneous Markov chain with state space $\Delta = \{\mathbf{i} = (i_1, i_2, i_3, i_4), i_v$ non-negative integers summing to N} and probability law

$$P_{\mathbf{i},\mathbf{j}} = \frac{N!}{j_1!j_2!j_3!j_4!}p_1^{j_1}p_2^{j_2}p_3^{j_3}p_4^{j_4} \tag{2.5}$$

for the transition from state $\overline{\mathbf{i}} = (i_1, i_2, i_3, i_4)$ to $\overline{\mathbf{j}} = (j_1, j_2, j_3, j_4)$; $\overline{\mathbf{i}}, \overline{\mathbf{j}} \in \Delta$.

The mean changes over successive generations with a prescribed current state $\mathbf{i}$ reduces to that of (5.1). The formulation of the stochastic model through (5.2) is named the Wright–Fisher type construction as they introduced sampling effects in such a multinomial guise.

In view of the unidirectional mutation process, it is clear that there exists a unique absorbing state $(N, 0, 0, 0) = \boldsymbol{\alpha}$ where the whole population consists of AB individuals. The stochastic variations in gamete numbers compel ultimate fixation of the AB haplotype (absorption into $\boldsymbol{\alpha}$) from any initial state.

There are two principal random variables associated with the stochastic process of importance to our investigations.

Define

$$\tau(\mathbf{i}) = \text{as the time elapsed until formation of the first double mutant.} \tag{5.3}$$

More specifically, $\tau(\mathbf{i})$ counts the number of elapsed generations until the i_1 component of the state variable first becomes positive. Thus, this random variable is a first passage time to that set of states $\mathbf{i} = (i_1, i_2, i_3, i_4) \in \Delta$ where $i_1 > 0$. We have exhibited the dependence of $\tau(\mathbf{i})$ on the initial state $\mathbf{i}$, it necessarily also is a function of the parameters r, μ and N. For ease of notation the dependence on these latter variables has been suppressed.

Define the random variable

$T(\mathbf{i})$ as the time elapsed until absorption into state $\boldsymbol{\alpha} = (N, 0, 0, 0)$
(i.e. fixation of the AB gamete) $\tag{5.4}$

where $\mathbf{i}$ denotes the initial state. Since population size is finite and all states, apart from $\boldsymbol{\alpha}$, communicate, the random variables $\tau(\mathbf{i})$ and $T(\mathbf{i})$ are well defined.

Bodmer (1970) stressed that the important time epoch relevant to the incorporation of the AB gamete in the presence of finite population size was the value of $\tau(\mathbf{i})$. Roughly, he argued that once a double mutant or a level of a few percent was reached, the environmental background is markedly altered, so that AB becomes endowed with a large selective advantage, subsequently causing its frequency to increase rapidly. For example, a possibility is that the mating preferences of the double mutant type are strongly assortative (and not random) quickly increasing its frequency in the population compared to the other types. Other authors insist that the small population size effect is maintained for substantially many generations and establishment of a double mutant requires essentially total incorporation, implying that $T(\mathbf{i})$ is the pertinent variable. It is possible to introduce a family of random variables measuring intermediate stages of incorporation. Let $T(\mathbf{i}; \gamma)$ (γ fixed, $0 < \gamma \le 1$) be the time when the population process (with initial state $\mathbf{i}$) first achieves a state where the number of AB gametes exceeds γN. Thus $T(\mathbf{i}; \frac{1}{2})$ indicates the time where incorporation is essentially half accomplished.

For purposes of comparing the rates of formation of the AB gamete in asexual (recombination fraction $r = 0$) and sexual ($r > 0$) populations influenced by sampling fluctuations, it would be valuable to determine the distribution functions or at least the first and/or second moments of $\tau(\mathbf{i})$ and $T(\mathbf{i})$. We concentrate on computing the expectations of $T(\mathbf{i})$ and $\tau(\mathbf{i})$ written $E(\mathbf{i})$ and $F(\mathbf{i})$ respectively. Based on familiar probabilistic reasoning the vector $E(\mathbf{i})$ can be formally calculated as the solution of the system of linear equations

$$E(\mathbf{i}) = 1 + \sum_{\mathbf{j} \in \Delta} p_{\mathbf{i}\mathbf{j}} E(\mathbf{j}) \qquad \mathbf{i} \in \Delta \tag{5.5}$$

where $p_{\mathbf{i}\mathbf{j}}$ is defined in (5.2) and Δ denotes the state space. For the special state $\boldsymbol{\alpha} = (N, 0, 0, 0)$, we necessarily prescribe the value $E(\boldsymbol{\alpha}) = 0$ since zero time is required to fixation of AB if the population composition is all AB.

In principle, the computation of $E(\mathbf{i})$ is done by inverting the system (5.5); this does not seem practicable. The case $N = 2$ of (5.5) involves 10 variables but using certain symmetries, these can be reduced to 6 unknowns.

$$\xi_1 = E(0, 0, 0, 2) \qquad\qquad \xi_2 = E(0, 0, 1, 1) + E(0, 1, 0, 1)$$

$$\xi_3 = E(0, 0, 2, 0) + E(0, 2, 0, 0) \qquad \xi_4 = E(0, 1, 1, 0)$$

$$\xi_5 = E(1, 0, 0, 1) \qquad\qquad \xi_6 = E(1, 1, 0, 0) + E(1, 0, 1, 0)$$

$$\xi_7 = E(2, 0, 0, 0) = 0.$$

A judicious calculation commencing from (5.5), labelling

$$\frac{\partial \xi_1}{\partial r} = \eta_1, \qquad \frac{\partial \xi_2}{\partial r} = \eta_2, \qquad \frac{\partial \xi_4}{\partial r} + \frac{\partial \xi_5}{\partial r} = \eta_3,$$

yields the equations

$$\eta_1[1 - (1 - \mu)^4] - \eta_2 4(1 - \mu)^3\mu - \eta_3 2\mu^2(1 - \mu)^2 = 0$$

$$\eta_1(-1)\frac{(1 - \mu)^4}{4} + \eta_2[1 - \tfrac{1}{2}(1 - \mu)^3(1 + 2\mu)] - \eta_3\tfrac{1}{2}\mu(1 + \mu)(1 - \mu)^2 = 0$$

$$\eta_1(-1)\frac{(1 - \mu)^4}{4} + \eta_2[(-1)\mu(1 - \mu)^3] + \eta_3\left[1 - \frac{(1 - \mu)^2(1 + \mu)^2}{4}\right]$$

$$= [4 - 2(1 - \mu)^4(1 - r)^2]^{-1}. \quad (5.6)$$

It is easy to infer from these equations that $\eta_1 > 0$, $\eta_2 > 0$ and $\eta_3 > 0$. Furthermore, we can deduce that all ξ_i increase with r except ξ_4 corresponding to the state variable $(0, 1, 1, 0)$ which has a *negative linkage disequilibrium value*.

A computational solution of (5.5) for $N = 3$ (here already 8 variables are present), and for all $N \le 10$ and Monte Carlo simulations for $N = 100$ and $N = 1000$ reveals the result that the expected time until fixation, commencing from the initial state $(0, 0, 0, N)$ is *monotone increasing* as a function of increasing recombination rate r (see Table 1(a)). *Thus, even with no selection differences among the gamete types, but provided the population size is finite, increasing recombination has an increasing deterrent effect on progress towards the total incorporation of the double mutant gamete.* More explicitly, the larger the recombination pressure the lengthier the expected time until fixation of AB. Recall however from Section 2, equation (2.9), that for *the case of no selection differences and the same mutation pattern, but with infinite population size, recombination has no influence on the rate of fixation if the initial frequency state has $D = 0$ value* (zero linkage disequilibrium). These striking comparisons suggest the implication that the larger the population size the smaller the deterrent effect due to increased recombination at least with respect to the objective of total incorporation of the AB gamete. This agrees somewhat with the view held by Crow and Kimura and reaffirmed by Maynard Smith (1971), concerning the effects and influences of population size on evolving sexual as against asexual populations; but it should be emphasized that these qualitative findings rest decisively on the prescription of the initial state being $(0, 0, 0, N)$. The identical conclusions appear to persist for any initial state specification $\mathbf{i}^0 = (i_1^0, i_2^0, i_3^0, i_4^0)$ satisfying $D^{(0)} \ge 0$. In line with Theorem 1, Section 2, it is confirmed by numerical computation for our stochastic model that where $D^{(0)} < 0$, the opposite fact ensues; namely increased recombination speeds the process for total incorporation. It further appears in these cases that increasing population number works concordantly with increasing recombination rate.

TABLE 1(a). Values of $E(0,0,0,N)$, the expected time for total fixation with initial state $\alpha = (0,0,0,N)$

	μ \ r	0·0	0·001	0·01	0·1	0·2	0·4	0·5
$N=2$	10^{-1}	$1{\cdot}6284158098^{1}$	$1{\cdot}6284224538^{1}$	$1{\cdot}6284816626^{1}$	$1{\cdot}6290196053^{1}$	$1{\cdot}6295162038^{1}$	$1{\cdot}6302564242^{1}$	$1{\cdot}6305238248^{1}$
	10^{-3}	$1{\cdot}5017438978^{3}$	$1{\cdot}5017439007^{3}$	$1{\cdot}5017439268^{3}$	$1{\cdot}5017441537^{3}$	$1{\cdot}5017443481^{3}$	$1{\cdot}5017446139^{3}$	$1{\cdot}5017447036^{3}$
	10^{-5}	$1{\cdot}5000174994^{5}$	$1{\cdot}5000174994^{5}$	$1{\cdot}5000174994^{5}$	$1{\cdot}5000174994^{5}$	$1{\cdot}5000174994^{5}$	$1{\cdot}5000174995^{5}$	$1{\cdot}5000174995^{5}$
	10^{-8}	$1{\cdot}5000000175^{8}$	$1{\cdot}5000000175^{8}$	$1{\cdot}5000000175^{8}$	$1{\cdot}5000000175^{8}$	$1{\cdot}5000000175^{8}$	$1{\cdot}5000000175^{8}$	$1{\cdot}5000000175^{8}$
$N=3$	10^{-1}	$1{\cdot}7516728359^{1}$	$1{\cdot}7516892750^{1}$	$1{\cdot}7518350238^{1}$	$1{\cdot}7530986133^{1}$	$1{\cdot}7541712235^{1}$	$1{\cdot}7556072313^{1}$	$1{\cdot}7560771543^{1}$
	10^{-3}	$1{\cdot}5035909586^{3}$	$1{\cdot}5035909731^{3}$	$1{\cdot}5035911003^{3}$	$1{\cdot}5035920916^{3}$	$1{\cdot}5035927990^{3}$	$1{\cdot}5035935886^{3}$	$1{\cdot}5035938122^{3}$
	10^{-5}	$1{\cdot}5000360698^{5}$	$1{\cdot}5000360698^{5}$	$1{\cdot}5000360698^{5}$	$1{\cdot}5000360699^{5}$	$1{\cdot}5000360700^{5}$	$1{\cdot}5000360701^{5}$	$1{\cdot}5000360701^{5}$
	10^{-8}	$1{\cdot}5000000361^{8}$	$1{\cdot}5000000361^{8}$	$1{\cdot}5000000361^{8}$	$1{\cdot}5000000361^{8}$	$1{\cdot}5000000361^{8}$	$1{\cdot}5000000361^{8}$	$1{\cdot}5000000361^{8}$
$N=4$	10^{-1}	$1{\cdot}8611137571^{1}$	$1{\cdot}8611385795^{1}$	$1{\cdot}8613576320^{1}$	$1{\cdot}8631792481^{1}$	$1{\cdot}8646139578^{1}$	$1{\cdot}8663614104^{1}$	$1{\cdot}8668843371^{1}$
	10^{-3}	$1{\cdot}5054975652^{3}$	$1{\cdot}5054976053^{3}$	$1{\cdot}5054979521^{3}$	$1{\cdot}5055003835^{3}$	$1{\cdot}5055018457^{3}$	$1{\cdot}5055032164^{3}$	$1{\cdot}5055035496^{3}$
	10^{-5}	$1{\cdot}5000552899^{5}$	$1{\cdot}5000552899^{5}$	$1{\cdot}5000552899^{5}$	$1{\cdot}5000552902^{5}$	$1{\cdot}5000552903^{5}$	$1{\cdot}5000552905^{5}$	$1{\cdot}5000552905^{5}$
	10^{-8}	$1{\cdot}5000000553^{8}$	$1{\cdot}5000000553^{8}$	$1{\cdot}5000000553^{8}$	$1{\cdot}5000000553^{8}$	$1{\cdot}5000000553^{8}$	$1{\cdot}5000000553^{8}$	$1{\cdot}5000000553^{8}$
$N=6$	10^{-1}	$2{\cdot}0459018565^{1}$	$2{\cdot}0459387105^{1}$	$2{\cdot}0462617411^{1}$	$2{\cdot}0487949811^{1}$	$2{\cdot}0505950801^{1}$	$2{\cdot}0525388057^{1}$	$2{\cdot}0530599751^{1}$
	10^{-3}	$1{\cdot}5093283613^{3}$	$1{\cdot}5093285149^{3}$	$1{\cdot}5093298077^{3}$	$1{\cdot}5093373555^{3}$	$1{\cdot}5093407930^{3}$	$1{\cdot}5093433308^{3}$	$2{\cdot}5093438396^{3}$
	10^{-5}	$1{\cdot}5000940520^{5}$	$1{\cdot}5000940521^{5}$	$1{\cdot}5000940522^{5}$	$1{\cdot}5000940530^{5}$	$1{\cdot}5000940534^{5}$	$1{\cdot}5000940536^{5}$	$1{\cdot}5000940537^{5}$
	10^{-8}	$1{\cdot}5000000941^{8}$	$1{\cdot}5000000941^{8}$	$1{\cdot}5000000941^{8}$	$1{\cdot}5000000941^{8}$	$1{\cdot}5000000941^{8}$	$1{\cdot}5000000941^{8}$	$1{\cdot}5000000941^{8}$
$N=8$	10^{-1}	$2{\cdot}1982482813^{1}$	$2{\cdot}1982920888^{1}$	$2{\cdot}1986741884^{1}$	$2{\cdot}2015503235^{1}$	$2{\cdot}2034558768^{1}$	$2{\cdot}2053638337^{1}$	$2{\cdot}2058430409^{1}$
	10^{-3}	$1{\cdot}5131730025^{3}$	$1{\cdot}5131733835^{3}$	$1{\cdot}5131765098^{3}$	$1{\cdot}5131921203^{3}$	$1{\cdot}5131978055^{3}$	$1{\cdot}5132037853^{3}$	$1{\cdot}5132020138^{3}$
	10^{-5}	$1{\cdot}5001331483^{5}$	$1{\cdot}5001331483^{5}$	$1{\cdot}5001331487^{5}$	$1{\cdot}5001331509^{5}$	$1{\cdot}5001331509^{5}$	$1{\cdot}5001331514^{5}$	$1{\cdot}5001331514^{5}$
	10^{-8}	$1{\cdot}5000001332^{8}$	$1{\cdot}5000001332^{8}$	$1{\cdot}5000001332^{8}$	$1{\cdot}5000001332^{8}$	$1{\cdot}5000001332^{8}$	$1{\cdot}5000001332^{8}$	$1{\cdot}5000001332^{8}$

TABLE 1(b). Values of $E(0, 1, 0, N - 1)$

N	μ \ r	0	0·1	0·5
2	10^{-3}	$1·2518689588^3$	$1·2518691721^3$	$1·2518696302^3$
	10^{-5}	$1·2500187494^5$	$1·2500187494^5$	$1·2500187495^5$
	10^{-8}	$1·2500000188^8$	$1·2500000188^8$	$1·2500000188^8$
5	10^{-3}	$1·4074598089^3$	$1·4074651105^3$	$1·4074697960^3$
	10^{-5}	$1·4000751145^5$	$1·4000751151^5$	$1·4000751156^5$
	10^{-8}	$1·4000000751^8$	$1·4000000751^8$	$1·4000000751^8$
8	10^{-3}	$1·4507042596^3$	$1·4507232195^3$	$1·4507330356^3$
	10^{-5}	$1·4376334608^5$	$1·4376334628^5$	$1·4376334639^5$
	10^{-8}	$1·4375001335^8$	$1·4375001335^8$	$1·4375001335^8$

TABLE 1(c). Values of $E(0, 1, 1, N - 2)$

N	μ \ r	0	0·1	0·5
2	10^{-3}	$1·0019940140^3$	$9·7933576205^2$	$9·1881094452^2$
	10^{-5}	$1·0000199994^5$	$9·7729341871^4$	$9·1668816859^4$
	10^{-8}	$1·0000000200^8$	$9·7727274797^7$	$9·1668816859^7$
5	10^{-3}	$1·3075098086^3$	$1·3004226955^3$	$1·2909328220^3$
	10^{-5}	$1·3000756145^5$	$1·2929332701^5$	$1·2834097567^5$
	10^{-8}	$1·3000000756^8$	$1·2928572190^8$	$1·2833334098^8$
8	10^{-3}	$1·3882354825^3$	$1·3846163252^2$	$1·3813744112^3$
	10^{-5}	$1·3751337733^5$	$1·3714576976^5$	$1·3681898901^5$
	10^{-8}	$1·3750001338^8$	$1·3713236636^8$	$1·3680556899^8$

Note. (i) The accuracy of the tabulations are to the order 10^{-11}; (ii) The end exponents in each entry indicates a multiplying factor of the power of 10. For example interpret $1·5093298097^3$ = $1·5093298097 \times 10^3$; (iii) Each unit corresponds to a generation.

Implications of numerical calculations

(a) The data shows that $E(0, 0, 0, N)$, the expected time until fixation of AB where initially the population consists of all "*ab*" is increasing (slightly but definitely) with increasing recombination fraction for each finite N. We have checked this property for $N \leq 20$ (tabulations displayed go to $N \leq 8$) and by Monte Carlo for $N = 10^2$ and 10^3. This fact was rigorously demonstrated for $N = 2$, consult (5.6).

(b) As $N \to \infty$, the general theory of Section 2 tells us that recombination will exercise diminishing influence (for $N = \infty$, no effect) on the rate of incorporation. This property is confirmed by the tables. Note that already for $N = 8, \mu = 10^{-8}$, we see no effect in $E(0, 0, 0, N)$ up to accuracy of 10^{-12}. Computer runs refining the accuracy up to 10^{-17} barely start to indicate the increasing property of $E(0, 0, 0, N)$.

(c) The deterrent effect (expressed through lengthier expected times of fixation) of increasing recombination in the presence of finite population size is weakened as $N \to \infty$ and also increased with diminished mutation rate. These properties are drawn from the tables.

(d) The numerical results also indicate that $E(0, 1, 0, N - 1)$ also increases with increasing r. Note that the D value of $(0, 1, 0, N - 1)$ is also zero. It is likely that for any critical state $\mathbf{i}$ where $D_{\mathbf{i}} \geq 0$, the function $E_N(\mathbf{i})$ is increasing with increasing recombination.

(e) It appears that the effects of increasing recombination is very slight. Nevertheless there is a discernable trend indicating that increased recombination slows the progress towards full fixation. We would expect in the presence of multiplicative or supermultiplicative fitnesses the retardation effects of increased recombination to be sharpened.

For μ very small, N large but $N\mu$ of moderate magnitude recombination appears to have a more pronounced expression of impeding the fixation rate.

(f) For the initial state $(0, 1, 1, N - 2)$ exhibiting one single mutant of each kind and the rest all wild type, we see that increased recombination markedly diminishes the fixation time. This was the initial state chosen by Bodmer (c.f. Section 2 of this paper). In agreement with Bodmer (1971), increased population size diminishes the relative advantage of each unit of increased recombination.

The initial state $(0, 1, 1, N - 2)$ has a negative linkage disequilibrium. To repeat, Table 1(c) indicates for each N that the value of $E(0, 1, 1, N - 2)$ is emphatically decreased as r increases.

Diffusion approximation

The Markov chain process (5.2) can be approximated by a diffusion process letting $N \to \infty$ in a suitable manner, namely, $N\mu \to \theta, Nr \to \rho, 0 < \theta, \rho < \infty$ (e.g., consult Kimura (1964) or Crow and Kimura (1970, Chapters 8 and 9) and Karlin and McGregor (1964)). The state space of the continuous process can be identified with the simplex in two dimensions $S = \{(x, y); 0 \leq x, y; x + y \leq 1\}$ where the variable $x = i_1/N$ is the proportion of the double mutants in the population and $y = (i_2 + i_3)/N$ is the proportion of single mutants. The reduction to two variables is legitimate owing to the equality in the mutation rates at the two loci and the symmetry that this entails for the numbers of Ab and aB individuals, where the initial state consists of equal numbers of Ab and aB individuals. The quantities $\theta = N\mu$ and $\rho = Nr$ represent the average intensity of mutation and the average recombination force of the total population, respectively. It can be proved that the expected time $E(x, y)$ until the population comprises only double mutants ($x = 1$), when the initial state is (x, y), is a solution of the differential equation

$$\tfrac{1}{2}x(1 - x)\frac{\partial^2 E}{\partial x^2} - xy\frac{\partial^2 E}{\partial x \partial y} + \tfrac{1}{2}y(1 - y)\frac{\partial^2 E}{\partial y^2} + (\theta y - \rho D)\frac{\partial E}{\partial y}$$

$$+ [\theta(2 - 2x - 3y) + 2\rho D]\frac{\partial E}{\partial y} + 1 = 0 \qquad (5.7)$$

where

$$D = D(x, y) = x(1 - x - 2y) - \frac{y^2}{4}.$$

Auxiliary conditions appended in order to characterize the desired solution $E(x, y)$ uniquely are the properties of non-negativity and obvious monotonicities of $E(x, y)$, finiteness of $E(0, 0)$ and the partial derivative of $(0, 0)$. We have not succeeded in solving (5.7), or in securing reasonable approximations.

The associated diffusion process for evaluating the expected time $\phi(x)$ to fixation for a simple haploid population with unidirectional mutation takes place on the segment $0 \le x \le 1$ where $1 - x$ represents the frequency of the mutant type. The appropriate differential equation satisfied by $\phi(x)$ is

$$\tfrac{1}{2}x(1 - x)\frac{\partial \phi}{\partial x} - \gamma x\frac{\partial \phi}{\partial x} + 1 = 0, \qquad \phi(0) = 0, \qquad \phi(1) < \infty, \qquad (5.8)$$

In order to determine ϕ uniquely, we can require that $\phi(x)$ be the smallest positive solution of (5.8) or alternatively $\phi'(1) < \infty$. The required explicit solution is

$$\phi(x) = 2 \int_0^x \frac{1}{(1 - \eta)^{2\gamma}}\left(\int_\eta^1 \frac{(1 - \xi)^{2\gamma - 1}}{\xi}\, d\xi\right) d\eta \qquad (5.9)$$

Continuous time stochastic model
Another common method for constructing a stochastic version of the model of Section 4 apart from that of Wright–Fisher, (5.2), is through the method of birth–death events. This we will refer to as the Moran type formulation (c.f. Moran, 1962).

The state space of the process is as before

$$\Delta = \{\mathbf{i}; \mathbf{i} = (i_1, i_2, i_3, i_4)\}.$$

Changes of state can be induced by birth, death, mutation and recombination events. Let λ be the rate of events, i.e. the probability of an event generating a change of state, during the time epoch $(t, t + dt)$ is $\lambda \, dt + o(dt)$. Transitions for birth–death and mutation events occur by choosing an individual to die (all possibilities equally likely) and then choosing an individual to replicate from among AB, Ab, aB, ab types with probabilities i_1/N, i_2/N, i_3/N, i_4/N respectively. Following birth, a mutation occurrence can alter the type of the new offspring. Define the quantities

$$q_1(\mathbf{i}) = \frac{i_1}{N} + \mu \frac{(i_2 + i_3)}{N} + \mu^2 \frac{i_4}{N}$$

$$q_2(\mathbf{i}) = (1 - \mu)\frac{i_2}{N} + \mu(1 - \mu)\frac{i_4}{N},$$

$$q_3(\mathbf{i}) = (1 - \mu)\frac{i_3}{N} + \mu(1 - \mu)\frac{i_4}{N},$$

$$q_4(\mathbf{i}) = (1 - \mu)^2 \frac{i_4}{N}.$$

$$(5.10)$$

There are 12 possibilities for birth–death mutation events where one type increases its number and a second type diminishes its number; these are

$$\mathbf{i} = (i_1, i_2, i_3, i_4) \to (i_1 + 1, i_2 - 1, i_3, i_4) \text{ with probability } \frac{i_2}{N} q_1(\mathbf{i})$$

$$\to (i_1 + 1, i_2, i_3 - 1, i_4) \text{ with probability } \frac{i_3}{N} q_1(\mathbf{i})$$

$$\to (i_1 + 1, i_2, i_3, i_4 - 1) \text{ with probability } \frac{i_4}{N} q_1(\mathbf{i})$$

$$\to (i_1 - 1, i_2 + 1, i_3, i_4) \text{ with probability } \frac{i_1}{N} q_2(\mathbf{i})$$

$$\vdots \qquad\qquad\qquad \vdots$$

$$\to (i_1, i_2, i_3 - 1, i_4 + 1) \text{ with probability } \frac{i_3}{N} q_4(\mathbf{i})$$

The two possible recombination events have the representation

$$(i_1, i_2, i_3, i_4) \rightarrow (i_1 + 1, i_2 - 1, i_3 - 1, i_4 + 1) \text{ with probability } \frac{i_2 i_3}{N^2} r$$

$$\rightarrow (i_1 - 1, i_2 + 1, i_3 + 1, i_4 - 1) \text{ with probability } \frac{i_1 i_4}{N^2} r.$$

During the infinitesimal period $(t, t + dt)$ no change of state occurs with probability

$$1 - \lambda \left[\left(1 - \frac{i_1}{N}\right) q_1 + \left(1 - \frac{i_2}{N}\right) q_2 + \left(1 - \frac{i_3}{N}\right) q_3 + \left(1 - \frac{i_4}{N}\right) q_4 \right.$$
$$\left. + r \frac{i_2 i_3}{N^2} + r \frac{i_1 i_4}{N^2} \right] dt + o(dt)$$

(where we write for brevity $q_i = q_i(\mathbf{i})$, $i = 1, 2, 3, 4$).

In this continuous time model the expected time $E(\mathbf{i})$ until total fixation satisfies the system of equations

$$\left[\left(1 - \frac{i_1}{N}\right) q_1 + \left(1 - \frac{i_2}{N}\right) q_2 + \left(1 - \frac{i_3}{N}\right) q_3 + \left(1 - \frac{i_4}{N}\right) q_4 + r \frac{i_2 i_3}{N^2} + \frac{i_1 i_4}{N^2} r \right] E(\mathbf{i})$$

$$= \frac{1}{\lambda} + \frac{i_2}{N} q_1 E(i_1 + 1, i_2 - 1, i_3, i_4) + \frac{i_3}{N} q_1 E(i_1 + 1, i_2, i_3 - 1, i_4)$$

$$+ \frac{i_4}{N} q_1 E(i_1 + 1, i_2, i_3, i_4 - 1)$$

$$+ \frac{i_1}{N} q_2 E(i_1 - 1, i_2 + 1, i_3, i_4) + \dots \qquad (5.11a)$$

$$\vdots$$

$$+ \dots \frac{i_2}{N} q_4 E(i_1, i_2 - 1, i_3, i_4 + 1) + \frac{i_3}{N} q_4 E(i_1, i_2, i_3 - 1, i_4 + 1)$$

$$+ \frac{i_1 i_4}{N^2} r E(i_1 - 1, i_2 + 1, i_3 + 1, i_4 - 1)$$

$$+ \frac{i_2 i_3}{N^2} r E(i_1 + 1, i_2 - 1, i_3 - 1, i_4 + 1)$$

The absorbing state $\boldsymbol{\alpha} = (N, 0, 0, 0)$ necessarily has $E(\boldsymbol{\alpha}) = 0$.

By scaling λ, μ and r properly relative to N, so that specifically $Nr \rightarrow \rho$, $\lambda \sim N^2$, $N\mu \rightarrow \theta$, $0 < \rho, \theta < \infty$ we can pass from (5.11a) to a limiting differential equation identical to (5.7).

Results from computer runs to evaluate $E(\mathbf{i})$ from the system (5.10) are recorded in Tables 2.

TABLE 2(a). Value of $E(0, 0, 0, N)$ for continuous time model

μ \ r		0·0	0·001	0·01	0·1	0·2	0·4	0·5
$N = 2$	10^{-1}	$3\cdot1824769433^1$	$3\cdot1824769433^1$	$3\cdot1824769433^1$	$3\cdot1824769433^1$	$3\cdot1824769433^1$	$3\cdot1824769433^1$	$3\cdot1824769433^1$
	10^{-3}	$3\cdot0019980030^3$	$3\cdot0019980030^3$	$3\cdot0019980030^3$	$3\cdot0019980030^3$	$3\cdot0019980030^3$	$3\cdot0019980030^3$	$3\cdot0019980030^3$
	10^{-5}	$3\cdot0000199998^5$	$3\cdot0000199998^5$	$3\cdot0000199998^5$	$3\cdot0000199998^5$	$3\cdot0000199998^5$	$3\cdot0000199998^5$	$3\cdot0000199998^5$
	10^{-8}	$3\cdot0000002000^8$	$3\cdot0000002000^8$	$3\cdot0000002000^8$	$3\cdot0000002000^8$	$3\cdot0000002000^8$	$3\cdot0000002000^8$	$3\cdot0000002000^8$
$N = 3$	10^{-1}	$5\cdot0223694565^1$	$5\cdot0223746292^1$	$5\cdot0224208994^1$	$5\cdot0228578928^1$	$5\cdot0232945276^1$	$5\cdot0240426721^1$	$5\cdot0243657583^1$
	10^{-3}	$4\cdot5059902780^3$	$4\cdot5059902795^3$	$4\cdot5059902927^3$	$4\cdot5059904159^3$	$4\cdot5059905354^3$	$4\cdot5059907351^3$	$4\cdot5059908187^3$
	10^{-5}	$4\cdot5000599990^5$	$4\cdot5000599990^5$	$4\cdot5000599990^5$	$4\cdot5000599990^5$	$4\cdot5000599990^5$	$4\cdot5000599991^5$	$4\cdot5000599991^5$
	10^{-8}	$4\cdot5000000600^8$	$4\cdot5000000600^8$	$4\cdot5000000600^8$	$4\cdot5000000600^8$	$4\cdot5000000600^8$	$4\cdot5000000600^8$	$4\cdot5000000600^8$
$N = 4$	10^{-1}	$7\cdot0006050507^1$	$7\cdot0006258603^1$	$7\cdot0008113949^1$	$7\cdot0025086393^1$	$7\cdot0041114561^1$	$7\cdot0066641805^1$	$7\cdot0076979583^1$
	10^{-3}	$6\cdot0119730035^3$	$6\cdot0119730123^3$	$6\cdot0119730907^3$	$6\cdot0119737816^3$	$6\cdot0119743945^3$	$6\cdot0119253012^3$	$6\cdot0119756468^3$
	10^{-5}	$6\cdot0001199972^5$	$6\cdot0001199972^5$	$6\cdot0001199972^5$	$6\cdot0001199973^5$	$6\cdot0001199973^5$	$6\cdot0001199974^5$	$6\cdot0001199975^5$
	10^{-8}	$6\cdot0000001198^8$	$6\cdot0000001198^8$	$6\cdot0000001198^8$	$6\cdot0000001198^8$	$6\cdot0000001198^8$	$6\cdot0000001198^8$	$6\cdot0000001198^8$
$N = 6$	10^{-1}	$1\cdot1314314829^2$	$1\cdot1314402054^2$	$1\cdot1315175264^2$	$1\cdot1321892708^2$	$1\cdot1327692244^2$	$1\cdot1336000278^2$	$1\cdot1339069075^2$
	10^{-3}	$9\cdot0298945943^3$	$9\cdot0298946669^3$	$9\cdot0298953034^3$	$9\cdot0299003455^3$	$9\cdot0299041042^3$	$9\cdot0299087222^3$	$9\cdot0299102406^3$
	10^{-5}	$9\cdot0002999889^5$	$9\cdot0002999889^5$	$9\cdot0002999890^5$	$9\cdot0002999895^5$	$9\cdot0002999898^5$	$9\cdot0002999903^5$	$9\cdot0002999904^5$
	10^{-8}	$9\cdot0000002994^8$	$9\cdot0000002994^8$	$9\cdot0000002994^8$	$9\cdot0000002994^8$	$9\cdot0000002994^8$	$9\cdot0000002994^8$	$9\cdot0000002994^8$
$N = 8$	10^{-1}	$1\cdot6031615830^2$	$1\cdot6031808001^2$	$1\cdot6033503723^2$	$1\cdot6047663934^2$	$1\cdot6091110601^2$	$1\cdot6074290615^2$	$1\cdot6079560909^2$
	10^{-3}	$1\cdot2055732547^4$	$1\cdot2055732828^4$	$1\cdot2055735259^4$	$1\cdot2055752689^4$	$1\cdot2055763833^4$	$1\cdot2055775691^4$	$1\cdot2055779189^4$
	10^{-5}	$1\cdot2000559972^6$	$1\cdot2000559972^6$	$1\cdot2000559972^6$	$1\cdot2000559974^6$	$1\cdot2000559975^6$	$1\cdot2000559976^6$	$1\cdot2000559976^6$
	10^{-8}	$1\cdot2000000559^9$	$1\cdot2000000559^9$	$1\cdot2000000559^9$	$1\cdot2000000559^9$	$1\cdot2000000559^9$	$1\cdot2000000559^9$	$1\cdot2000000559^9$

TABLE 2(b). Value of $E(0, 1, 0, N - 1)$

N	μ \ r	0	0·1	0·5
2	10^{-3}	$2 \cdot 5019980030^3$	$2 \cdot 501998030^3$	$2 \cdot 501998030^3$
	10^{-5}	$2 \cdot 5000199998^5$	$2 \cdot 5000199998^5$	$2 \cdot 5000199998^5$
	10^{-8}	$2 \cdot 50000000200^8$	$2 \cdot 50000000200^8$	$2 \cdot 50000000200^8$
5	10^{-3}	$7 \cdot 0199423718^3$	$7 \cdot 0199448126^3$	$7 \cdot 0199497223^3$
	10^{-5}	$7 \cdot 0001999940^5$	$7 \cdot 0001999942^5$	$7 \cdot 0001999947^5$
	10^{-8}	$7 \cdot 0000001998^8$	$7 \cdot 0000001998^8$	$7 \cdot 0000001998^8$
8	10^{-3}	$1 \cdot 1555732547^4$	$1 \cdot 1555752689^4$	$1 \cdot 1555779189^4$
	10^{-5}	$1 \cdot 1500559972^6$	$1 \cdot 1500559974^6$	$1 \cdot 1500559976^6$
	10^{-8}	$1 \cdot 15000000559^9$	$1 \cdot 15000000559^9$	$1 \cdot 15000000559^9$

TABLE 2(c). Value of $E(0, 1, 1, N - 1)$

N	μ \ r	0	0·1	0·5
2	10^{-3}	$2 \cdot 0019980020^3$	$2 \cdot 0019980020^3$	$2 \cdot 0019980020^3$
	10^{-5}	$2 \cdot 0000199998^5$	$2 \cdot 0000199998^5$	$2 \cdot 0000199998^5$
	10^{-8}	$2 \cdot 0000000200^8$	$2 \cdot 0000000200^8$	$2 \cdot 0000000200^8$
5	10^{-3}	$6 \cdot 5199423605^3$	$6 \cdot 5000133456^3$	$6 \cdot 4645415448^3$
	10^{-5}	$6 \cdot 5001999940^5$	$6 \cdot 4802006831^5$	$6 \cdot 4446459267^3$
	10^{-8}	$6 \cdot 5000001998^8$	$6 \cdot 4800002005^8$	$6 \cdot 4444446457^8$
8	10^{-3}	$1 \cdot 1055732514^4$	$1 \cdot 1037995502^4$	$1 \cdot 1014294325^4$
	10^{-5}	$1 \cdot 1000559972^6$	$1 \cdot 0982703841^6$	$1 \cdot 0958895146^6$
	10^{-8}	$1 \cdot 1000000559^9$	$1 \cdot 0982143417^9$	$1 \cdot 0958333894^9$

Note. (i) The accuracy of the tabulations are to the order 10^{-11}. (ii) The end exponents in each entry indicates a multiplying factor of the power of 10. For example interpret $1 \cdot 5093298097^3$ $= 1 \cdot 5093298097 \times 10^3$. (iii) Each unit of time corresponds to the expected time for a birth–death mutation event.

Numerical implications of Table 2

These calculations reaffirm the qualitative findings extracted from Table 1. Thus the effects of finite population size, whether the process occurs continuously in time or over discrete generations, are no different.

It can be proved analytically that for $N = 2$ only, in the continuous model the values of $E(0, 0, 0, N)$ and generally $E(i_1, i_2, i_3, i_4)$ where $D(\mathbf{i}) = 0$ are independent of r.

Expected first formation time of double mutant type

We return now to the Wright–Fisher type model (5.2). The expectations of $\tau(\mathbf{i})$, (the expected times to formation of the first double mutant) denoted by $F(\mathbf{i})$ also satisfy the system of linear equations (5.5) but with a different set of boundary conditions. The obvious interpretations require that here $F(\mathbf{i}) = 0$ for all $\mathbf{i} = (i_1, i_2, i_3, i_4)$ where $i_1 \geq 1$. It can be checked for $N = 2$ that

$$F(\mathbf{i}) \text{ is a decreasing function of } r \text{ for every } \mathbf{i}, \qquad (5.11)$$

i.e. increased recombination shortens the expected time to the formation of the first double mutation. An intuitive justification is available. Whatever the possibilities for creating the AB type by mutation, added recombination can only increase the contingencies of its first manifestation. The breaking apart by recombination of advantageous double mutants does not contribute to the evaluation of the first formation time of the haplotype AB. Undoubtedly, the fact of (5.11) applies for all N.

To sum up, the preceding discussion points up the following two contrasting features of the effects of recombination. (We assume here the initial state $(0, 0, 0, N)$.)

1. Increased recombination diminishes the expected time of first formation of the favoured double mutant.

2. Increased recombination increases the expected time of total fixation of the AB gamete. With increased population size the disadvantages of more recombination are relaxed to the extent that as $N \to \infty$ the effects of recombination is eliminated.

Under certain biological conditions it is clear that the mean first passage time is of most relevance while for others, the total fixation time is of paramount importance. It is likely that the estimation of the "advantage" of recombination on the mean time for the population to reach a level of say 5, 50 or 90% incorporation are also of major evolutionary significance. In view of the divergent facts of (1) and (2), it is surmised that the merits of increased recombination on a goal of 50% incorporation is exceptionally sensitive to initial conditions; this probably leads to no clear overall principles or conclusions.

Orders of magnitude of expected times to fixation and first formation of double mutations.

It is worth featuring the orders of magnitude of time involved (expressed in terms of μ and N) for the occurrence of first formation and also for total fixation of the AB gamete. The order of magnitude as a function of population size N and mutation μ has the form

$$F(\mathbf{i}) = \frac{c(\mathbf{i})}{\mu\sqrt{N}} \qquad (5.12)$$

where $c(\mathbf{i})$ is a constant dependent on $\mathbf{i}$ only through the frequencies (i_1/N, i_2/N, i_3/N, i_4/N). In particular, if $\mu \sim \theta/N$, (mutation rate is of order of the reciprocal of population size) the expected duration until the first formation of the AB gamete is of the order $c \cdot \sqrt{N}$. Section 6 discusses the formal validations of these estimates. On the other hand note that if μ is constant and of order greatly exceeding $1/\sqrt{N}$ then the creation of the double mutant occurs quite quickly, in practically zero time duration, as $N \to \infty$.

It is noteworthy that for small mutation rate $\mu = 1/N$ the requisite time until formation of the first AB gametes is of substantial magnitude ($1/\mu\sqrt{N}$ generations). For example with $N = 10^4$ and $\mu \sim 10^{-5}$ then $F(\mathbf{i})$ involves approximately 1000 generations. (See also Tables I and II.)

It is of interest to contrast these magnitudes with the order of magnitude of the mean time until total fixation of AB, still emphasizing the dependence on population size. It can be established (see Section 6 below) that

$$E(\mathbf{i}) \sim \frac{d(\mathbf{i}) \log N}{-\log(1 - \mu)} \qquad (5.13)$$

provided μ is not too small compared with $1/N$. The interpretation of the constant $d(\mathbf{i})$ parallels that of $c(\mathbf{i})$ in (5.12).

When the mutation rate is small of order $\mu = \theta/N$, θ moderate, we find that

$$E(\mathbf{i}) = d^*(\mathbf{i})N \qquad (5.14)$$

or

$$E(\mathbf{i}) = d^*(\mathbf{i})N \log N \qquad (5.15)$$

depending on the mating mechanism. In fact in a case of a selfing population where lines of descent are kept separated (5.15) is relevant, while for a random mating population (5.14) appears to apply (see Section 6).

On the other hand, if μ is a positive constant which is not too small, then the expected number of generations for the creation of the first double mutants is negligible of order $c/\mu\sqrt{N}$, while the expected number of generations until fixation of the double mutant is still large as $N \to \infty$, namely

$$\frac{d \log N}{-\log(1 - \mu)} \sim d \log N.$$

The above discussion underscores the sensitivity of any analysis to the assumptions placed on the magnitude of mutation rate in relation with the order of population size.

In the next section we take up the determination of the time magnitudes for achieving a pth order mutant form corresponding to p separate loci. Maynard Smith (1971) discussed parts of this problem in an intriguing manner, employing rough and ready methods.

6. First Passage and Fixation Times for pth Order Mutants

The model investigated hitherto was structured on a two-locus character with mutant forms to be established at each locus. Consider now the extension of the model to that of a p-locus character with unidirectional mutation events occurring at each locus and where the pth order mutant carries highest fitness. It would be of considerable interest to calculate the expected time until first formation of a pth order mutant and also the expected time until fixation of the population as pth order mutants. The analysis of the appropriate p-locus stochastic model involving recombination in its general formulation appears at the present time prohibitive. To achieve some insight, we examine two simpler versions of a related model.

MODEL I. (Selfing lines). Consider a population of N individuals. In each generation each individual bears a single progeny which may independently with probability μ undergo a mutation change and with probability $1 - \mu$ retain its parental expression. A resulting mutant offspring is called a first order mutant. The population reproduces itself in this manner over successive generations so that first order mutants can produce second order mutants, second order mutants can change into third order mutants, etc.

We calculate now the distribution of the time $\tau_p^{(N)}$ until the first manifestation of a pth order mutant. (Note that we display the dependence of $\tau_p^{(N)}$ on population size.) To this end, conceive of reproduction and the mutation process as continuing forever. Certainly each line of descent will ultimately accumulate loci achieving any level (order) of mutation. Since each parent produces a single offspring we can unambiguously keep the lines of descent separated. Let $T_i, i = 1, 2, \ldots, N$ denote the minimum number of generations required until the line of descent of the ith individual is a pth order mutant. Evidently

$$\Pr\{\tau_p^{(N)} > m\} = \Pr\{\min(T_1, T_2, \ldots, T_N) > m\}$$
$$= \Pr\{T_1 > m\}^N = \left[\sum_{k=0}^{p-1} \binom{m}{k}\mu^k(1 - \mu)^{m-k}\right]^N. \tag{6.1}$$

Set $\mu = \theta/N$, $m = xN^{1-1/p}$ and using the asymptotic relation

$$\log\left(\sum_{k=0}^{p-1} \frac{\xi^k}{k^\beta}\right) \sim \left(\xi - \frac{\xi^p}{p^\beta}\right) \quad \text{as } \xi \to 0, \quad p \ge 2.$$

we can deduce the limit law

$$\lim_{N \to \infty} \Pr\left\{\frac{\tau_p^{(N)}}{N^{1-1/p}} \ge x\right\} = e^{-x^p\theta^p/p!}.$$

It can also be proved that the moments of the random variable $\tau_p^{(N)}/N^{1-(1/p)}$ also converge to the moments of the distribution $1 - e^{-u^p\theta^p/p!}$.

In particular, we have

$$E(\tau_p^{(N)}) \sim N^{1-(1/p)} \frac{p^{(1/p)-1}\Gamma(1/p)}{\theta} \quad \text{as } N \to \infty.$$

Thus, as claimed in Section 5 when $\mu \sim 1/N$ the expected time until formation of the first double mutant is of order $c\sqrt{N}$. More generally, the expected time until first formation of a pth order mutant is of order $cN^{1-(1/p)}$.

Consider next the random variable $T_p^{(N)}$ as the time when all lines of descent are at least pth order mutants. It follows from the definitions that

$$\Pr\{T_p^{(N)} \leq m\} = P\{\max(T_1, T_2, \ldots, T_N) \leq m\}$$

$$= \left[1 - \sum_{k=0}^{p-1} \binom{m}{k} \mu^k (1-\mu)^{m-k}\right]^N. \tag{6.2}$$

Limit laws for maxima of independent random variables are classical. The proper normalization is determined by the following rule. For arbitrary x determine $m = m(x)$ as the closest positive integer solution of the equation

$$\sum_{k=0}^{p-1} \binom{m}{k} \mu^k (1-\mu)^{m-k} = \frac{x}{N}.$$

It is necessary to determine m quite finely in order to extract a limit law from (6.2); precisely

$$m = \frac{\log N}{\gamma} + \frac{((p-1)!)\log\log N}{\gamma} - \frac{\log x}{\gamma} + c_p$$

with $\gamma = -\log(1-\mu)$, where c_p is an appropriate constant depending on p and μ. Taking $y = -\log x$, we can then establish

$$\lim_{N\to\infty} \Pr\left\{T_p^N - \frac{\log N}{\gamma} - \frac{(p-1)!\log\log N}{\gamma} - c_p \leq \frac{y}{\gamma}\right\}$$

$$= e^{-e^{-y}}.$$

It can also be proved that

$$E(T_p^{(N)}) \sim \frac{\log N}{\gamma} + \frac{(p-1)!\log\log N}{\gamma} + c_p + 0\left(\frac{1}{N}\right)$$

where $\gamma = -\log(1-\mu)$.
  For

$$\mu \sim \frac{1}{N}$$

we have

$$E(T_p^{(N)}) \sim N \log N + (p - 1)!N \log \log N. \qquad (6.3)$$

Note that the order of magnitude of time to achieve a pth order mutant is essentially independent of p. The level p only influences the lower order terms like $[(p - 1)! \log \log N]$.

MODEL II. Birth–Death events—Random mating. The preceding analysis was based on a model where pure lines were maintained: once a mutant was born, all succeeding generations in his line of descent preserved it or yielded progeny from one of his descendants expressing a higher order mutant form. We now formulate a birth and death model to allow the possibility that a mutant may be formed but then dies before his line produces double mutants.

Consider a population of N individuals. Birth–death events occur continuously in time subject to the following rules. With chance $\lambda\, dt$ a birth–death event occurs during the time interval $(t, t + dt)$ (λ is usually a function of population size measuring the intensity of matings and associated birth–death occurrences). An individual is chosen at random to be replaced (die) and an individual from the same pool is chosen at random to serve as parent producing a new offspring. After birth, the offspring may change its form by mutation to that of a higher order mutant.

An individual is referred to as an rth order mutant if among his ancestors there occurred exactly r mutation events.

Consider the simplest situation where all N parents are 0-mutants and suppose we wish to compute the expected time until the population consists entirely of one or higher order mutants. Let ϕ_k be the expected time for the population to achieve a state containing no 0-mutant types where the initial population is comprised of k, 0-mutant types and $N - k$, higher than 0-mutant types. By the nature of the process, and a standard renewal argument, we derive the equations

$$\phi_k = \frac{1}{\lambda} + \left(\frac{k}{N}\right)\left(\frac{k}{N}\mu + \frac{N - k}{N}\right)\phi_{k-1} + \left(\frac{N - k}{N}\right)\frac{k}{N}(1 - \mu)\phi_{k+1}$$
$$+ \left\{\frac{N - k}{N}\left[\frac{N - k}{N} + \frac{k}{N}\mu\right] + \frac{k}{N}\frac{k}{N}(1 - \mu)\right\}\phi_k \qquad (6.4)$$

with $\phi_0 = 0$ by definition.

Let us assume $\lambda \sim N^2$ (rate of events taken proportional to N^2; this order of contacts between individuals partly reflects the phenomena of random mating), $k/N = x$ and $\mu = \theta/N$ and let $N \to \infty$ in (6.4). The existence of the limits

$$\phi_k = \phi(Nx) \to \psi(x)$$

$$N(\phi_k - \phi_{k-1}) \to 2\psi'(x)$$

and

$$N^2(\phi_{k+1} + \phi_{k-1} - 2\phi_k) \to \psi''(x)$$

can be proved and then the system of equation (6.4) passes into the differential equation

$$0 = 1 - \theta x \psi'(x) + x(1 - x)\psi''(x). \tag{6.5}$$

The ancillary requirements added to (6.5) that will characterize uniquely the desired solution of (6.5) are: $\psi'(1) < \infty$, $\psi(0) = 0$ and ψ is monotone increasing. Comparing to (5.9) we secure the formula

$$\psi(x) = \int_0^x \frac{1}{(1 - \eta)^\theta} \left(\int_\eta^1 \frac{(1 - \xi)^{\theta-1}}{\xi} \, d\xi \right) d\eta \tag{6.6}$$

Translating these results back to the original model we find that starting with a population of N 0-mutants, the expected time until the population are all at least 1-mutants is asymptotically $\sim N\psi(1)$. Equivalently, a unit of time in the limit process associated with (6.5) involves approximately N generations.

In the context of Model II we tackle next the problem of evaluating the expected time until the *first* 2nd order mutant is formed. Consider a parent population comprised of k mutants of order one and $N - k$ zero order mutants. Let F_k be the expected time until formation of the first double mutant (2nd order mutant). We readily derive the equations

$$F_k = \frac{1}{\lambda} + \lambda_k F_{k+1} + \mu_k F_{k-1} + \nu_k F_k, \qquad 1 \le k \le N - 1 \tag{6.7}$$

where

$$\lambda_k = \left(\frac{N - k}{N} \right) \left[\frac{N - k}{N} \mu + \frac{k}{N} (1 - \mu) \right]$$

$$\mu_k = \frac{k}{N} \frac{N - k}{N} (1 - \mu)$$

$$\nu_k = \left(\frac{N - k}{N} \right) \left[\frac{N - k}{N} (1 - \mu) \right] + \frac{k}{N} \left(\frac{N - k}{N} \mu + \frac{k}{N} (1 - \mu) \right)$$

Note that

$$\lambda_k + \mu_k + \nu_k = 1 - \frac{k}{N}\mu,$$

and $(k/N)\mu$ is exactly the probability that a birth yields a second order mutant.

Let $\mu = \dfrac{\theta}{N}$, $k = x\sqrt{N}$; the correct scaling of time is now $\lambda \sim N^{3/2}$.　(6.8)

(If k is larger than order $x\sqrt{N}$ then double mutants are practically formed instantly). Note that the average time between events is $1/\lambda$ and consequently the mean length of time for the changeover of a complete generation is $1/\lambda \cdot N = 1/\sqrt{N}$. It follows that the mean number of generations until a first double mutant is created will be of the order

$$\frac{F(0)}{1/\sqrt{N}} = \sqrt{N}F(0) \tag{6.9}$$

As $N \to \infty$ maintaining (6.8), the limit

$$F_k \to u(x) \tag{6.10}$$

exists and $u(x)$ satisfies

$$xu''(x) + \theta u'(x) - \theta x u(x) + 1 = 0 \tag{6.11}$$

subject to the boundary conditions

$$u(0) < \infty, \qquad u(\infty) = 0, \qquad u(x) \text{ is bounded and decreasing.} \tag{6.12}$$

Careful analysis of (6.11) yields the precise formula

$$u(x) = Cv_1(x) - \frac{1}{W(v_1, v_2)} \int_0^x [v_1(x)v_2(t) - v_1(t)v_2(x)]t^{\theta-1} \, dt \tag{6.13}$$

where

$$C = \frac{1}{W(v_1, v_2)} \int_0^\infty t^{(\theta-1)/2} v_2(t) \, dt$$

and

$$v_1(t) = t^{(1-\theta)/2} I_{(\theta-1)/2}(t\sqrt{\theta}), \qquad v_2(t) = t^{(1-\theta)/2} K_{(\theta-1)/2}(t\sqrt{\theta}),$$

I_α is the standard Bessel function with imaginary argument and K_α is the solution of Bessel's equation of the second kind; $W(v_1, v_2)$ is the Wronskian of the indicated functions.

On account of (6.9)–(6.13) we find that the expected number of generations until formation of the first double mutant is of the order

$$\sqrt{N}u(0) \quad \text{(it is readily checked that } 0 < u(0) < \infty) \quad (6.14)$$

provided the original population makeup was entirely 0-mutant individuals.

Sum up: The preceding two calculations based on Model II strongly suggest that the orders of magnitude indicated in (5.14) and (5.15) of Section 5 are undoubtedly correct for the mutation-recombination genetic model. Exact calculations are as yet intractable. That is the effects of mating, birth, death and recombination tends to complicate the exact determinations but apparently will not alter orders of magnitude. On the basis of the above discussion, we would also expect that the expected time until first formation of a pth order mutant is $\sim cN^{1-(1/p)}$.

7. Summary and Discussion

We present a succinct summary of our main qualitative findings. More complete biological discussions on several facets of the problems concerning the rate of incorporation of favourable mutations is mostly contained in the body of the text. Of special interest are the numerical tabulations of Section 5 for finite population models.

1. *Deterministic model with all mutation events favourable* (supermultiplicative or multiplicative viabilities):

We pointed out in Section 2 that recombination is advantageous in accelerating the incorporation of the most favoured double mutant only under the condition that the initial population state exhibits negative linkage disequilibrium. In these circumstances small mutation rates work better in consonance with recombination in giving the "advantage" to sexual evolution. If the linkage disequilibrium value $D^{(0)}$ is positive then increasing recombination decidedly retards the process of incorporation. Most likely, in the case of all favourable mutation forms, the natural initial state has $D^{(0)} < 0$. Biologic justifications and interpretations are indicated in our discussion of Section 2 sometimes leading to $D^{(0)} < 0$ and in other circumstances to $D^{(0)} = 0$.

The above facts were deduced from the analysis of a two-locus haploid model. We do not know whether a study of recombination effectiveness in a diploid setting may result in different conclusions. In this vein it is important to realize that a stable polymorphism perpetuated by balance of selection forces is non-existent in a haploid model. On the other hand, diploid populations maintain a wide variety of polymorphic representations attributable to balancing viability selection among the genotypes.

2. *Deterministic case with favourable double mutations but singly deleterious mutants.*

We indicated in Section 3 that in this case it is possible for the recombination mechanism to establish a polymorphism and the nature of this equilibrium state is such that the wild type is most abundant while the other haplotypes are represented with positive but small frequency. In a haploid two-locus selection model under the condition where the forms Ab and aB have fitness disadvantage to ab (the wild type) and AB is the most fit type, the pure population consisting only of the ab gamete presents a stable state provided recombination is large enough. If now undirectional mutation $a \rightarrow A, b \rightarrow B$ is introduced into the system then a mutation selection balance is developed but a sufficiently large recombination force is essential for the stability of the polymorphism. The foregoing result testifies to the greater importance of recombination as a mechanism for maintaining potential and actual variability rather than for expressing advantages for the efficient incorporation of favoured double mutants. It is possible that the notion of the "advantage" of recombination should be phrased in terms not of the speed of incorporation of mutants but in terms of the variability it permits.

3. *Continuous time deterministic formulations.*

The discussion of Section 4 points out that whether the model is set up as a discrete time (non-overlapping generations) situation or whether the model is formulated with changes occurring continuously in time (overlapping generations), the qualitative conclusions of Section 2 appear to remain unaltered.

4. *Effects of finite population size and relative recombination rates on the rate of incorporation*

(a) In view of the diverse effects of recombination on infinite populations (see (i) above) especially in its sensitivity to the character of the initial population state, we cannot expect a consistent relationship between recombination strength and the rate of incorporation of double mutants in the case of large population size N.

(b) The authors Crow and Kimura (1965), Maynard Smith (1971) and Felsenstein (1973) in attempting to take account of "genetic drift" (sampling fluctuations) in evaluating the impact of recombination on the rate of incorporation, proceed through a strange mixture of deterministic and stochastic analyses. To quote from Felsenstein, "The arguments involve a large number of crude approximations". Formulas derived in one locus theory are used freely for multi-locus phenomena. Expectations of products of dependent random variables are multiplied freely, and similarly. These practices create doubts on the validity of several of their claims.

Two proper stochastic versions of the standard deterministic model of Section 2 (with finite population size) are formulated in Section 5. The first non-overlapping generation stochastic model is constructed in the spirit of the Wright–Fisher sampling procedure. The second stochastic model along the lines of the Moran construction allows overlapping generation transitions among states and the process unfolds in continuous time. Most of the conclusions are robust with respect to the continuous time as against the discrete time stochastic formulation.

There are two basic random variables or expectations of relevance to the problem; (i) *the expected time until first formation* of a double mutant labeled $F(\mathbf{i})$ and *the expected time until total fixation of AB* denoted by $E(\mathbf{i})$, where $\mathbf{i} = (i_1, i_2, i_3, i_4)$ indicates the initial state of the population where i_1, i_2, i_3, i_4 $(\Sigma i_v = N)$ are the numbers of AB, Ab, aB, ab gametes respectively. Concentrating on the initial state $\mathbf{w} = (0, 0, 0, N)$ such that the population consists initially of the haplotype ab only, we inferred the following contrasting facts:

(i) For fixed population size N, $F(\mathbf{w})$ is a decreasing function of r, (the recombination fraction); while

(ii) $E(\mathbf{w})$ is an increasing function of r.

Thus, in a finite population, without selection effects, increasing recombination is on an average advantageous in speeding the time of first formation of the double mutant but has the opposite effect of impeding the time of total incorporation.

In other words, sex (recombination) is advantageous in speeding the time of first formation of the double mutant; but it later breaks apart the favoured gamete types tending to stretch the mean total fixation time beyond that which would obtain if no recombination mechanism was operating. These assertions are perhaps surprising since in the infinite population model where no selection differences operate, recombination exerts absolutely no effect on the rate of incorporation (c.f. the discussion with (2.9)).

The above divergence of results leads to the following anomaly. If the important variable is actually *the expected time $T_\gamma(\mathbf{i})$ until the population first achieves a level of γ percentage* incorporation (say, $\gamma = 5$ or 50 or 80% rather than 100% or first formation) we could expect no simple consistent relationship pertaining to the influence of the strength of recombination on the values of $T_\gamma(\mathbf{i})$.

The assertions of (i) and (ii) apply only for an initial state $\mathbf{i} = (i_1, i_2, i_3, i_4)$ of non-negative linkage disequilibrium. *For an initial negative linkage disequilibrium the corresponding conclusion for (ii) seems to be reversed.*

The above discussion emphasizes the delicacy of the concept of "incorporation" especially in the presence of genetic drift factors. With respect to the controversy on the relative merits of small or large population size (see the introduction) we find that increased population size can only enhance the advantage of one additional unit of recombination in the following sense. Let $E_N(\mathbf{w}, r)$ denote the expected time of total incorporation of AB where the population size is N and recombination fraction is r and $\mathbf{w}$ is the initial state, $\mathbf{w} = (0, 0, 0, N)$. We have the inequality

$$E_N(\mathbf{w}, r) - E_N(\mathbf{w}, r^*) > 0 \quad \text{for all } r > r^* > 0 \quad \text{(see Section 5).} \quad (7.1)$$

The quantity in (7.1) goes to zero as $N \to \infty$ showing that the retardation effect of increased recombination on the rate of total incorporation is diminished with increasing population size. The result is opposite if the initial state has $D^{(0)} < 0$ as is the case with $(0, 1, 1, N - 2)$.

5. Orders of magnitude of expected times to total fixation

The main results highlighted in Section 6 can be summarized as follows. If mutation rate is of the order of the reciprocal of population size then the expected time until the first formation of a double mutant is of the order $c_2\sqrt{N}$ and to the first formation of a pth order mutant of order $c_p N^{1 - (1/p)}$. These estimates apply both for random mating and selfing processes.

The estimates of the expected time until total fixation is more sensitive to the underlying mating system. Thus, we found that for selfing systems the expected time until total fixation of pth order mutants is of the order $N \log N + (p - 1)! N \log \log N$ where the dependence on p is clearly reflected only in lower order terms. For a randomly mating population (model II of Section 6) we can infer that this same expected time is of the order $c_p N$ where the constant c_p appears to depend on p in a complicated fashion.

If μ is of larger order than $1/N$ then the expected time until fixation behaves like $c \cdot \log N/\mu$. The expected time until first formation of a pth order mutant is $(\mu N^{1/p})^{-1}$. Again we see the sensitivity of the estimates to the assumptions made in comparing the order of $1/\mu$ and the population size.

A valid objection to much of the discussion of Sections 5 and 6 is that differential selection forces have been left out of the analysis. This criticism is indeed justified. Nevertheless, our qualitative findings undoubtedly apply to cases of small selection effects. The correct treatment of a proper stochastic version of the model involving a grade of selection parameters remains a challenging problem. We believe that the orders of magnitude deduced are probably correct for the complex stochastic model involving recombination as well.

6. Future prospects

A proper understanding of the role of recombination in the evolutionary process cannot be discussed only in the context of the problem of incorporation of favourable mutations. There is a booming literature underlining the importance and effects of recombination for explaining multi-locus equilibria theory, characterization of modifier genes, the nature of polygenic inheritance, the structure of special mating systems, problems related to sex determination and other problems.

For further perspectives on some of these problem areas we refer to Bodmer and Felsenstein (1967), Franklin and Lewontin (1970), Karlin and Feldman (1970), Feldman and Balkau (1973), Karlin and McGregor (1973), Karlin and Lieberman (1974) and to references cited therein.

Acknowledgments

I want to express my great indebtedness to Professor J. McGregor with whom I discussed some of the developments of Sections 5 and 6; and my gratitude to Professors W. Bodmer, J. Gani, J. Maynard Smith, M. Feldman and Ilan Eshel for valuable criticisms of the manuscript. I am also thankful to Boris Shkoller for helping with the computer runs.

REFERENCES

BODMER, W. F. (1970). The evolutionary significance of recombination in prokaryotes. *In* "Prokaryotic and Eukaryotic cells." Symposia of the Society for General Microbiology No. XX, 279–294.

BODMER, W. F. AND FELSENSTEIN, J. (1967). Linkage and selection: Theoretical analysis of the deterministic two locus random mating model. *Genetics* **57**, 237–265.

CREW, F. A. E. (1965). "Sex Determination." Dover Pub. Inc., New York.

CROW, J. F. AND KIMURA, M. (1965). Evolution in sexual and asexual populations. *Amer. Natur.* **99**, 439–450.

CROW, J. F. AND KIMURA, M. (1969). Evolution in sexual and asexual populations: a reply. *Amer. Natur.* **103**, 89–91.

CROW, J. F. AND KIMURA, M. (1970). "An Introduction to Population Genetics Theory." Harper and Row, New York.

ESHEL, I. AND FELDMAN, M. W. (1970). On the evolutionary effect of recombination. *Theor. Pop. Biol.* **1**, 88–100.

FELDMAN, M. (1971). Equilibrium studies of two locus haploid populations with recombination. *Theor. Pop. Biol.* **2**, 299–318.

FELDMAN, M. AND BALKAU, B. (1973). Some results in the theory of three gene loci. Proc. Symp. Pop. Dynamics, Madison, Wisconsin.

FELSENSTEIN, J. (1973). The evolutionary advantage of recombination. (In preparation.)

FISHER, R. A. (1930). "The Genetical Theory of Natural Selection." p. 291 (2nd Ed. 1958). Dover Publications, New York.

FRANKLIN, I. AND LEWONTIN, R. (1970). Is the gene the unit of selection. *Genetics* **65**, 707–734.

KARLIN, S. (1972). Notes on Mathematical Genetics, Weizmann Institute of Science, Rehovot, Israel.

KARLIN, S. AND FELDMAN, M. W. (1970). Linkage and selection: two locus symmetric viability model. *Theor. Pop. Biol.* **1**, 39–71.

KARLIN, S. AND LIEBERMAN, U. (1974). Equilibrium Theory for Multi-locus selection models, to be published.

KARLIN, S. AND MCGREGOR, J. (1964). On some stochastic models in genetics. *In* "Stochastic Models in Medicine and Biology," (J. Gurland, ed.), pp. 245–279. University of Wisconsin Press, Madison, Wisconsin.

KARLIN, S. AND MCGREGOR, J. (1971). On mutation selection balance for two-locus haploid and diploid population. *Theor. Pop. Biol.* **2**, 60–70.

KARLIN, S. AND MCGREGOR, J. (1973). Towards a theory of the evolution of modifier genes. *Theor. Pop. Biol.* (To appear.)

KIMURA, M. (1964). Diffusion models in population genetics. *J. App. Prob.* **1**, 177–232.

KIMURA, M. (1969). The length of time required for a selectively neutral mutant to reach fixation through random frequency drift in a finite population. *Genet. Res.*

MAYNARD SMITH, J. (1968). Evolution in sexual and asexual populations. *Amer. Natur.* **102**, 469–473.

MAYNARD SMITH, J. (1971) What use is sex? *J. Theor. Biol.* **30**, 319–335.

MORAN, P. A. P. (1962). "The Statistical Processes of Evolutionary Theory." The Clarendon Press, Oxford.

MULLER, H. J. (1932). Some genetic aspects of sex. *Amer. Natur.* **66**, 118–138.

MULLER, H. J. (1958). Evolution by mutation. *Bull. Amer. Math. Soc.* **64**, 137–160.

MULLER, H. J. (1964). The relation of recombination to mutational advance. *Mutation Research* **1**, 2–9.

10. Stochastic Processes in Artificial Selection

A. ROBERTSON

*Department of Genetics, University of Edinburgh,
Edinburgh, Scotland*

The importance of stochastic variation in gene frequency as a cause of evolutionary change has been the subject of great controversy, not least here in Oxford. The argument began some forty years ago and has recently started up again with new adversaries. In an evolutionary context, the importance of "genetic drift" was stressed mainly by Wright. His views were often presented as being much more naïve than they in fact were. Looking on natural selection as a hill-climbing exercise in n dimensions (the allele frequencies at n loci), he argued that, if in the selection process there was much interaction between genes at different loci, it was quite possible that populations could find themselves stuck at a "local peak". Though there were higher peaks in the neighbourhood, these could never be reached by a large population because the first steps would have to be down hill. Wright (1948) then suggested that chance changes in gene frequencies, arising from temporary restrictions in population size, could well dislodge the population and move it into a region from which it could ascend to an even higher peak. More recently, arising from the fascinating evidence now being put before us on the changes of amino acid sequences in proteins during the course of evolution, the argument has restarted in a simpler form (King and Jukes, 1969). The adherents of "non-Darwinian evolution" then hold that a large fraction of the evolutionary changes in amino-acid sequence in a particular peptide chain have occurred by chance rather than by natural selection.

Artificial Selection

In the case of artificial selection, there can be little doubt about the importance of chance changes in gene frequency. In laboratory selection experiments, the number of animals chosen as parents for the next generation is usually less than fifty and, even in large populations of domestic animals, the effective size, as measured in terms of the variance of chance changes in gene frequency from generation to generation, may be very small. In a dairy cattle population of, say, 100,000 cows, all may be bred to less than twenty bulls using artificial insemination. If the chance variation is large, there is then a finite probability that a gene with a positive effect on the character under selection may very well be lost by chance before selection has had the opportunity to increase its frequency. The possibility of loss will of course be greater for genes which are initially rare or which have only a small effect on the character being selected for.

In almost all treatments of this problem, it has been assumed that the genes are acting additively on the character under selection, in the sense that the expected mean value of an individual with a given genotype can be arrived at by summing the contributions of the separate genes which it carries. The basic equation for the chance of fixation of a gene was derived by Kimura (1957) in what was primarily an evolutionary context. The concept of the "gene frequency distribution", first introduced by Wright, then becomes of great value; this is the distribution of the numbers of populations whose gene frequencies lie within a specific range. It is therefore possible to describe the change of the gene frequency distribution in continuous time with two terms: one involving the steady change in gene frequency due to artificial selection and the other to changes due to stochastic variation, analogous to a diffusion equation in physics. Even though the equation is not explicitly soluble, such an approach may be useful in that one can by examination of it at least establish the form of the solution. In the case of an additive gene with a selective advantage s (in the sense that the expected relative numbers of the gene and its alternative allele in the subsequent generation are $1 + s$ times the value in this) it becomes clear by an examination of the differential equation that the process can be described by three parameters: q the initial frequency of the desirable alternative, Ns, where N is the effective number of parents used, and time measured by t/N. The equation for the change of the gene frequency distribution $\phi(q, t)$ with time is

$$\frac{\partial \phi}{\partial t} = \frac{\partial^2}{\partial q^2} \frac{(\phi q(1 - q))}{4N} - \frac{\partial}{\partial q} \frac{(sq(1 - q)\phi)}{2}.$$

Kimura (1957) gave an explicit expression for the probability $u(q_0)$ that the desirable alternative with initial frequency q_0 would be fixed in the population and this is the basis of all work on the limits to artificial selection in finite populations. It is

$$u(q_0) = \frac{1 - e^{-2Nsq_0}}{1 - e^{-2Ns}}.$$

It remains to relate the selective advantage of a gene with its effect on the character under selection. Haldane (1931) many years ago showed that, under artificial selection for a metric character, there was a proportional relationship between the effect of a gene of the character (expressed in terms of the standard deviation of the character) and its consequent selective advantage, the coefficient of proportionality being equal to the intensity of selection, i, if this is defined as the superiority of selected parents above the mean of the group from which they were chosen, divided by the standard deviation of the character. Converting the chance of fixation of the desirable allele into the expected change in the character under selection at the limit and expressing the genetic variation in terms of the total genetic variance, Kimura's equation can then be transformed into one relating the probable change in the character to various parameters describing the selection process. This can easily be extended to several such loci affecting the character, under the assumption that they are segregating independently in the population, i.e. within the population there is no association between genes at different loci within individuals. This is an important limitation to which we shall return later. The theory then predicts the expected change at the limit (expressed in relation to the initial genetic standard deviation) in terms of only two combinations of parameters, one being the product of the population size N, the intensity of selection, i, and the proportion, h, which the genetic variation made up of the total and the second being the initial frequency, q, of the desirable allele, assumed equal at all loci. An interesting observation was that, for a constant genetic variance in the population, the predicted advance tends to a limiting value as the number of loci affecting the character increases. This finite predicted change is made up of infinitely small changes at an infinite number of loci. In the limiting case, it can be predicted that the advance at the limit will be $2N$ times the response in the first generation, that the rate of response will fall off exponentially with time and that one half the total advance will be achieved within $1\cdot4\,N$ generations. This situation may never be approached in practice but the results are nevertheless important in describing the limiting case.

The Problem of Linkage

This simple treatment only applies on the assumption of no association between genes at different loci within individuals. Unfortunately the organization of the genetic material into a small number of chromosomes provides an immediate cause of such association. The importance of linkage in artificial selection has been stressed for many years, notably by Mather, but arguments as to the magnitude of the effects have always had to be qualitative only. In fact it now becomes clear that the effect of linkage can

only be understood properly in relation to the population size. This determines the time scale of an artificial selection process; the crucial problem
is whether the negative associations built up by selection can be disentangled
by crossing over before they are fixed by chance. This interaction between
population size and linkage is apparent from an examination of the differential
equations which describe the situation with two loci in which the recombination fraction between the loci (r) appears multiplied by N. Hill and Robertson
(1966), attempted a treatment of two linked loci under selection by a combination of algebra and computer simulation and a general solution was arrived
at. Not all of the results have been published because they did not appear
to lead to an extension to any number of loci.

In the following we will discuss the problem of a large number of loci
with linkage, with which we have been much concerned (Robertson, 1970),
not so much because of its intrinsic interest but as an example of the use
of computer simulation in the unravelling and understanding of a complex
situation. In this simplest form in which we assume n equivalent loci, all
with the same effect on the character under selection, the same initial gene
frequency and assume them to be equally spaced along the chromosome
with no initial association between alleles at different loci, the statement of
the problem requires eight parameters. We are concerned with a complex
system with stochastic variation imposed. The rules are known for the
parts of this system; the problem is to understand its behaviour as a whole.
What exactly do we mean when we say that we "understand" a system?
We mean that we have arrived at a description of it in terms of a few parameter
combinations with the minimum of interactions between them. This is on the
whole how science progresses—by the introduction of conceptual frameworks
within which a "simple" description can be given in the sense of not having
to specify interactions. In analysing this complex system, we were trying to
discover a pattern within the results which would lead to such a "re-para-
meterization". This approach is essentially that followed by an experimenter
rather than a statistician. The statistician, given a problem in eight dimensions
may be tempted to do a 3^8 factorial experiment, describing the results in
terms of main effects and interactions and leave it at that. On the other hand
an experimenter searching for a pattern usually does not design his experiments in any depth. He may, for the examination of some parts of the system,
carrying out a fully balanced design, but by and large the experiment that
he does next week is conditional on the results that he has got this week.
As a consequence the application of the standard kind of statistical analysis
to the whole set of results is completely out of the question. The basic
conditions are just not satisfied. If they are, this often happens after the
fact, in that, having decided what the pattern really is, the experimenter then
designs a balanced experiment so that he can write a neat and elegant paper
to demonstrate it.

It should be stressed that in such computer simulation work, one should use the computer only when it is unavoidable. If limiting cases can be dealt with by algebra, then this should certainly be done. This will give firm points at the edges of the space and may also suggest what parameter combinations are likely to be useful. In my own case I found myself treating the computer results exactly as I would if I were an experimental scientist, struggling to find a pattern amongst them. I did as much algebra as I could for special cases and in one particular instance, having seen a pattern in the results and having checked over a wider range of parameter combinations that this pattern really held, I then managed to show theoretically that this is precisely what would have been expected. As a result I arrived at a reasonable description of the results within 5 parameter combinations—one of these, involving the product of population size, intensity of selection and heritability, was the same as in the case of no association, and the others were the number of loci, the initial frequency of the desirable allele at these loci, the product of population size and the length of the single chromosome on which the loci were assumed to be situated (the latter describing the extent of recombination in the system) and finally the time in generations divided by the effective population size. It further became clear that as the number of loci increased, a limiting situation was reached in which the results were independent not only of the number of loci but also of the initial gene frequency. Thus I had arrived at a limiting theory with a very large number of loci which needed only three parameter combinations. I would then claim to have understood the system.

The computer programs used in this work treat each locus separately, the alternative alleles being represented by either 0 or 1. If there is a limiting situation with an infinite number of loci, can this be simulated on a computer? After many tribulations, I finally answered this during a visit to Professor Karlin at the Weizmann Institute. Each chromosome was treated as being composed of an infinite number of small segments, each segment contributing equally to the overall variation between chromosomes in the effect on the selected character, which was of course normally distributed as the limiting case of the binomial distribution. Each individual had an expected value equal to the sum of its two chromosomal values and, in the production of new gametes by crossing over between the two parental chromosomes carried by an individual, a value had to be attached to the parental chromosomes at the crossover position, knowing only the position and the total effect on the chromosome. The program is very consuming in time and space in that it involved keeping a library of the value of all initial chromosomes at any points on them at which crossing over had occurred. Thus it was much less expensive to run a programme with 50 loci with an initial frequency of the desirable allele of 0·5 rather than to use my new continuous program with an infinite number of loci. The results may be illustrated by

Table 1 in which it is assumed that the loci concerned are evenly distributed on a single chromosome with length 0·50, i.e. there is a probability of 0·5 δl that a cross over will occur in any small length of the chromosome making up a proportion δl of the total. It is assumed that the heritability and the selection process is not altered. The following points may help understanding of the table:

TABLE 1. The Mean and Standard Deviation of Response to 20 Generations of Selection in Terms of the Initial Genetic Standard Deviation. Ten individuals were selected out of 20 with heritability 0·25 and chromosome length 0·50 cM. Each value is based on 30 replicates.

| | Initial gene frequency | | | | | |
| | 0·2 | | 0·5 | | 0·8 | |
No. of loci	Mean	S.D.	Mean	S.D.	Mean	S.D.
	3·53	0·98	1·60	0	1·00	0
5	4·19	1·25	2·75	0·52	1·58	0
20	4·00	1·25	3·01	0·74	2·31	0·69
50	3·78	0·97	3·14	1·10	2·78	0·71
∞			Mean = 2·86		S.D. = 0·97	

(i) The lower the initial gene frequency, the larger the response. There is an absolute limit to response, at which all desirable alleles are fixed and when there is no variance between replicates. This is more likely to be reached when the initial gene frequency is high.

(ii) The absolute limit is smaller and more likely to be reached, when the number of loci is small. At high gene frequencies, the response then increases as the number of loci increases.

(iii) The standard deviation of the response is of course small if the absolute limit is almost certainly reached. Thus at high gene frequencies the standard deviation increases with number of loci. Otherwise the variation between replicates will mainly be determined by the mean gene frequency at the end of the experiment. The variation will be at a maximum when this is in the region of 0·5. This accounts for the rise and subsequent decline in the standard deviation when the initial frequency is 0·2.

(iv) The effect of initial gene frequency clearly declines as the number of loci increases.

The final line of the table gives the results from the new simulation program with a conceptually infinite number of loci. I then realized that just as one feels that one has solved a problem by arriving at an algebraic solution so too one may have vindicated a conceptual approach merely by being able to write a simulation program which embodies it. So I have succeeded in reducing the problem to a limiting situation of only three parameters and can justifiably say that I understand something of the problem of linkage in quantitative selection with many linked loci.

REFERENCES

HALDANE, J. B. S. (1931). A mathematical theory of natural and artificial selection. Part VII. Selection intensity as a function of mortality rate. *Proc. Camb. Phil. Soc.* **27**, 137–142.

HILL, W. G. AND ROBERTSON, A. (1966). The effect of linkage on limits to artificial selection. *Genet. Res.* **8**, 269–284.

KING, J. L. AND JUKES, T. H. (1969). Non-Darwinian evolution: random fixation of selectively neutral mutations. *Science* **164**, 788–798.

KIMURA, M. (1957). Some problems of stochastic processes in genetics. *Ann. Math. Stat.* **28**, 882–901.

ROBERTSON, A. (1970). A theory of limits in artificial selection with many linked loci. *In* "Mathematical Topics in Population Genetics." (K. Kojima, ed.). Springer-Verlag, Berlin.

WRIGHT, S. (1948). On the roles of directed and random changes in gene frequency in the genetics of populations. *Evolution* **2**, 279–294.

11. The Dynamics of Migration and Selection in Human Populations

R. W. HIORNS

Department of Biomathematics,
University of Oxford, Oxford, England

There are occasions in biological research when the phenomena of prime importance, as judged by criteria presented from theoretical analysis, prove to be particularly difficult to investigate in practical situations, thereby providing a considerable obstacle to the understanding of these phenomena and the validation of their theoretical and generalized conceptual basis. When this happens, and when the theoretical analysis is in terms of mathematical models, the difficulties may arise at several stages in the model building process. At one extreme, the models may develop from little more than biologically intuitive statements rather than statements based upon comprehensive observation, and then no amount of mathematical sophistication will assist in relating ideas to the actual biological mechanisms. On the other hand, observations may be plentiful and summary statements may be attempted in the form of mathematical models, but the difficulties may lie in the complexity of these models and the intractability of the mathematical analysis. In simplified terms, these obstacles to understanding may be classified in terms of the failure on the part of the biologist to undertake measurements which would allow verification of some mathematical model established by using theoretical arguments or, alternatively, the inability of the mathematician to produce sufficiently realistic models and closed form solutions for them which may be applied using existing observations.

Turning to a specific phenomenon, Dobzhansky (1972) has deplored the state of our knowledge concerning natural selection. Although it is accepted that various forms of natural selection act on human populations, only, as he says, "in a few exceptional instances has conclusive direct evidence become available". After a typically careful and illuminating analysis of the various forms of selection which could operate he concludes that much more observation is needed, expensive though this would undoubtedly be, and that the state of our knowledge concerning natural selection over the past few decades has advanced almost entirely at the level of conceptualization and theoretical analysis rather than through the collection of factual data. There seems little doubt that Dobzhansky places the responsibility for our further understanding of natural selection squarely with the human biologists, the situation thus corresponding to the first category outlined above.

In this contribution, the problems which confront a human biologist are considered in four situations in which migration and/or some form of selection are thought to be acting; this consideration will be in terms of the appropriate mathematical models and their eventual application. The first situation concerns a single population and the problem of detecting a balanced polymorphism by the restrictive techniques of sampling which are available to a field investigator in "primitive" populations.

The increasing availability of data on migration, which is generally not difficult to collect, has prompted several recent theoretical investigations (see e.g. Hiorns *et al.*, 1969) and those concerning migration and selection are discussed in the second example with differential selection acting in a system of populations described by a stepping-stone model. From particular numerical solutions, some indication is given of the difficulties which are likely to arise when making statistical inference from observations.

The estimation of migration or intermixture rates is the third problem considered and some of the methodology for this is reviewed. Once again, the presence of selection in the populations concerned has effects upon certain estimators of the intermixture rates and the possibility arises for the detection of this selection.

From the range of mechanisms which embrace selective migration between populations and assortative mating within populations, one particular model for a form of selective migration is used as the fourth example to illustrate the difficulty of applying such models to measures of polygenic characters in human populations. This differs from the other examples in that mating is assumed to be non-random.

By consideration of these four situations, some assessment may be made of the degree to which the mathematician or the biologist is likely to be responsible for the major obstacle to understanding of the various phenomena considered. From the rather specific nature of the problems chosen here, it would be surprising if a clear conclusion were possible and if it were it would then be presumptuous in the extreme to suppose that this conclusion would apply generally. However, it is hoped that some slight insight will be gained into the critical relationship between the two disciplines of mathematics and biology.

1. Natural Selection in a Single Population

In order to understand the reason for the lack of knowledge concerning natural selection in human populations, it is necessary to appreciate the sampling aspects of field investigations and the difficulties they present to the human biologist. There are, of course, several forms of selection which may be acting but only one of these, post-natal somatic selection, can be chosen for illustration here. By observing a cohort of individuals born at the same time and by measuring gene frequencies (or alternatively, phenotype or genotype frequencies) of the survivors at various ages, any selection can be directly estimated. This longitudinal technique is clearly not appropriate to those field situations which can only be visited once. Paradoxically, in these situations are found 'primitive' populations in which natural selection through mortality may well be expected to be acting more strongly than in many other human populations. In other words, where selection is operating most strongly, it may be most difficult to detect. How difficult this is can be explored in terms of an alternative technique, in which a cross-sectional sample of age groups is taken and a comparison is made of genotype frequencies in the groups. Some of these groups will deviate from Hardy–Weinberg equilibrium and such deviations may be used to estimate selection coefficients, assuming random mating.

Two sampling schemes suggest themselves, providing two tests for selection. These are depicted in Fig. 1 and conceptually they involve three groups of individuals: in the first group are the younger children whose ages are less than some specified age at which selection may be thought to be operating; in the second group are the older children and adults whose ages are greater than this specified age; and in the third group are the younger children whose parents form the second group. For the purpose of constructing matched samples, younger and older sibs can be taken for comparisons of the first and second groups whilst parents and offspring may be matched to compare the second and third groups. Taking account of whether selection has or has not acted upon the groups, these comparisons may be termed, for convenience, "pre/post" and "post/pre", respectively (Hiorns and Harrison, 1970).

One selective mechanism of some interest is the maintenance of a balanced polymorphism by means of selection acting against the homozygotes. If the symbols s and S are used to denote the selection acting against the recessive and dominant homozygotes, respectively, then the frequencies of the recessive homozygote in each of the three groups will be q^2,

$$q^2(1 - s)/(1 - Sp^2 - sq^2) \quad \text{and} \quad q^2(1 - sq)^2/(1 - Sp^2 - sq^2)^2.$$

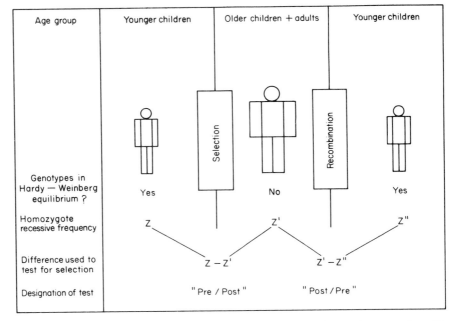

FIG. 1. The three types of individual used in the cross-sectional sample are shown together with their frequency notation and the differences used in the tests.

For convenience, let these three frequencies be represented by z, z', and z''. In order to carry out the required tests for selection, which, in effect, seek to establish whether the pairs z and z' or z' and z'' exhibit differences, the covariances of these pairs of frequencies will be required. After some algebra, the required results are:

$$\text{cov}\,(z, z') = q^2(1 - q)(1 + 3q)(1 - s)/\{4(1 - Sp^2 - sq^2)\}$$

$$\text{cov}\,(z', z'') = pq^3(1 - s)(1 - sq)(1 - Sp)/\{q(1 - sq)\}$$

in which $q = S/(s + S)$. Furthermore, if u is the value of a variable following a standard normal distribution which is exceeded with some specified probability, say 0·05 (for a 5 % significance level in the appropriate one-sided test), the same size, n, representing the number of matched pairs in the "pre/post" sample, may be computed from the expression

$$n = u^2\{z(1 - z) + z'(1 - z') - 2\,\text{cov}\,(z, z')\}/(z - z')^2,$$

and the expression for the "post/pre" sample sizes has a similar form.

On the assumption that mating is random, these sample sizes, which are based upon the deviation from Hardy–Weinberg equilibrium arising from certain specific values for the selection coefficients, will provide samples which are just large enough to detect the deviation caused by the amounts of selection specified at the appropriate significance level for the test. However, it is apparent from Fig. 2 that even for quite substantial selection

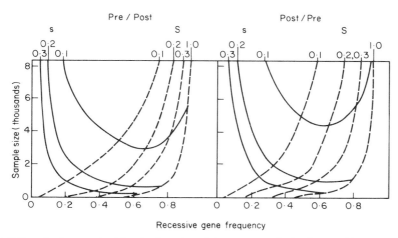

FIG. 2. Sample sizes required for the detection of the amounts of selection specified by s and S which maintain a balanced polymorphism, by means of the two comparisons "pre/post" and "post/pre".

coefficients of 0·1 or 0·2, the sample sizes required would have to be of the order of several thousands in order to establish the existence of selection using a 5 % significance level for the test concerned. These sizes have to be considered in relation to existing data which, because of the practical problems and the expense of collecting and analysing blood samples, are typically only of the order of a few hundred.

The conclusion we can reach from this illustration, and from a detailed consideration of certain other selection mechanisms (Hiorns and Harrison, 1970), is that although many investigators have collected data in the hope of detecting selection in certain primitive populations, it seems probable that the samples collected would have been too small, probably by a factor of at least 10, to detect selection of the reasonably expected type and intensity even in the event that this was in fact operating in the populations. The above illustration has concentrated on situations where matched samples can be obtained but this is by no means always the case. If this cannot be done, unmatched samples, in the form of two random samples, are taken from the two age groups being compared, but the covariances between relatives do not apply, and the samples required are consequently higher by a thousand or two than in the case of matched samples. From these remarks it will be seen that matched samples provide a greater ability to detect selection over a wider range of gene frequency values or selection intensities for a given amount of resources for the collection and analysis and although this principle is not unknown to field workers, the illustrations given here may emphasize its importance in quantitative terms.

2. Natural Selection and Migration in a System of Populations

In the example described above, the effects of natural selection within a single generation or, at most, over the period of one generation were considered. At this point, several populations are introduced with migration between them, but the interest will remain in the short term behaviour, though this will usually be more than a single generation, and sometimes considerably more (Hiorns and Harrison, 1972).

Let a single locus two allele (Aa) system with dominance be described by the frequency q of the dominant A gene at this locus. If the selection coefficients corresponding to each of the three genotypes are represented by the elements of the vector $\mathbf{s}$ so that $f(q, \mathbf{s})$ is the frequency of the allele after selection and if a simple stepping stone migration pattern is assumed with $(k + 1)$ populations in a linear array, the frequencies in the populations after one generation are

$$q'_0 = (1 - m)f(q_0, \mathbf{s}_0) + mf(q_1, \mathbf{s}_1)$$

$$q'_1 = mf(q_{i-1}, \mathbf{s}_{i-1}) + (1 - 2m)f(q_i, \mathbf{s}_i) + mf(q_{i+1}, \mathbf{s}_{i+1})$$

$$(\text{for } i = 1, 2, \ldots, k - 1.)$$

$$q'_k = mf(q_{k-1}, \mathbf{s}_{k-1}) + (1 - m)f(q_k, \mathbf{s}_k).$$

There is no general solution to this set of non-linear equations, but frequencies may be computed directly. The progress towards an equilibrium in two situations is of interest:

(1) When selection opposes the gene flow by migration;
(2) When selection reinforces this gene flow.

In the first situation, when selection opposes gene flow by migration, the behaviour of the system is illustrated in Fig. 3, in which values of $k = 9$, $m = 0 \cdot 1$ and $s = 0 \cdot 01$ were used to obtain a numerical solution of the equations, with initially $q_0 = 1$ and $q_i = 0$ $(i > 0)$. In these computations, the particular form of selection assumed involves relative fitnesses for the genotypes $AA : Aa : aa$ of $1 : 1 : 1 - s$ in population 0 and $1 - s : 1 - s : 1$ in the remaining populations. This represents selection against the recessive gene a in the first population and for this gene in all others. From Fig. 3 a cline of gene frequencies may be seen to emerge after a relatively short

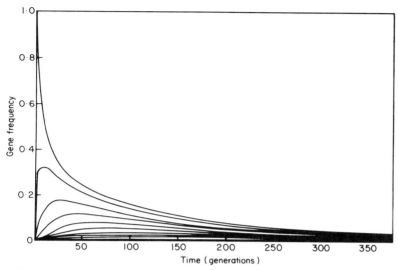

FIG. 3. For values $k = 9$, $m = 0.1$ and $s = 0.01$, in the model described in the text, the frequency of the dominant gene over time in each terminal population where initially this gene is represented in all individuals of one terminal population. This gene is selected for in the terminal population and against in all other populations where the recessive homozygote is selected for. The selection therefore *opposes* the gene flow promoted by migration.

time whereas the progress towards an equilibrium level is slow, requiring several hundred generations before the changes in frequency of the gene are of the same order of magnitude as would be expected to arise solely by mutation. As the effects of drift and mutation were excluded from the model, the validity of the model is confined to the period of major changes in gene frequency levels, which is anyway long before the frequency approaches unity and fixation. However, the length of this period indicates that real populations may well be currently at that stage of development where the dispersal of a locally advantageous gene may be in a dynamic state. If this is the case, and if the time since the arrival of the gene at the outset is unknown, estimation regarding the level of the selection operating based upon a cross-sectional sample, would prove impossible even if the migration flow is accurately known.

When selection reinforces the flow, in the second situation, a similar computation is made using the model with, in all populations, the same selection mechanism as in the population 0 in the first situation. The result of this is that the gene is now favoured everywhere and being initially possessed by all individuals of the first population and by none elsewhere, the spread of the gene and its fixation may be expected to be rapid. The fact that migration and selection are forces acting together to bring about homogeneity of the populations as well as complete representation does, in the event, cause an early convergence of the populations in the system. Their progress together towards fixation, however, proves to be rather slow, and a lengthy period of time is required. This is illustrated for migration $m = 0.1$ and selection $s = 0.01$ in Fig. 4. Once more the remarks made concerning

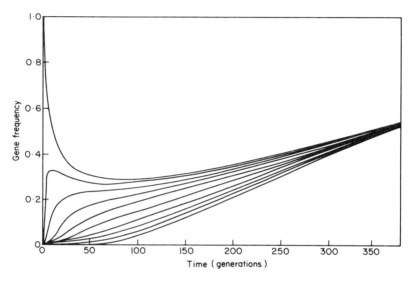

FIG. 4. The frequency of the dominant gene over time for the same system and values of the migration rate and selection coefficient as depicted in Fig. 3. However, here all populations select for the gene so that the selection here *reinforces* the gene flow promoted by migration.

drift and mutation apply but the model would seem to be valid over quite long time periods. The dynamic state of the system would again make the detection of selection difficult, in the absence of accurate information concerning the time of onset and amounts of migration.

Another aspect of this situation is possibly informative, and this is to assume that the selection *mechanism* itself is unknown. For example, if the two forms of selection operating in the opposing and reinforcing models are seen as alternative interpretations of a particular set of observed frequencies, how probable is it that they would be correctly designated? The short term behaviour for both models (as before for $k = 9$, $m = 0.1$, $s = 0.01$) as shown in Fig. 5 leads to a cline of a similar nature for ten or twenty generations and it would be impossible to decide which of the models was more appropriate from a cross-sectional sample of the populations at such a time. In other words, these (or indeed *other*) selection mechanisms

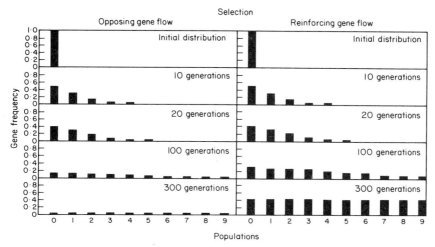

FIG. 5. The dispersion of a gene with time by migration in the presence of selection. Here the frequency of the dominant gene is plotted against the positions of the populations in the linear array at several times. The values of k, m and s are as for Figs. 3 and 4.

may describe the observations equally well and estimates of selection coefficients for a particular mechanism may therefore be meaningless.

Some comparable results have been derived by Karlin and Macgregor (1972) for other simple specific forms of selection, and these authors mention briefly, without detailed argument, one point which can be clearly seen from the opposing situation above, namely, that clines of gene frequency do not necessarily imply different selection pressures in all populations concerned. Following other work by Moran (1959) and Maynard Smith (1966) who discuss the fixation of alternative genes in two niches brought about by differential selection, Karlin and MacGregor (1972) have shown that in the particular case of disruptive selection, which acts in two populations in favour of one homozygote in one population and of the other in the other, a stable polymorphism may be maintained if the gene flow through migration is "slight" between the populations. However, since they do not discuss the time which would be needed for such a stable polymorphism to be established and since this time in the case of "slight" migration would be very considerable (and certainly from the above results this would be expected to be of the order of several thousands of generations), it remains doubtful whether their findings relate to present day polymorphisms in man. These latter authors concede, nevertheless, that if some form of assortative mating were present this would permit larger migration rates to operate to produce a stable polymorphism and this would somewhat reduce the time and make such systems more probable.

3. The Estimation of Migration or Admixture Rates for Intermixing Populations

Certain situations exist in which populations intermix over a period of several generations, thus producing hybrid populations. There are occasions when it is meaningful to seek to determine the rates of intermixing from knowledge of the gene pools in the various populations before and after this intermixture. Such knowledge over a time period of several generations is unlikely to be available and an alternative method usually adopted is to seek present-day populations which have contributed to admixture without having been affected by it. These populations may then be compared with other present-day hybrid populations.

In the simplest situation, there are two parental populations which contribute to a third (hybrid) population and the frequency of a gene at a single two allele locus may be observed in these three populations, as q_1, q_2 and q_0; the estimator which suggests itself for the proportion m_1 of individuals in the third population who derive from the first, takes the form

$$m_1 = \frac{q_0 - q_1}{q_2 - q_1}$$

and similarly for m_2 (Glass and Li, 1953). Where the frequencies in the contributing populations are well separated and all the sample values used are based upon large samples, no difficulty arises because the q_3 value will then be expected to be intermediate between the values of q_1 and q_2. When this is not the case, it is not possible to estimate m since, because of the unacceptability of migration rates which are negative or which exceed unity, the information content on this parameter in such samples will be nil.

Unfortunately, actual gene frequency data which are employed in this estimation process tend to be based upon small samples, usually concerning only a few hundred individuals, and the difficulty outlined above is not a trivial one. However, such data are never confined to observations at a single gene locus. Invariably, as many as ten or more genetic systems are available and if the migration rates themselves are of prime importance the combination of the information on the parameters from the several loci is required in the estimation process (Roberts and Hiorns, 1962). Where all available loci are included and where sample sizes are quite small, there would seem to be some advantage in including the constraints mentioned above in the estimation process. The technique available which provides for non-negative parameter estimates achieves a minimization of absolute deviations by the technique of linear programming instead of the more conventional least squares technique (Hiorns, 1964). However, this kind of technique, unlike the constrained maximum likelihood method, does not provide standard errors for estimates and is of use mainly for comparative purposes for indicating the variability of estimates with different criteria and constraints.

For the general problem of a number of parental populations, two methods of maximum likelihood estimation may be distinguished for the cases in which all population sample sizes are of the same order of magnitude or for the case in which the parental populations are based upon much larger samples and therefore have gene pools which may be assumed to be exactly known. The former type, devised by the present author (Hiorns, 1964), uses a model in the form of a linear functional relationship incorporating the admixture rates and the gene frequencies. This leads to a log likelihood with the structure, ignoring a constant term,

$$-\sum_j \sum_k w_{jk}\left(q_{0jk} - \sum_i m_i q_{ijk}\right)^2 - \lambda\left(\sum_i m_i - 1\right)$$

in which the w_{jk} are weights involving the admixture rates m_i from the n contributing populations $i = 1, 2, \ldots, n$ and the variances σ^2_{ijk} of the observed gene frequencies q_{ijk} of the kth allele at the jth locus in the ith population. Here q_{0jk} and σ^2_{0jk} refer to the intermixed or receiving population, λ is a Lagrangian parameter, and the w_{jk} are defined by $w_{jk} = 1/(\sigma^2_{0jk} + m_i^2 \sigma^2_{0ij})$. In practice, a two stage iterative process is generally effective in achieving these estimates, first obtaining approximate m_i values with unit weights then calculating weight values from these m_i values and proceeding in this alternating manner. The least squares structure of the estimation problem here, which so simplifies the first stage of the iterative process, relies upon an assumption of normality in constructing the likelihood function. What this in effect requires is that the multinomial distributions which strictly underlie the gene frequencies do not produce values which are close to zero or 1. Although this may seem restrictive, since such extreme values are common, particularly at loci with multiple alleles, the procedure described does provide values which are roughly comparable with other techniques.

The second type of estimation referred to employs the likelihood function incorporating the multinomial distribution (Krieger et al., 1965). Here it is assumed that the migrants who form the new hybrid population are randomly sampled, but in unknown proportions, from contributing populations whose gene pools are exactly known. This assumption makes the estimation simpler, although an iterative technique is still required, as was demonstrated by Elston (1971) and this may be seen from the structure of the log likelihood which is

$$\text{constant} + \sum_j \sum_k n_{jk} \log p_{jk}$$

in which

$$p_{jk} = \sum_{i=1}^{n} m_i q_{ijk}$$

and $n_{jk}/n_0 = q_{0jk}$ where n_0 is the total size of the hybrid population, n_{jk} individuals having the kth allele at the jth locus and the m_i, q_{0jk} and q_{ijk} are as defined previously.

TABLE 1. Gene frequencies for Nordestino and parental populations

Locus	Allele	African	Indian	Portuguese	Nordestino
Rhesus	cDe	0·635	0	0·046	0·233
	CDe	0·091	0·517	0·421	0·368
	cDE	0·049	0·483	0·101	0·116
	cde	0·179	0	0·398	0·257
	Cde	0·046	0	0·034	0·026
	N	(900)	(238)	(3000)	(288)
PTC	T	0·804	1·000	0·510	0·517
	t	0·196	0	0·490	0·483
	N	(280)	(86)	(618)	(296)

	Estimated intermixture rates			Method source
1. Maximum likelihood (linear functional relationship)	0·2967 ±0·0156	0·0393 ±0·0142	0·6641 ±0·0281	(Hiorns, 1964)
2. Linear programming	0·3240	0·0834	0·5926	(Hiorns, 1964)
3. Likelihood with more systems but exact parent values	0·2170	0·0707	0·7123	(Elston, 1971)
4. Least squares with more systems	0·2583	0·0920	0·6497	(Roberts and Hiorns, 1965)

Table 1 shows some results of using the various techniques for two genetic systems observed in the hybrid Nordestino populations of São Paulo, Brazil and the parental contributing African, Indian and Portuguese populations (Data source: Saldanha, 1962). The general conclusion from seeing these estimated rates which would be drawn by a human biologist is that they are not substantially different and almost any of these sets of values would be acceptable to him. This is despite the fact that, in some cases, more systems than can conveniently be reproduced here were used, and also that the first maximum likelihood method described may be inappropriate because of the rather low frequencies. However, the standard errors for that method which are shown following the $\pm$ symbols are in general accordance with the variety of estimated rates from these methods. Elston (1971) also gives a maximum likelihood method based upon phenotype frequencies but the results do not seem to be very different from methods using gene frequencies.

In discussing the application of models for intermixture certain other considerations are necessary. The choice of parental populations and representative samples are of critical importance in determining the admixture rates. The prevalence of genetic variation within these populations may make this choice difficult. However, a much more critical choice may concern the set of loci which are used in the analysis. In the absence of natural selection, the migrants carry with them these representative samples from the parental populations and if the numbers of migrants were large enough, migration rates determined from gene frequencies at any set of loci would be equivalent to those determined from those at any other set.

The importance of this feature of intermixture situations now emerges, for if differential selection acts with respect to the parental populations at a particular locus, the migration rates determined from gene frequencies at that locus will be out of line with those rates determined using loci where there is no such selection. This provides, in principle, another method for detecting selection and may be seen as an exciting by-product of the estimation of admixture rates. It has been applied by Workman *et al.*, (1963). Their results of estimating admixture rates for American Negroes in Claxton, Georgia in terms of White Americans in Claxton and West African Negroes show that whereas most systems indicate a rate of 0·1 White American and 0·9 African Negro, some systems show a much higher white contribution. It is concluded that these latter systems (G6PD, Hp_1, Tfd and Hb), were subject to selection for the 'white' alleles at the loci concerned when the original West African Negroes migrated into the American environment. A detailed discussion of these and other results together with the admixture rates and their standard errors appears in Cavalli–Sforza and Bodmer (1971, Chapter 8).

In these applications, only two contributing populations were used and the difficulty of using more populations and multiple loci would be considerable if selection were incorporated into a comprehensive model. The approach of using the model without selection separately, i.e. a simpler model to a simpler data structure, does illustrate that exploring inconsistencies in parameter estimates which arise can be more informative even if the inference is on a piecemeal basis.

4. Selective Migration and Population Differences

Whereas much of the theoretical analysis of the dynamics of human population has been in terms of an assumption of random (i.e. non-selective) migration, it is nevertheless true that much evidence exists that migrants rarely constitute a representative sample of the populations which they leave. Undoubtedly, this assumption, though erroneous, does prove useful in providing an indication, in gross terms, of the relationship between migration and natural selection and of admixture rates, at least for polymorphic systems. In the case of polygenic characters, selective migration needs to be taken into account.

In an attempt to balance the description of models, a model for selective migration will be described below. To date there appears to have been little work on this important topic although there is a small stock of models dealing with that other non-random phenomenon, assortative mating (Crow and Felstenstein, 1968; Karlin, 1972). It is important to distinguish between the models for selective migration, assortative mating and artificial selection. It will suffice here to appreciate that if populations are defined in different ways what is an assortatively mating population could become several distinct randomly mating ones with the aid of the device of selective migration. This seems to be the case, for example, with respect to social class structures within certain populations (Hiorns, et al., 1973). But there are characters which are employed as a basis for assortative mating which would not conceivably be acceptable descriptors of population structures. In contrast with artificial selection, in which selection is seeking to enhance specific characteristics by selective breeding, the model to be discussed differs from those of Robertson (1973, this volume) in that, for human populations within a closed conserving system, no genes are lost by the selective migration, only moved into different populations. The important consequence is that the extreme phenotypic values which may be achieved are limited by the feedback of balancing genes if migration is of a fixed amount each generation.

To illustrate this point, consider a single population in which some polygenic character has a normal distribution, and is dependent upon genes at L independently assorting loci each having an additive effect a upon the phenotypic value of the character. Suppose that this population splits into two equal populations and that migration takes place each generation, with the fraction m being selected as the individuals having the highest phenotypic values from one population being moved to the other population to be replaced by an equal number but having the lowest phenotypic values in that population. In this way, one "higher" and one "lower" population emerges and if the same form of migration takes place each generation, the difference between the mean phenotypic values in the populations will increase and eventually stabilize. Provided there is some amount of migration, the situation will not stabilize to one when all individuals in one population have a higher phenotypic value than those in the other, because of the forced movement of "high" individuals from the "higher" population in order to meet the constant migration requirement. For this reason the populations will not achieve fixation by the action of this form of migration, (very much to the contrary, in fact) and only by the effects of small population drift in the usual way.

To make this model more realistic in terms of human populations, the selective migration process can be obstructed by an "environmental component" of variance, V, which may be defined in terms of a random perturbation of the phenotypic value of the individuals selected for movement.

With these assumptions it may be shown that the separation at equilibrium leads to a difference in phenotypic value between the populations equal to

$$4La\left[\frac{2 + k^2 - 2(1 + k^2 V)^{\frac{1}{2}}}{4(4(1 - V) + k^2)}\right]^{\frac{1}{2}}$$

where $k = az/m$ and z is the ordinate from a normal distribution corresponding to a tail area probability m. (Kempton, 1971).

Figure 6 shows the relation between the environmental component of variance, the number of loci and the distance between two populations in units of the combined population standard deviation. From this graph for any observed difference in a phenotypic character, which is known to depend upon a number of independently assorting loci, the environmental component may be determined. Such an observable character is IQ and although the number of independently assorting loci is, of course, unknown, it may well be that this number is between 1 and 100. Before considering sample values of this character, a weakness of the model must be indicated. This concerns the fact that no account is taken of the change in the phenotypic value of the character which arises following the migration of particular individuals into different environments. The so-called "environmental variance" in this model has only served to obscure the proper identification of individuals for the purpose of selective migration. However, the neglected effect here would, if included in an improved model, increase the population difference predicted by such a model. In other words, the existing model produces for any given difference a value for the environmental component which is too low, and this value could be used as a minimum estimator of the true environmental component of variance.

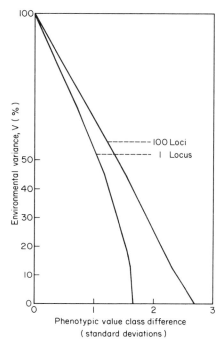

FIG. 6. The environmental component of variance V for some phenotypic character is plotted against the difference between two classes expressed as a multiple of the population standard deviation of the character value. The two lines represent phenotypes dependent upon genes at 1 or 100 independently assorting loci, respectively.

The data summarized in Table 2 is from two sources. Burt's (1961) IQ scores for parents of children in a London borough are regrouped from six social classes into two, together with more recent data collected this year in Charlton-on-Otmoor near Oxford, similarly grouped. Also shown are standardized predicted values for Charlton based on the Burt data but using the rather different social class composition of the Oxfordshire village. Ignoring the effects of the different IQ tests which were used in these two situations, the difference between these standardized values indicates that there are fundamental differences between the people within the social class groups quite apart from the different social class distributions. On close examination, such differences no doubt arise out of the very small number (3 out of 1000) of social class I individuals in the Burt data, the decline in social class V in recent times and the exclusion of housewives from the Charlton data. According to which set of data is used, the mean difference in IQ between two classes, one upper, one lower, seems to be either one or two standard deviations. From Fig. 6 the Charlton data indicate that the

TABLE 2. A comparison of two samples of IQ measured in the social classes

Social class	Burt (1961) sample				Charlton-on-Otmoor (1972) sample		
	number	IQ class mean	number	mean	number	Standardized predicted	observed
I	3	139·7 ⎱			4 ⎱		
II	31	130·6 ⎬	34	131·4	37 ⎬ 41	131·2	111·5
III non-manual	122	115·9 ⎰			57		
III manual	258	108·2 ⎱			35 ⎱		
IV	325	97·8 ⎬	966	99·4	⎬ 96	101·2	102·8
V	261	84·9 ⎰			4 ⎰		
	1000				137		
			d = difference = 32·0			29·9	8·7
			s = standard deviation = 15·0				11·6

IQ mean scores for social class groups in two samples: a London borough and Charlton-on-Otmoor. The standardized predicted values for Charlton are based upon the London borough observed values but adjusted to have the Charlton social class composition.

environmental component of variance is rather more than 0·55 if a single locus is involved, or more than 0·65 if 100 loci. In a similar way the Burt data gives corresponding minimum values of zero and 0·30. With such different results, arising as much out of the very different observed social class values in the two cases as out of an inadequate model, there is probably not much to be gained at this juncture by seeking further sophistication of models for use with such data.

5. Conclusions

It is not easy to assess the state of understanding of basic phenomena in relation to four specific problems with particular reference to the criterion of whether more complex mathematical models or more biological observation would be profitable. No clear conclusion is possible, as in the case of the simplest model situation, selection against a recessive gene in a single population, the difficulty of the problem would disappear if much larger samples could be obtained at less expense or if the constraints upon the sampling method could be removed. There is little need for further mathematical analysis in such a situation. However, the other three examples presented involve the inter-relationship between some forms of migration and selection and the development of population systems over time. The changes in genetic constitution caused by these factors are considerable and there is clearly more room for mathematical modelling, both deterministic and stochastic in approach, as well as a great need for more observation of the genetic variability, the quantity and *character* of the migration between populations and, although this was not considered in the above examples, the possibly non-random nature of the mating systems themselves, particularly with regard to polygenic characters. Although the conclusions are not clear they do indicate in some way that an understanding of migration and selection phenomena will be advanced by more mathematical analysis and more collection of data but the existing states of the problem areas at the present time may be indicative of where effort is needed to restore some degree of balance.

Perhaps more relevant to the actual development of understanding of biological phenomena will be the state of the two technologies, on the one hand the biologists have increased their ability to describe genetic systems and their biochemical bases (Cavalli-Sforza and Bodmer, 1971, Chapter 5), whilst on the other hand the mathematicians have developed the electronic computer as a tool to do algebraic analysis in non-numeric form (Barton *et al.*, 1970), to compute statistical distributions of numerically computed solutions (Dempster, 1969) as well as to compute these solutions in an optimal manner (Fletcher, 1971), and to perform complex simulations of biological systems.

REFERENCES

BARTON, D., BOURNE, S. R. AND HORTON, J. R. (1970). The structure of the Cambridge algebra system. *Computer Journal* **13**, 243–249.

BURT, C. (1961). Intelligence and social mobility. *Brit. Jour. Statist. Psychol.* **14**, 3–24.

CAVALLI-SFORZA, L. L. AND BODMER, W. F. (1971). "The Genetics of Human Populations." Freeman, San Francisco.

CROW, J. F. AND FELSENSTEIN, J. (1968). The effect of assortative mating on the genetic composition of a population. *Eugen. Quart.* **15**, 85–97.

DEMPSTER, M. A. H. (1969). Distributions in intervals and linear programming. *In* "Interval Analysis." (E. Hansen, ed.). Proc. Oxford University Computing Laboratory Symposium. Oxford University Press, London.

DOBZHANSKY, T. (1972). Natural selection in mankind. *In* "The Structure of Human populations." (G. A. Harrison and A. J. Boyce, eds.). Oxford University Press, London.

ELSTON, R. C. (1971). The estimation of admixture in racial hybrids. *Ann. of Hum. Genet.* **35**, 9–17.

FLETCHER, R. (1971). (Ed.). "Optimization" I.M.A. symposium. Academic Press, London.

GLASS, B. AND LI, C. C. (1953). The dynamics of racial intermixture, an analysis based on the American Negro. *Amer. Jour. Hum. Genet.* **5**, 1–20.

HIORNS, R. W. (1964). Statistical estimation for asymptotic regression models and the derivation of such models in a new theory for racial intermixture. Unpublished Ph.D. thesis, University of Edinburgh.

HIORNS, R. W., HARRISON, G. A., BOYCE, A. J. AND KÜCHEMANN, C. F. (1969). A mathematical analysis of the effects of movement on relatedness between populations. *Ann. Hum. Genet.* **37**, 237–250.

HIORNS, R. W. AND HARRISON, G. A. (1970). Sampling for the detection of natural selection by age group genetic differences. *Human Biology* **42**, 53–64.

HIORNS, R. W. AND HARRISON, G. A. (1972). Gene flow under selection and its evolutionary implications. Unpublished.

HIORNS, R. W., HARRISON, G. A. AND KÜCHEMANN, C. F. (1973). Factors affecting the genetic structure of populations: an urban/rural contrast in Britain. *In* "Genetic Variation in Britain." (D. F. Roberts and E. Sunderland, eds.), Taylor and Francis, London.

KARLIN, S. (1972). Some mathematical models of population genetics. *Amer. Math. Monthly* **79**, 699–739.

KARLIN, S. AND MACGREGOR, J. (1972). Application of method of small parameters to multi-niche population genetic models. *Theor. Pop. Biol.* **3**, 186–209.

KARLIN, S. AND MACGREGOR, J. (1972). Polymorphisms for genetic and ecological systems with weak coupling. *Theor. Pop. Biol.* **3**, 210–238.

KEMPTON, R. A. (1971). Differences in genetic composition between populations experiencing selecting migration. *Ann. Hum. Genet.* **35**, 25–33.

KREIGER, H., MORTON, N. E., MI, M. P., AZEVEDO, E., FREIRE-MAIA, A. AND YASUDA, N. (1965). Racial admixture in North Eastern Brazil. *Ann. Hum. Genet.* **29**, 113–125.

MAYNARD SMITH, J. (1966). Sympatric speciation. *Amer. Natur.* **100**, 637–650.

MORAN, P. A. P. (1959). The theory of some genetical effects of population subdivision. *Aust. J. Biol. Sci.* **12**, 109–116.

ROBERTS, D. F. AND HIORNS, R. W. (1962). The dynamics of racial intermixture. *Amer. Jour. Hum. Genet.* **14**, 261–277.

ROBERTS, D. F. AND HIORNS, R. W. (1965). Method of analysis of the genetic composition of a hybrid population. *Human Biology* **37**, 38–43.

ROBERTSON, A. (1973). Stochastic processes in artificial selection. *In* "The Mathematical Theory of the Dynamics of Biological Populations." (M. S. Bartlett and R. W. Hiorns, eds.). Academic Press, London.

SALDANHA, P. H. (1962). Race intermixture among North Eastern Brazilian populations. *American Anthropologist* **64** 751–759.

WORKMAN, P. L., BLUMBERG, B. S. AND COOPER, A. J. (1963). Selection, gene mutation and polymorphic stability in a U.S. White and Negro population. *Amer. J. Hum. Genet.* **15**, 429–437.

Part IV

Estimation and Simulation Problems

12. Commonsense Estimates from Capture-Recapture Studies

R. M. CORMACK

*Department of Statistics, University of Edinburgh,
Edinburgh, Scotland**

Many of the papers in this volume are concerned with the process of describing the development of an animal population by a mathematical model. The properties of such a model can then be derived, either by elegant mathematics or equally elegant computer simulation, in order to describe the future state of the population in terms of certain initial boundary conditions. The model becomes of scientific value when such predictions can be tested, which requires in turn that the mathematical symbols can be replaced by numbers. The parameters of the model must be estimated from data of a type that a biologist can collect about the population he is studying.

Ideally the mathematician would like the biologist to provide a continuous record of the whole population over some reasonable period of time. But only rarely, if at all, is the complete population observable or a continuous monitoring possible. Sample observations at discrete points S_i in time form the usual data.

First let us specify the model for an open population subject to birth and death, and possibly immigration and emigration. We shall describe death and emigration by a set of parameters φ_i, the probability that an individual alive in the population at the time of the ith sample S_i is still alive and in the population at the time of the $(i + 1)$th sample S_{i+1}. We shall describe birth and immigration by a set of parameters B_i, the number of individuals in the population at S_{i+1} that were not in the population at S_i. These changes affect N_i, the size of the population at S_i. Let us suppose that this population contains a subpopulation of M_i identifiable individuals which are subject to the same model with respect to their leaving the population, but to which there can be no additions without the experimenter's knowledge. In practice

* Present address: Department of Statistics, University of St. Andrews, St. Andrews, Scotland.

this implies that the M_i individuals have been marked by the experimenter. It also imposes the requirement that an identifiable individual does not emigrate temporarily to return at a later sampling period. This population is then sampled in such a way that each individual independently has probability p_i of being recorded in S_i. Observations n_i, of which m_i belong to the identifiable subpopulation, result. We shall suppose here that all n_i individuals are returned immediately to the subpopulation of identifiable individuals. Modifications to this scheme make no difference in principle to what follows, but merely complicate the notation unnecessarily.

This is a standard biological procedure, known as capture-recapture sampling or a multiple-recapture census. Under the assumptions stated, or with slight modification to the specification of immigration, maximum likelihood estimators have been obtained, after lengthy algebra, by Seber (1965) and by Jolly (1965). Let us see here what can be achieved by use of only minimal algebra, by the basic procedure of estimating a probability by the corresponding proportion, incorporating the commonsense idea that the largest relevant group of individuals should form the base for the proportion.

Suppose first that we know M_i. At S_i this is the largest known group, of which a proportion m_i/M_i are obtained in S_i.

$$\therefore \quad \hat{p}_i = m_i/M_i \tag{1}$$

Now, of the population N_i alive at S_i, n_i are observed in S_i.

$$\therefore \quad n_i/\hat{N}_i = \hat{p}_i$$

or

$$\hat{N}_i = n_i/\hat{p}_i = n_i M_i/m_i. \tag{2}$$

To estimate φ_i we require the largest known group alive immediately after S_i whose survivors at the time of S_{i+1} are also known. This is clearly the $(M_i + n_i - m_i)$ individuals in the identifiable subpopulation after S_i, of which M_{i+1} are alive at S_{i+1}.

$$\therefore \quad \hat{\varphi}_i = M_{i+1}/(M_i + n_i - m_i). \tag{3}$$

Applying this as a proportion to the whole population, N_{i+1} comprises those of N_i still alive, together with the B_i additions.

$$\therefore \quad \hat{B}_i = \hat{N}_{i+1} - \hat{N}_i \hat{\varphi}_i. \tag{4}$$

Unfortunately, however, in an open population we do not know M_i since the additions to the identifiable subpopulation have been subject to death and emigration. We do know m_i, the number of them observed in S_i. We must estimate $(M_i - m_i)$. We seek the largest known set of individuals known to be a subset of $(M_i - m_i)$—those that are observed after S_i, since those not so observed cannot be known to be alive at S_i. Denote the number of such individuals by z_i. Also n_i is the largest group of individuals known to be alive at S_i and hence comparable with $(M_i - m_i)$: denote by r_i the number of the n_i observed after S_i.

$$\therefore \quad z_i/(\hat{M}_i - m_i) = r_i/n_i$$

or

$$\hat{M}_i = m_i + \frac{z_i n_i}{r_i}. \tag{5}$$

The estimates which result from replacing M_i in (1) to (4) by $\hat{M}_i$ from (5) are those derived by Seber and by Jolly.

No Death or Emigration

If there is assumed to be no death in the population during the experiment there is no need to estimate M_i by (5): M_i is known to be the total number of individuals added to the identifiable subpopulation by the experimenter prior to S_i. The other equations are unaltered and again give the known maximum likelihood estimates.

No Birth or Immigration

In a population with no birth, estimation of M_i, p_i and N_i by (5), (1) and (2) is unaltered. But when we now ask for the largest group alive at S_i whose survivors at S_{i+1} are known, the answer is now N_i whereas in the more general model N_i could not be used since the subset of N_i alive at S_{i+1} cannot be distinguished from the B_i (unknown) new arrivals into the population. Thus in this case φ_i is estimated by N_{i+1}/N_i.

Closed Population

With a closed population, this type of argument is less convincing, although it still leads to maximum likelihood estimates. The parameters to be estimated are N and the set of p_i. The largest known subset of N is the set of all individuals M_{k+1} sampled throughout the complete experiment. Again equating proportion to probability to obtain estimates:

$$M_{k+1}/\hat{N} = 1 - \prod_{i=0}^{k} (1 - \hat{p}_i).$$

At S_i, n_i of the N individuals are observed (each with probability p_i).

$$\therefore \quad \hat{p}_i = n_i/\hat{N}.$$

These equations were first given by Darroch (1958).

Limitations of the Commonsense Approach

This discussion of the closed population illustrates the limitations of our commonsense approach. When the model describes the population as changing in a very general way we can see fairly clearly which groups of individuals contain information about a particular parameter, and what maximal subset should be considered. With a less general model with fewer parameters, each parameter will have an influence on more, often overlapping, sets of individuals, and it is less easy to be convinced that the largest relevant sets are being considered. In statistical terms (which I am trying to avoid as this is essentially an exercise in showing biologists the basic principles underlying fairly complicated statistical manipulation) the sufficient statistics are "obvious" for the general model, less so for the restricted case.

With an intermediate model the difficulties are perhaps more clear. Suppose that it is known that between S_j and S_{j+1} neither death nor emigration occurs, so that φ_j is 1. Then it is known that $M_{j+1} = (M_j + n_j - m_j)$. Using the same reasoning as led to (5), in order to estimate the size of the marked population we must add to the number *known* to be alive an estimate of those alive but not observed at either S_j or S_{j+1}. Denote by m_j^* the number of the m_{j+1} marked individuals observed at S_{j+1} that were not observed at S_j. Then we know that $(m_j + m_j^*)$ individuals marked before S_j are alive at S_j (and at S_{j+1} since $\varphi = 1$). The largest known subset of $(M_j - m_j - m_j^*)$ are those individuals that are observed after S_{j+1}, and we obtain the estimate $\hat{M}_j$ by equating that proportion of such individuals to the proportion of the largest group known to be alive at S_{j+1} that are observed after S_{j+1}. Since $\varphi_j = 1$, all individuals observed at either S_j or S_{j+1} are known to be alive at S_{j+1} (though not necessarily at S_j because of new arrivals between S_j and S_{j+1}), a total of $n_j + (n_{j+1} - m_{j+1} + m_j^*)$ individuals. Taking z_{j+1}, r_{j+1} as before to denote numbers of these groups observed after S_{j+1}:

$$\frac{z_{j+1}}{(\hat{M}_j - m_j - m_j^*)} = \frac{r_{j+1}}{n_j + n_{j+1} - m_{j+1} + m_j^*}.$$

Estimates $\hat{N}_j$ and $\hat{N}_{j+1}$ follow from (2), and $\hat{B}_j = \hat{N}_{j+1} - \hat{N}_j$.

Type I Losses

The advantages of this commonsense approach becomes clearer in the analysis of models with an increased, rather than a restricted, number of parameters compared with the homogeneous open model discussed earlier. If the population consists of mutually exclusive subpopulations (e.g. different sexes) governed by different sets of parameters, then, as Jolly (1965) pointed out, the parameters for each subpopulation can be estimated separately by the method described earlier. In many cases, however, the model may specify subpopulations with some parameters in common.

Consider the following example. The behaviour of an individual may reasonably be assumed to be affected by the act of first capture and marking.

With fish particularly the initial handling may cause what are known to fisheries biologists as Type I losses, a temporary reduction in survival over the first period after marking. At S_i we must distinguish two subpopulations of marked individuals, M_i' newly marked at S_{i-1} and M_i^* marked before S_{i-1}. The simplest model has survival parameters φ_{i-1}', φ_{i-1}^* for these two groups, φ_{i-1}^* applying also to unmarked individuals, while all other parameters remain common to all individuals. Both groups $(M_i' - m_i')$ and $(M_i^* - m_i^*)$ consist entirely of animals marked before S_i and thus have survival φ_i^*. Hence the group of animals known to be alive at S_i that are used for comparison (with both $(M_i' - m_i')$ and $(M_i^* - m_i^*)$) must also consist only of animals marked before S_i, that is, m_i ($= m_i' + m_i^*$) rather than n_i as previously. If r_i^+ of the m_i, z_i' of the $(M_i' - m_i')$, z_i^* of the $(M_i^* - m_i^*)$, are observed after S_i,

$$\hat{M}_i' = m_i' + z_i' m_i / r_i^+$$

$$\hat{M}_i^* = m_i^* + z_i^* m_i / r_i^+.$$

Survival estimates for the two groups are then given by:

$$\hat{\varphi}_i' = \hat{M}_{i+1}'/(n_i - m_i)$$

$$\hat{\varphi}_i^* = \hat{M}_{i+1}^*/\hat{M}_i$$

where $\hat{M}_i = \hat{M}_i' + \hat{M}_i^*$.

This model specifies that all marked individuals M_i are subject to the same probability p_i of being observed.

$$\therefore \quad \hat{p}_i = m_i/\hat{M}_i \quad \text{and} \quad \hat{N}_i = n_i/\hat{p}_i.$$

These estimates are identical to maximum likelihood estimates derived by Robson (1969) on the basis of a hypergeometric model.

An alternative assumption could be that the change in behaviour caused by initial marking alters both an individual's survival and its probability of being observed in the following sample. Since the estimates $\hat{M}_i'$, $\hat{M}_i^*$, $\hat{\varphi}_i'$, $\hat{\varphi}_i^*$ all depend only on individuals marked prior to the preceding sample they remain unchanged under this model. But we must estimate probabilities of capture separately from the two subpopulations:

$$\hat{p}_i' = m_i'/\hat{M}_i', \qquad \hat{p}_i^* = m_i^*/\hat{M}_i^*$$

and, since p_i' applies to marked individuals only, so that $n_i' = m_i'$,

$$\hat{N}_i = \hat{N}_i' + \hat{N}_i^* = \hat{M}_i' + (n_i - m_i')/\hat{p}_i^*.$$

Example

To demonstrate these commonsense estimates, we shall use data from a long-term study of fulmars (*Fulmarus glacialis*) at their breeding site, collected by G. M. Dunnet and his colleagues, part of which was analysed by Dunnet *et al.* (1963). In this study of adult birds initial marking was such that birds in subsequent years could be individually identified in flight without physical recapture. Thus marked and unmarked birds do not have the same p_i, and hence N_i cannot be estimated. This does not affect the validity of derivation of estimates of M_i, φ_i and a p_i appropriate to marked birds. The behaviour of the two sexes is sufficiently different for a model with separate sets of parameters for each sex to be tried. Unfortunately, sex identification is difficult even in the hand, but a discriminant function based on bill dimensions misclassifies only about 5% of individuals (Dunnet and Anderson, 1961). Some individuals are recorded as of unknown sex either because the measurements were not taken or because the indicated sex implied that two individuals of the same sex were successfully breeding.

Estimates of survival φ_i and of sampling intensity p_i for various groups of animals are given in Tables 1 and 2, obtained from basic equations (1), (3) and (5). The apparent paradox that birds of unknown sex have shorter lifespan than either males or females is explained by them not living long enough to be measured. Various properties of estimates obtained from capture–recapture data can be observed from these tables, and will be fully discussed elsewhere. Here let us merely illustrate the degree of modification imposed by different models by brief study of the data on males only (Table 3).

In the interest of tidiness, for example, we may not be satisfied with impossible survival estimates greater than unity. In the absence of a general theory of maximum likelihood estimation of parameters with inequality restraints, we might use a model in which φ_i is assumed to be 1 if $\hat{\varphi}_i \geq 1$. This is certainly a commonsense procedure, and we can here forget the distributional maze which will result. Data from the two periods between which no death occurs are then amalgamated. Over the first period, this gives $M_1 = 3$, $\hat{\varphi}_2$ increasing from

$$\frac{20 \cdot 53}{21 \cdot 11} = 0 \cdot 9725 \text{ to } \frac{20 \cdot 53}{21} = 0 \cdot 9776$$

to compensate for the reduction from $\hat{\varphi}_1 = 1 \cdot 0367$ to $\varphi_1 = 1$. The alteration to $\hat{M}_5$ caused by the assumption that $\hat{\varphi}_5 = 1$ makes $\hat{\varphi}_4$ in turn greater than 1, so that we must assume $\varphi_4 = \varphi_5 = 1$. Survival estimates in the first six periods change from

	1·04	0·97	0·94	1·00	1·03	0·94	(as in Table 3)
to	1·00	0·98	0·95	1·00	1·00	0·98,	

TABLE 1. Estimates of annual survival for subpopulations of fulmars

	All	Males	Females	Unknown	1951	1955	Marked 1960	1963	1966
1950–51	0.99	1.04	1.11	0.82					
1–2	0.90	0.97	0.90	0.83	0.89				
2–3	0.96	0.94	1.00	0.97	0.94				
3–4	0.99	1.00	1.16	0.92	1.04				
4–5	0.94	1.03	0.77	0.96	0.95				
5–6	0.95	0.94	1.00	0.88	0.92	1.02			
6–7	0.97	0.92	0.96	1.13	0.96	0.92			
7–8	0.88	0.95	1.00	0.67	0.82	0.98			
8–9	0.92	0.98	0.97	0.73	0.89	0.93			
9–60	0.96	0.94	0.97	1.00	0.96	0.99			
60–1	0.93	0.97	0.92	0.79	0.96	0.87	0.93		
1–2	0.88	0.93	0.99	0.58	0.82	0.88	0.96		
2–3	0.92	0.97	0.88	0.91	0.92	0.98	0.94		
3–4	0.94	0.93	0.97	0.88	0.91	0.94	1.01	0.96	
4–5	0.97	0.98	0.98	0.92	1.01	1.01	0.86	0.95	
5–6	0.97	0.97	0.95	1.09	1.01	0.85	0.93	1.02	
6–7	0.97	0.95	1.00	0.90	0.86	0.93	1.01	0.95	1.01
7–8	0.95	0.96	0.94	0.95	0.86	0.86	0.97	1.04	0.94
8–9	0.93	0.97	0.89	0.92	0.95	0.92	0.91	0.93	1.08
9–70	0.95	0.94	0.96	0.96	0.85	1.02	0.92	0.89	0.97
Expected Lifespan	16.2 ±1.4	25.0 ±4.3	22.3 ±4.2	8.1 ±1.1	11.9 ±1.9	15.5 ±3.0	17.2 ±5.4	25.9 ±12.6	398.0 ±9130
No. of birds	442	165	190	87	66	51	25	36	20

TABLE 2. Estimated probability of capture for subpopulations of fulmars

	All	Males	Females	Unknown	1951	1955	Marked 1958	1963	1967
1951	0·30	0·32	0	0·61					
2	0·51	0·54	0·50	0·52	0·53				
3	0·47	0·62	0·42	0·39	0·51				
4	0·56	0·73	0·40	0·55	0·58				
5	0·67	0·64	0·70	0·67	0·63				
6	0·51	0·64	0·37	0·55	0·62	0·38			
7	0·68	0·72	0·71	0·56	0·71	0·69			
8	0·69	0·75	0·64	0·65	0·77	0·60			
9	0·37	0·58	0·24	0·29	0·49	0·37	0·29		
60	0·72	0·84	0·69	0·57	0·74	0·65	0·79		
1	0·86	0·86	0·85	0·94	0·81	0·88	0·83		
2	0·85	0·88	0·78	1·00	0·83	0·81	0·92		
3	0·90	0·92	0·92	0·78	0·91	0·83	0·73		
4	0·74	0·70	0·77	0·80	0·77	0·72	0·82	0·64	
5	0·69	0·74	0·68	0·40	0·76	0·72	0·46	0·73	
6	0·67	0·73	0·64	0·43	0·58	0·70	0·70	0·59	
7	0·81	0·89	0·77	0·55	0·94	0·75	0·70	0·75	
8	0·60	0·65	0·58	0·51	0·67	0·63	0·71	0·60	0·62
9	0·76	0·75	0·77	0·74	0·77	0·79	1·00	0·84	0·71
70	0·80	0·83	0·76	0·84	0·75	0·82	0·71	0·77	1·00

TABLE 3. Data and estimates for male fulmars

	$(n_i - m_i)$	m_i	n_i	r_i	z_i	$\hat{M}_i$	$\hat{\varphi}_i$	$\hat{\varphi}'_i$	$\hat{p}_i$	$\hat{p}'_i$
1950	3	0	3	3	0	0	1.04	1.00	—	—
1	18	1	19	18	2	3.1	0.97	0.94	0.32	0.33
2	7	11	18	17	9	20.5	0.94	0.86	0.54	0.59
3	0	16	16	16	9	25.0	1.00	—	0.62	0.50
4	1	19	20	20	7	26.0	1.03	1.00	0.73	—
5	14	18	32	29	9	27.9	0.94	1.07	0.64	1.00
6	6	25	31	28	13	39.2	0.92	1.03	0.64	0.47
7	3	30	33	31	11	41.7	0.95	1.00	0.72	0.48
8	9	32	41	39	10	42.5	0.98	0.81	0.75	0.33
9	3	29	32	30	20	50.3	0.94	1.00	0.58	0.41
60	13	42	55	53	8	50.3	0.97	0.93	0.84	1.00
1	2	53	55	51	8	61.6	0.93	1.00	0.86	0.91
2	9	52	61	59	7	59.2	0.97	0.78	0.88	1.00
3	12	61	73	67	5	66.4	0.93	0.93	0.92	1.00
4	10	51	61	59	21	72.7	0.98	1.03	0.70	0.63
5	8	60	68	65	20	80.9	0.97	1.02	0.74	0.40
6	11	63	74	70	22	86.1	0.95	1.01	0.73	0.61
7	7	82	89	84	10	92.6	0.96	0.86	0.89	0.90
8	11	62	73	69	32	95.8	0.97	0.83	0.65	0.83
9	8	78	86	76	24	105.2	0.94	0.80	0.75	0.89
70	10	87	97	64	12	105.2	—	—	0.83	0.48

a change which would be less trivial if the sampling intensity were less than in this study.

Of more potential interest is to look at the effect of initial marking, although the estimates $\hat{\varphi}_i'$ and $\hat{p}_i'$ in Table 3 do not suggest any obvious fallacies in the homogeneous model. The data are at one and the same time too limited, in that few new animals are introduced into the survey each year, and too extensive, in that the length of the study and its high sampling intensity combine to cause few animals to be lost to the study each year, for differences of the order which might occur to be obvious. Algebraic gymnastics are needed to derive any formal parametric test, and even a simple non-parametric test must be avoided because of the correlations between estimates for successive periods. An analysis of the complete data does give some suggestion that initial marking does affect the behaviour of an animal in the succeeding year, strangely not that its probability of survival is reduced but that its probability of being observed in the following year is reduced.

Acknowledgements

I am most grateful to my colleague, Mr A. Carothers, for the analyses given in Tables 1 and 2, and to N.E.R.C. for a research grant to study these problems.

REFERENCES

DARROCH, J. N. (1958). The Multiple-Recapture Census. I. Estimation of a Closed Population. *Biometrika* **45**, 343–359.

DUNNET, G. M. AND ANDERSON, A. (1961). A method for sexing living fulmars in the hand. *Bird Study* **8**, 119–126.

DUNNET, G. M., ANDERSON, A., AND CORMACK, R. M. (1963). A study of survival of adult fulmars with observations on the pre-laying exodus. *British Birds* **56**, 2–18.

JOLLY, G. M. (1965). Explicit estimates from capture-recapture data with both death and immigration—stochastic model. *Biometrika* **52**, 225–247.

ROBSON, D. S. (1969). Mark-recapture methods of population estimation. *In* "New Developments in Survey Sampling." (N. L. Johnson and H. Smith Jr., Eds.) Wiley, New York.

SEBER, G. A. F. (1965). A note on the multiple-recapture census. *Biometrika* **52**, 249–259.

13. An Evaluation of Two Capture-Recapture Models Using the Technique of Computer Simulation

J. A. Bishop

Department of Zoology, University of Liverpool, Liverpool, England

and

P. M. Sheppard

Department of Genetics, University of Liverpool, Liverpool, England

1. Introduction

Statisticians and population biologists have had an uneasy relationship in the development and application of models for use in capture-recapture studies. Biologists are frequently ignorant of mathematical notation and are depressingly uncritical of the models that they use, e.g. Manga (1972). There is little excuse for this since Cormack (1969) reviewed the statistics of capture-recapture in a lively and interesting way as well as providing (Cormack, 1973) a commonsense description of the popular stochastic model of Jolly (1965). Statisticians on the other hand are usually aware of the difficulties of studying animal populations. In their enthusiasm for the elegance and simplicity of Jolly's model they have tended to rule out some more restricted, but none-the-less useful, models on statistical rather than biological grounds.

We have investigated several populations of insects using capture-recapture methods (e.g. Bishop, 1972; Sheppard *et al.*, 1969). We became interested in the relative performances of Jolly's model and Fisher and Ford's (1947) deterministic model in dealing with the sometimes inadequate data that we obtained. To compare them a population was simulated in a computer, sampled, and the data analysed by the appropriate methods. The characteristics of the population were known so that we could assess the reliability of each model. Manly (1970) compared Fisher and Ford's, Jolly's and Manly and Parr's (1968) models in a similar way. His results were not useful to us since he samples his populations at an unrealistically high intensity (0·1–0·75).

Adequate estimates of population parameters will not be obtained if the data does not fulfil assumptions (listed by Southwood, 1966, p. 75) that are basic to all models. Roff (1973) stresses that, in particular, the assumption that all members of a population are equally likely to be captured is rarely, if ever, met. This can drastically affect estimates. Bishop and Bradley (1972) describe how taxi-cabs can be used as subjects for a population study. The basic assumption of equal catchability was not met in samples taken at the rank in Lime Street Station, (since it was found that only 62 of the 300 taxis in Liverpool regularly operate from there) though it was met in samples taken about 200 metres away in Lime Street. Table 1 compares the

TABLE 1. Estimates of the size of the population of taxi-cabs in Liverpool. Samples were taken at Lime Street Station taxi rank and about 200 metres away in Lime Street. In both cases samples were taken between 1430 and 1530 hrs. (Jolly's model used in estimation)

	Lime Street Liverpool	Taxi rank Lime Street Station
Wed.	208·4	97·1
Thurs.	336·3	98·0
Fri.	352·8	130·4
Sat.	286·9	178·8
Sun.	213·9	267·2
Mon.	230·7	90·5
Tues.	240·7	59·6
Mean	267·1	131·6

estimates of population size obtained at the two localities. The spectacular differences emphasize that any data for analysis must be taken from a population that is as homogeneous as possible. This requires separate analysis of data from subpopulations of each sex and of each instar or age class. These subgroups may further show genetic variability that alters their probability of capture (Bishop, 1973). Biological insight used to assemble organisms into homogeneous groups will greatly improve estimates of population parameters obtained from existing population models.

2. Methods

We simulated populations with known characteristics with an I.C.L. KDF 9 computer. Given proportions of these were sampled, the individuals concerned were marked and released into the parent population so that they were available for later recapture. The data obtained were analysed by Fisher and Ford's and Jolly's models. Ten populations with every combination of the following conditions were simulated:

(i) Probability of survival from sampling occasion n to occasion $n + 1$ 0·5, 0·9.

(ii) Size of population 200, 1000, 3000.

(iii) Number of occasions (separated by equal time intervals) on which population was sampled 10 and 20.

(iv) Proportion of population in sample on each occasion 0·05, 0·09, 0·12.

Numbers were kept constant during each simulation. Animals that died as a result of an unfavourable encounter with a pseudo-random number were replaced by unmarked animals.

On some sampling occasions no individual was recaptured. In such cases one recapture is assumed (i.e. it is acknowledged that the population is finite) and the corresponding estimate is calculated. With Fisher and Ford's model two estimates of population size are obtained. Firstly, there is the standard uncorrected estimate and secondly there is one subject to Bailey's (1951, 1952) correction. The uncorrected maximum likelihood estimate $\hat{N} = Mn/m$ has a positive bias of about $1/m$. Bailey's correction attempts to counteract this bias by adding one to the capture and to the recapture figures ($\hat{N} = M(n + 1)/(m + 1)$). The results of a similar correction applied in Jolly's model will be discussed elsewhere.

The mean population size and survival rate for each simulation has been calculated in those cases where there are "daily" estimates. These have been used to calculate overall means and standard errors for the results for the 10 simulations with constant parameters. It is only possible to present broad trends on this basis and a detailed examination of results will be presented in a later paper.

The conditions in the simulated populations were such that there was no undue bias against the Fisher and Ford model. Particularly important was the assumption of a constant probability of survival which had an approximately binomial variance. Later it will be necessary to investigate situations in which the population departs further from the assumptions of the Fisher and Ford model.

3. Results

(i) Estimates of population size

The results appear in Table 2. Jolly's model gave poor estimates where a low proportion (0·05) of populations with survival rates of 0·5 were sampled. This is not surprising since in these circumstances few recaptures were made and little information was available. Fisher and Ford's model provided better estimates because the calculation of a single constant survival rate and its subsequent use in calculating population estimates is less subject to sampling error. Where more information from recaptures was available

TABLE 2. Estimates of overall mean population size ($\pm$standard errors) for series of 10 computer simulated populations of known (indicated) characteristics. Jolly's and Fisher and Ford's (with and without Bailey's correction) models used for analysis

(a) Actual population size 200

Survival rate	Jolly	Fisher–Ford	Fisher–Ford (Bailey's correction)	Sampling occasions	% Population sampled
	Estimates of Population				
	37 ± 5	177 ± 43	170 ± 24	20	5
	39 ± 5	207 ± 61	137 ± 37	10	
0·5	214 ± 18	242 ± 25	190 ± 19	20	9
	225 ± 30	334 ± 69	251 ± 49	10	
	307 ± 28	284 ± 28	231 ± 22	20	12
	234 ± 33	241 ± 46	200 ± 36	10	
	233 ± 10	246 ± 22	206 ± 16	20	5
	162 ± 13	221 ± 30	174 ± 22	10	
0·9	253 ± 17	226 ± 13	213 ± 12	20	9
	268 ± 16	228 ± 17	205 ± 14	10	
	225 ± 11	213 ± 9	206 ± 9	20	12
	268 ± 23	221 ± 11	206 ± 10	10	

(b) Actual population size 1000

Survival rate	Jolly	Fisher–Ford (U)	Fisher–Ford (C)	Sampling occasions	% Population sampled
	Estimates of Population				
	1669 ± 143	1420 ± 136	1143 ± 100	20	5
	1205 ± 154	1089 ± 143	867 ± 110	10	
0·5	1243 ± 55	1039 ± 39	961 ± 35	20	9
	1252 ± 127	1115 ± 129	1023 ± 108	10	
	1137 ± 44	1119 ± 46	1076 ± 44	20	12
	1260 ± 112	1198 ± 81	1139 ± 82	10	
	1220 ± 41	1192 ± 39	1141 ± 36	20	5
	1267 ± 53	1183 ± 48	1096 ± 45	10	
0·9	1051 ± 24	1008 ± 25	997 ± 24	20	9
	1046 ± 37	1006 ± 28	989 ± 26	10	
	1025 ± 9	1017 ± 6	1011 ± 5	20	12
	1021 ± 27	1008 ± 25	998 ± 25	10	

(c) Actual population size 3000

| Survival rate | Estimates of Population | | | Sampling occasions | % Population sampled |
	Jolly	Fisher–Ford (U)	Fisher–Ford (C)		
	4358 ± 379	3517 ± 272	3214 ± 241	20	5
	5727 ± 889	4062 ± 346	3580 ± 259	10	
0·5	3106 ± 66	2987 ± 71	2924 ± 69	20	9
	2948 ± 164	2825 ± 138	2757 ± 133	10	
	3114 ± 50	3072 ± 54	3036 ± 53	20	12
	3378 ± 137	3287 ± 124	3242 ± 122	10	
	3083 ± 62	2974 ± 49	2927 ± 47	20	5
	3156 ± 117	3138 ± 133	3062 ± 129	10	
0·9	3017 ± 31	2997 ± 42	2986 ± 42	20	9
	3135 ± 56	3141 ± 71	3120 ± 71	10	
	3021 ± 13	3008 ± 17	3003 ± 17	20	12
	3057 ± 24	3017 ± 30	3007 ± 30	10	

because survival rate was higher and/or more of the population was sampled there was much less difference between the performance of the two models. There was also less variation in the 10 estimates of mean population size as indicated by the reduction in size of standard errors in Table 2.

Jolly's model almost always overestimated population size and Fisher and Ford's model frequently did. When Bailey's correction is applied to the latter there is as expected a reduction in the estimated population size. Table 2 shows that there is sometimes overcorrection.

There is little difference between the means and standard errors obtained in those populations sampled on 10 and 20 occasions when few recaptures were made. When the numbers of recaptures increases the size of the standard errors for the populations sampled on 20 occasions drops, as one would expect, far more rapidly than those for the populations sampled on 10 occasions.

(ii) Estimates of survival rate
The results appear in Table 3. The overall pattern is similar to that described for population size. Unweighted estimates from Jolly's model were less accurate than those of Fisher and Ford's model when little recapture information was available but the difference is less apparent where the survival rate was higher or more of the population was captured.

Jolly's model consistently and considerably overestimates survival rate at population sizes 200 and 1000. No correction factor has, to our knowledge,

TABLE 3. Estimates of overall mean survival rate ($\pm$ standard errors) for series of 10 computer simulated populations of known (indicated) characteristics. Jolly's and Fisher and Ford's models used for analysis

(a) Population size 200

Actual survival rate	Jolly estimate	Fisher–Ford estimate	Sampling occasions	% Population sampled
	0.339 ± 0.037	0.426 ± 0.039	20	5
	0.357 ± 0.030	0.479 ± 0.061	10	
0.5	0.594 ± 0.037	0.508 ± 0.026	20	9
	0.678 ± 0.074	0.572 ± 0.049	10	
	0.644 ± 0.033	0.497 ± 0.022	20	12
	0.614 ± 0.071	0.491 ± 0.040	10	
	1.083 ± 0.030	0.895 ± 0.010	20	5
	0.894 ± 0.053	0.862 ± 0.044	10	
0.9	1.031 ± 0.023	0.898 ± 0.007	20	9
	1.084 ± 0.038	0.870 ± 0.020	10	
	0.978 ± 0.024	0.893 ± 0.006	20	12
	1.027 ± 0.040	0.883 ± 0.018	10	

(b) Population size 1000

Actual survival rate	Jolly estimate	Fisher–Ford estimate	Sampling occasions	% Population sampled
	0.632 ± 0.022	0.516 ± 0.012	20	5
	0.553 ± 0.034	0.440 ± 0.031	10	
0.5	0.595 ± 0.019	0.473 ± 0.009	20	9
	0.597 ± 0.028	0.484 ± 0.016	10	
	0.527 ± 0.011	0.497 ± 0.009	20	12
	0.580 ± 0.036	0.535 ± 0.018	10	
	0.966 ± 0.012	0.907 ± 0.004	20	5
	1.090 ± 0.039	0.915 ± 0.016	10	
0.9	1.066 ± 0.049	0.895 ± 0.003	20	9
	0.936 ± 0.021	0.900 ± 0.008	10	
	0.905 ± 0.007	0.899 ± 0.002	20	12
	0.927 ± 0.009	0.900 ± 0.008	10	

(c) Population size 3000

Actual survival rate	Jolly estimate	Fisher–Ford estimate	Sampling occasions	% Population sampled
	0·594 ± 0·027	0·482 ± 0·018	20	
	0·683 ± 0·047	0·511 ± 0·021	10	5
0·5	0·512 ± 0·007	0·488 ± 0·004	20	
	0·501 ± 0·016	0·473 ± 0·012	10	9
	0·505 ± 0·006	0·498 ± 0·006	20	
	0·520 ± 0·012	0·502 ± 0·011	10	12
	0·910 ± 0·006	0·891 ± 0·003	20	
	0·906 ± 0·013	0·888 ± 0·012	10	5
0·9	0·903 ± 0·003	0·896 ± 0·003	20	
	0·911 ± 0·006	0·906 ± 0·006	10	9
	0·904 ± 0·002	0·899 ± 0·001	20	
	0·917 ± 0·010	0·902 ± 0·002	10	12

been developed to deal with this, perhaps because survival was of subsidiary interest to the model. In some studies of animal populations survival rate is a more important quantity than population size. For example the peppered moth *Biston betularia* is of great interest to students of evolution since it exhibits the phenomenon of industrial melanism (Kettlewell, 1955; 1956). In many areas populations of the species are polymorphic and individuals of both the black and white typical and the black *carbonaria* form occur. Release-recapture experiments can be performed to determine the survival rates of these two types of moth. We simulated a series of populations similar to those described in Section 2 but more in keeping with the known ecology of the species (Clarke and Sheppard, 1966; Bishop, 1972). Two populations of 150 and 300 individuals were simulated where the typical and *carbonaria* forms were equally common. Survival rates of 0·3, 0·4 and 0·5 were used and 0·2 and 0·3 of the population was sampled for a period of 15 days. Results appear in Fig. 1. The actual survival rates are plotted on the abscissas and the corresponding estimated values are plotted on the ordinates. If there was an exact correspondence between the actual and the estimated rates then the estimates should fall along the diagonal lines in each graph. The values obtained from Fisher and Ford's model do approach this ideal. In only one case does the 95% confidence limit of a mean estimate not include the actual survival rate. The values obtained from Jolly's model almost always significantly overestimate the actual survival rate. The bias is so great that it is clear that the model must be used with great caution in studies where survival rates are important.

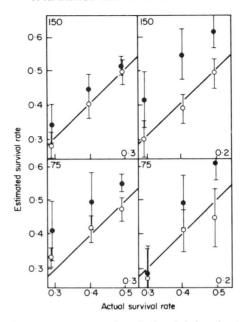

Fig. 1. A comparison between parameters of survival and their estimates obtained by Jolly's model (closed circles) and Fisher and Ford's model (open circles). In each case 95% confidence limits of the means of 10 simulations with constant parameters are shown. Figures in the upper left hand corners are the population sizes and in the lower right hand corners are the proportions of the populations that were sampled on each occasion (see text for further details).

5. Summary and Conclusions

The fashionable Jolly's model is statistically elegant and gives reliable estimates of population size and when adequate recapture data is available. This is usually achieved when about 0·09 or more of a population with a survival rate of 0·5 is sampled. In practice it is frequently difficult to obtain sufficient data particularly when the basic assumptions of capture-recapture models have been satisfied by making samples as homogeneous as possible. Biological common sense and statistical expertise indicates that data for the sexes and age classes should be analysed separately. Here the more restricted Fisher and Ford model can be useful particularly if there is evidence that survival rate remains relatively constant while the population is under study. The constancy of the survival rate can be checked during the analysis of data (Fisher and Ford, 1947).

Jolly's model seriously overestimates survival rate and should not in its present form be used in studies where this quantity is important. We believe, with other ecologists, that Jolly's method is useful and important. There are several situations in which its short comings are obvious but where Fisher and Ford's model will give reliable estimates. Computer programs for Jolly's method are widely available (White, 1971a; 1971b; Davies, 1971) and we take this opportunity of presenting a program for Fisher and Ford's method. This program has a built in check of the assumption of constant survival rate.

REFERENCES

BAILEY, N. T. J. (1951). On estimating the size of mobile populations from recapture data. *Biometrika* **38**, 293–306.
BAILEY, N. T. J. (1952). Improvements in the interpretation of recapture data. *J. Anim. Ecol.* **21**, 120–127.
BISHOP, J. A. (1972). An experimental study of the cline of industrial melanism in *Biston betularia* (L.) (Lepidoptera) between urban Liverpool and rural North Wales. *J. Anim. Ecol.* **41**, 209–243.
BISHOP, J. A. (1973). The proper study of populations. *In* "Insects: Studies in Population Management." *Mem. Ecol. Soc. Aust.* **1** (in press).
BISHOP, J. A. AND BRADLEY, J. S. (1972). Taxi-cabs as subjects for a population study. *J. Biol. Educ.* **6**, 227–231.
CLARKE, C. A. AND SHEPPARD, P. M. (1966). A local survey of the distribution of industrial melanic forms in the moth *Biston betularia* and estimates of the selective values of these in an industrial environment. *Proc. R. Soc. B.* **165**, 424–439.
CORMACK, R. M. (1969). The statistics of capture-recapture methods. *Oceanogr. Mar. Biol. Ann. Rev.* **6**, 455–506.
CORMACK, R. M. (1973). Commonsense estimates from capture-recapture studies. This volume. *In* "The Mathematical Theory of the Dynamics of Biological Populations." (M. S. Bartlett and R. W. Hiorns, eds.), Academic Press, London and New York.
DAVIES, R. G. (1971). "Computer Programming in Quantitative Biology." Academic Press, London and New York.
DOWDESWELL, W. H., FISHER, R. A. AND FORD, E. B. (1940). The quantitative study of populations in the Lepidoptera. I. *Polyommatus icarus* (Rott). *Ann Eugen* **10**, 123–136.
FISHER, R. A. AND FORD, E. B. (1947). The spread of a gene in natural conditions in a colony of the moth *Panaxia dominula* (L). *Heredity, Lond.* **1**, 143–174.
GASKELL, T. J. AND GEORGE, B. J. (1972). A Bayesian modification of the Lincoln Index. *J. appl. Ecol.* **9**, 377–384.
JOLLY, G. M. (1965). Explicit estimates from capture-recapture data with both death and immigration—stochastic model. *Biometrika* **52**, 225–247.
KETTLEWELL, H. B. D. (1965). Selection experiments on industrial melanism in the Lepidoptera. *Heredity, Lond.* **9**, 323–342.
KETTLEWELL, H. B. D. (1966). Further selection experiments on industrial melanism in the Lepidoptera. *Heredity, Lond.* **10**, 287–301.
MANGA, N. (1972). Population metabolism of *Nebria brevicollis* (F.) (Coleoptera: Carabidae). *Oecologia* (Berl.) **10**, 223–247.
MANLY, B. F. J. (1970). A simulation study of animal population estimation using the capture-recapture method. *J. appl. Ecol.* **7**, 13–40.
MANLY, B. F. J. AND PARR, M. J. (1968). A new method of estimating population size survivorship and birth rate from capture-recapture data. *Trans. Soc. Br. Ent.* **18**, 81–89.

ROFF, D. A. (1973). An examination of some statistical tests used in the analysis of mark-recapture data. *Oecologia* (Berl.) **12**, 35–54.
SHEPPARD, P. M., MACDONALD, W. W., TONN, R. J. AND GRAB, B. (1969). The dynamics of an adult population of *Aedes aegypti* in relation to dengue haemorrhagic fever in Bankok. *J. Anim. Ecol.* **38**, 661–702.
SOUTHWOOD, T. R. E. (1966). "Ecological Methods with Particular Reference to the Study of Insect Populations." Methuen, London.
WHITE, E. G. (1971a). A versatile FORTRAN computer program for the capture-recapture stochastic model of G. M. Jolly. *J. Fish Res. Bd. Canada* **28**, 443–446.
WHITE, E. G. (1971b). A computer program for capture-recapture studies of animal populations. A FORTRAN listing for the stochastic model of G. M. Jolly. Tussock Grasslands and Mountain Lands Institute, Lincoln College, Christchurch, N.Z. Special Publication No. 8, pp. 33.

Appendix

An ALGOL 60 Computer Program for the Analysis of Capture–Recapture Data using the Deterministic Model of Fisher and Ford.

R. J. White, Department of Zoology, University of Liverpool.
P. M. Sheppard, Department of Genetics, University of Liverpool.

The program listed uses the character representations and input–output routines of the Egdon System for the ICL KDF9. Slight modification may be needed for other machines.

The program is divided into three sections:
a short calling program,
procedure FFINP, which reads and stores the input, and
procedure FFORD, which performs the analysis and prints the results.

The advantages of this division are, first, the FFORD routine may be used alone within more complex programs, and second, the FFINP routine is easily modified to allow for other data formats, or input from magnetic tape or disc.

The first of the three samples of input given below is taken from Dowdeswell, Fisher and Ford (1940). The corresponding output is also given, for checking purposes.

The Fisher and Ford analysis requires to know the "total days survived" by all marks recaptured during the experiment. (See Fisher and Ford, 1947, for a discussion of the quantities involved.) This information can be presented to FFINP in three ways:

(i) As a complete trellis diagram, which specifies completely the numbers of marks recaptured on each occasion and the days survived since their release.

(ii) By a list of numbers of marks recaptured and the days survived by these marks, for each sampling occasion.

(iii) By a list of numbers of marks recaptured, and the value of the "total days survived". This latter quantity is the sum of the individual days survived by marks recaptured on each occasion as given in (ii).

Program input consists of:

1. A title used for labelling the output.

2. *m*, the duration of the experiment. The units of *m* are called days in the print-out, but need not necessarily correspond to 24 h periods. The rest of the input is read by FFINP.

3. *phi*. If $0.0 < phi < 1.0$, it is treated as the survival rate on which the analysis is to be based. Otherwise the survival rate is calculated from the data.

4. *tds*. This quantity determines the form of subsequent input. If -1, a trellis diagram is given as in (i) above. If 0, input follows the scheme of (ii) above. If greater than 0, its exact value is the total days survived, as in (iii) above.

5. Subsequent input consists of rows of data, each row pertaining to a particular sampling occasion. The first item in any row is the "date" of the sampling occasion. The program assumes that these "dates" progress chronologically in the same units as *m*, but they may take any integer values between -9999 and 99999. In the example given, 31 is equivalent to 31 August 1938; 33 is equivalent to 2 September. No data is given for day 32, 1 September.

6. The second column of figures, i.e. the items immediately following the dates, refer to the number of animals captured on that occasion.

7. If $tds \geq 0$, a column referring to the number of marks recaptured on each day is now included.

8. If $tds = 0$, a column referring to the number of days survived by all marks recaptured on each day is also included.

9. The next column of figures is the numbers of animals released on the days in question; it need not bear any relation to the numbers captured.

10. If $tds = -1$, the recaptures are then specified by means of the standard trellis diagram. Each row of figures refers to marks recaptured on the appropriate date. Columns refer to the date of the last release. Note that if days are omitted from the rows due to gaps in the data, the corresponding columns should also be omitted.

Further problems for analysis may follow, beginning with a new title.

The ouput is largely self-explanatory. Unless input type (iii) is being used, the values of expected and actual days survived by each day's recaptured marks is output under "Survival Estimation", together with the difference. These differences give an indication of the constancy of the survival rate over the period of the experiment.

Bailey's Correction is used in calculating the population estimates. It is possible to modify this part of FFORD if it is desired to use Gaskell and George's (1972) Bayesian formula. In addition to the raw estimates, smoothed values are given. These are third-order harmonic running means, except for the first and last. The figures for the gains and losses from the population (birth plus immigration and death plus emigration respectively) are based on the smoothed estimates. It is also possible to incorporate Sheppard's parameter m to allow for migration (see Sheppard et al., 1969).

PROGRAM

```
*ALGOL
'BEGIN INTEGER 'M, DATE1$ 'REAL 'PHI$
'COMMENT 'CALLS FFINP AND FFORD$
WRITETEXT(30,'('''('P')'POPULATION'**'ESTIMATION'**'BY
'**'DETERMINISTIC'**'MODEL'**'OF'**'FISHER'**'AND'**'FORD
'('C')'PROGRAM'*'BY'*'P.M.'*'SHEPPARD'*'AND'*'R.J.'*'WHITE'*'
(BISHOP'*'AND'*'SHEPPARD,'*'1972)'('2C')'')')$
HERE :COPYTEXT(20,30,'('$')')$ M= READ(20)$
'BEGIN INTEGERARRAY 'C, D, DAYS, DD(1:M)$
      FFINP(M,DATE1,PHI,C,DD,DAYS,D)$
      FFORD(M,DATE1,PHI,C,DD,DAYS,D)
'END 'OF ESTIMATIONS FOR THIS POPULATION$
'GOTO 'HERE
'END 'OF FISHER-FORD CALLING PROGRAM
*ALGOL
'PROCEDURE 'FFINP(M,DATE1,PHI,CAPS,RECAPS,DAYS,RELS)$
'COMMENT 'READS THE FISHER-FORD TRELLIS
DIAGRAM, OR A SUMMARY THEREOF. ON ENTRY
M IS THE DURATION OF THE EXPERIMENT$
'VALUE 'M$ 'INTEGER 'M, DATE1$ 'REAL 'PHI$
'INTEGERARRAY 'CAPS, RECAPS, DAYS, RELS$
'BEGIN INTEGER 'F, I, J, K, N, TDS$
'INTEGERARRAY 'DATE, SURV, RECOVS(1:M)$
PHI= READ(20)$ TDS= READ(20)$
DATE(1)= DATE1= READ(20)$ N= 1$
'FOR 'I= 1 'STEP '1 'UNTIL 'M 'DO'
'IF 'I 'EQ' DATE(N) - DATE1 + 1 'THEN'
'BEGIN 'CAPS(I)= READ(20)$
      RECAPS(I)= 'IF 'TDS 'GE' 0 'THEN 'READ(20) 'ELSE '0$
      DAYS(I)= 'IF 'TDS 'EQ' 0 'THEN 'READ(20) 'ELSE '0$
      RELS(I)= READ(20)$
      'COMMENT 'READ TRELLIS$
      RECOVS(N)= SURV(N)= 0$
      'IF 'TDS 'LT' 0 'THEN'
      'FOR 'J= 1 'STEP '1 'UNTIL 'N-1 'DO'
      'BEGIN 'K= READ(20)$
            RECAPS(I)= RECAPS(I) + K$
            RECOVS(J)= RECOVS(J) + K$
            K= K*(I + DATE1 - 1 - DATE(J))$
            DAYS(I)= DAYS(I) + K$
            SURV(J)= SURV(J) + K
      'END 'J$
      'COMMENT 'READ DATE OF NEXT ROW$
      'IF 'I 'LT' M 'THEN'
      'BEGIN 'N= N + 1$
            DATE(N)= READ(20)
      'END'
'END ELSE'
CAPS(I)= RECAPS(I)= DAYS(I)= RELS(I)= 0$
'IF 'TDS 'LT' 0 'THEN'
'BEGIN 'WRITETEXT(30,'(''('2C')'RECAPTURES'**'ANALYZED
      '**'BY'**'DAY'**'OF'**'RELEASE'*':('2C24S')'NUMBER
      '('5S')'SUBSEQUENTLY'('5S')'DAYS'('C11S')'DATE'('8S')'
      RELEASED'('5S')'RECOVERED'('5S')'SURVIVED'('2C')'')')$
      F= LAYOUT('('SS-NDDDDDDDDDD')')$
      'FOR 'I= 1 'STEP '1 'UNTIL 'N-1 'DO'
```

```
'BEGIN 'K= RELS(DATE(I) - DATE1 + 1)$
      'IF 'K 'GT' 0 'THEN'
'BEGIN 'WRITE(30,F,DATE(I))$
      WRITE(30,F,K)$
      'IF 'RECOVS(I) 'GT' 0 'THEN'
      'BEGIN 'WRITE(30,F,RECOVS(I))$
            WRITE(30,F,SURV(I))
      'END'$
      NEWLINE(30,1)
'ENDEND 'I
'END ELSE 'DAYS(I)= TDS
'END 'FFINP 18TH AUGUST 1972$
*ALGOL
'PROCEDURE 'FFORD(M,DATE1,PHI,CAPS,RECS,DAYS,RELS)$
'COMMENT 'ESTIMATES POPULATION PARAMETERS
ACCORDING TO THE FISHER-FORD RECIPE
:M IS DURATION OF THE EXPERIMENT
DATE1 IS DATE OF FIRST SAMPLING OCCASION
PHI IS SURVIVAL RATE (ON EXIT)
CAPS CONTAINS NO OF ANIMALS CAPTURED EACH DAY
RECS CONTAINS NO OF MARKS RECAPTURED EACH DAY
DAYS CONTAINS DAYS SURVIVED BY RECAPTURED MARKS
RELS CONTAINS NO OF ANIMALS RELEASED EACH DAY$
'VALUE 'M, DATE1$ 'INTEGER 'M, DATE1$ 'REAL 'PHI$
'INTEGERARRAY 'CAPS, RECS, DAYS, RELS$
'BEGIN ARRAY 'AMP(2:M)$
'INTEGER 'I, J, K, F1, F3, F4, F5, F6, NZERO$
'REAL 'LAST, DIFF, ETDS, TDS, PHI1, PHI2, EXP, TG, TL, G, L$
'PROCEDURE 'ESTDS(M,PHI,ETDS,AMP,AGE,RELS,RECAPS)$
'COMMENT 'THIS PROCEDURE CALCULATES THE ESTIMATED
TOTAL DAYS SURVIVED AND SETS UP AMP AND AGE$
'VALUE 'M, PHI$ 'INTEGER 'M$ 'REAL 'PHI, ETDS$
'ARRAY 'AMP, AGE$ 'INTEGERARRAY 'RELS, RECAPS$
'BEGIN INTEGER 'I$ 'REAL 'SUM$
      AMP(2)= RELS(1)*PHI$
      AGE(2)= 1.0$ ETDS= RECAPS(2)$
      'FOR 'I= 3 'STEP '1 'UNTIL 'M 'DO'
'BEGIN 'SUM= AMP(I-1) + RELS(I-1)$
      AMP(I)= SUM*PHI$
      AGE(I)= ((AGE(I-1)+1.0)*AMP(I-1) + RELS(I-1))/SUM$
      ETDS= ETDS + AGE(I)*RECAPS(I)
'ENDEND 'ESTDS 26TH AUG 1972$
F1= LAYOUT('('NDDDD')')$ F3= LAYOUT('('NDDDD.D')')$
F4= LAYOUT('('-NDDDDDD')')$ F5= LAYOUT('('NDDDDD.DD')')$
F6= LAYOUT('('+NDDD.DD')')$
'COMMENT 'THIS SECTION CALCULATES THE FISHER-FORD SURVIVAL
RATE AND THE NUMBERS OF MARKS AT LARGE ON EACH DAY$
'BEGIN ARRAY 'AGE(2:M)$
'IF 'DAYS(1) 'NE' 0 'THEN 'TDS= DAYS(1) 'ELSE'
'BEGIN 'TDS= 0.0$
      'FOR 'I= 2 'STEP '1 'UNTIL 'M 'DO'
      TDS= TDS + DAYS(I)
'END'$
'IF 'PHI 'GT' 0.0 'AND 'PHI 'LT' 1.0 'THEN'
ESTDS(M,PHI,ETDS,AMP,AGE,RELS,RECS) 'ELSE'
```

```
'BEGIN 'PHI= ETDS= 0.0$ PHI2= 1.0$
     'COMMENT 'CYCLE TO FIND SURVIVAL RATE PHI$
     'FOR 'I= 1 'STEP '1 'UNTIL '14 'DO'
     'BEGIN IF 'ETDS 'GT' TDS 'THEN'
          'BEGIN 'PHI2= PHI$
                PHI= (PHI + PHI1)/2.0
          'END ELSE'
          'BEGIN 'PHI1= PHI$
                PHI= (PHI + PHI2)/2.0
          'END'$
          ESTDS(M,PHI,ETDS,AMP,AGE,RELS,RECS)
     'END 'I
'END'$
WRITETEXT(30,'(''('2C')'SURVIVAL'**'ESTIMATION'('2C7S')'
MARKS-AT-LARGE'('11S')'AGE-FROM-RELEASE'***'---'*'
DAYS'*'SURVIVED'*'---'('C2S')'DAY'**'RELEASED'*'TOTAL'***'RECAPS
'**'EXPECTED'**'ACTUAL'***'EXPECTED'*'ACTUAL'*'DIFF.'('C')'')')$
'FOR 'I= 1 'STEP '1 'UNTIL 'M 'DO'
'BEGIN 'NEWLINE(30,1)$
     WRITE(30,F1,I+DATE1-1)$
     'IF 'RELS(I) 'GT' 0 'THEN'
     WRITE(30,F4,RELS(I)) 'ELSE 'SPACE(30,8)$
     'IF 'I 'GT' 1 'THEN'
     WRITE(30,F4,AMP(I)) 'ELSE 'SPACE(30,8)$
     'IF 'RECS(I) 'GT' 0 'THEN'
     WRITE(30,F4,RECS(I)) 'ELSE 'SPACE(30,8)$
     'IF 'I 'GT' 1 'THEN'
     WRITE(30,F5,AGE(I)) 'ELSE 'SPACE(30,9)$
     'IF 'DAYS(1)'EQ'0 'AND 'I'GT'1 'AND 'RECS(I)'GT'0 'THEN'
'BEGIN 'WRITE(30,F5,DAYS(I)/RECS(I))$
     EXP= AGE(I) * RECS(I)$
     WRITE(30,F5,EXP)$
     WRITE(30,F4,DAYS(I))$
     WRITE(30,F6,DAYS(I) - EXP)
'ENDEND 'OF SURVIVAL ESTIMATION TABLE$
WRITETEXT(30,'(''('2C')'POPULATION'*'ESTIMATES'*'FOR')')$
WRITE(30,F1,M)$
WRITETEXT(30,'(''*'DAYS'**'T.D.S.'*''EQ'')')$
WRITE(30,F5,TDS)$
WRITETEXT(30,'(''('C')'SURVIVAL'*'RATE'*''EQ'')')$
WRITE(30,LAYOUT('('ND.DDDDD')'),PHI)$
WRITETEXT(30,'(''***'ESTIMATED'*'T.D.S.'*''EQ'')')$
WRITE(30,F5,ETDS)
'END 'OF SURVIVAL ESTIMATION SECTION$
WRITETEXT(30,'(''('2C')'POPULATION'**'ESTIMATION
'('2C9S')'TOTAL'('11S')'MARKS-IN-POP'*****'POPULATION
'*'SIZE'('C2S')'DAY'**'CAPTURES'**'RECAPS'***'NO.'*'
PERCENT'('6S')'RAW'('5S')'SMOOTH'***'GAINS'*'LOSSES'('C')'')')$
'BEGIN ARRAY 'POP, IPOP, SPOP(2:M)$
'COMMENT 'CALCULATE AND STORE RAW AND SMOOTHED
POPULATION ESTIMATES USING BAILEYS CORRECTION$
K= NZERO= 0$
'FOR 'I= 2 'STEP '1 'UNTIL 'M 'DO'
'IF 'CAPS(I) 'GT' 0 'THEN'
```

```
'BEGIN IF 'K 'EQ' 0 'THEN 'K= I$
     POP(I)= IPOP(I)= AMP(I)*(CAPS(I)+1)/(RECS(I)+1)$
     'IF 'NZERO 'GT' 0 'THEN'
     'BEGIN 'DIFF= (POP(I) - IPOP(I-1))/NZERO$
          'FOR 'J= 1 'STEP '1 'UNTIL 'NZERO 'DO'
          IPOP(I-J)= IPOP(I-J) + DIFF*(NZERO-J+1)$
          NZERO= 0
     'END'
'END ELSE'
'BEGIN 'NZERO= NZERO + 1$
     POP(I)= 0.0$
     IPOP(I)= IPOP(I-1)
'END'$
SPOP(K)= 2.0/(1.0/IPOP(K)+1.0/IPOP(K+1))$
'FOR 'I= K+1 'STEP '1 'UNTIL 'M-1 'DO'
SPOP(I)= 3.0/(1.0/IPOP(I-1)+1.0/IPOP(I)+1.0/IPOP(I+1))$
SPOP(M)= 2.0/(1.0/IPOP(M-1)+1.0/IPOP(M))$
'COMMENT 'CALCULATE REMAINING QUANTITIES AND PRINT TABLE$
TG= TL= 0.0$
'FOR 'I= 2 'STEP '1 'UNTIL 'M 'DO'
'BEGIN 'NEWLINE(30,1)$
     WRITE(30,F1,I+DATE1-1)$
     'IF 'CAPS(I) 'GT' 0 'THEN'
     WRITE(30,F4,CAPS(I)) 'ELSE 'SPACE(30,8)$
     'IF 'RECS(I) 'GT' 0 'THEN'
     WRITE(30,F4,RECS(I)) 'ELSE 'SPACE(30,8)$
     WRITE(30,F4,AMP(I))$
     'IF 'RECS(I) 'GT' 0 'THEN'
     WRITE(30,F3,(RECS(I)+1)/(CAPS(I)+1)*100.0)
     'ELSE 'SPACE(30,7)$
     'IF 'POP(I) 'GT' 0.0 'THEN'
     'BEGIN IF 'RECS(I) 'EQ' 0 'THEN 'WRITETEXT(30,'('*')')
          'ELSE 'SPACE(30,2)$
          WRITE(30,F4,POP(I))$
          'IF 'RECS(I) 'EQ' 0 'THEN 'WRITETEXT(30,'('*')')
          'ELSE 'SPACE(30,2)
     'END ELSE 'SPACE(30,12)$
     'IF 'I 'GE' K 'THEN 'WRITE(30,F4,SPOP(I))$
     'IF 'I 'GT' K 'THEN'
     'BEGIN 'LAST= SPOP(I-1) * PHI$
          G= SPOP(I) - LAST$ TG= TG + G$
          L= SPOP(I-1) - LAST$ TL= TL + L$
          WRITE(30,F4,G)$
          WRITE(30,F4,L)
     'END'
'END 'I$
WRITETEXT(30,'('('2C')'TOTAL '**'GAINS'*''EQ'')',)$
WRITE(30,F4,TG)$
WRITETEXT(30,'('('C')'TOTAL '*'LOSSES'*''EQ'')')$
WRITE(30,F4,TL)$
GAP(30,1)
'ENDEND 'FFORD 22ND AUGUST 1972$
```

INPUT

```
POLYOMMATUS'**'ICARUS'**'TEAN'**'ISLES'**'OF'**'SCILLY'**'1938$

14$ -1$ -1$

26$40$40$
27$43$40$    5$
29$13$12$    0$  3$
30$52$50$    3$  8$  5$
31$56$51$    6$12$  6$15$
33$52$52$    4$10$  3$16$14$
34$50$50$    4$  5$  1$11$  5$14$
37$15$15$    1$  1$  1$  3$  1$  5$  5$
38$20$20$    1$  1$  2$  3$  2$  7$  8$  6$
39$ 7$ 0$    0$  0$  0$  2$  2$  4$  1$  0$  4

PANAXIA'**'DOMINULA'**'MARMONTSFLAT'**'WOOD'**'ALL'*'ADULTS'**'1972$

10$ -1$ 0$

1$  19$  0$  0$19$
2$  12$  0$  0$11$
3$   9$  1$  2$  9$
4$  12$  1$  1$  8$
5$  25$  4$  8$21$
6$  40$  2$  2$39$
7$  19$  1$  5$19$
8$  83$10$26$79$
9$  43$10$16$42$
10$61$ 8$12$57

PANAXIA'**'DOMINULA'**'COMFREY'**'COLONY'**'ALL'**'ADULTS'**'1972$

24$ -1$ 342$

1$   0$  0$  2$
2$   4$  1$  4$
3$   4$  1$  4$
4$  14$  5$14$
5$   4$  2$  4$
6$   3$  5$  3$
7$   1$  0$  1$
8$   6$  3$  6$
9$  10$  7$10$
10$ 7$  6$  7$
11$ 8$  2$  8$
12$ 9$  1$  9$
13$17$  5$17$
14$16$  8$16$
15$14$  8$14$
16$16$  5$16$
17$12$  6$12$
18$12$10$12$
19$ 8$  5$  8$
20$ 4$  7$  4$
21$10$11$10$
22$ 5$12$  5$
23$ 2$  3$  2$
24$ 2$  3$  0
```

OUTPUT

POPULATION ESTIMATION BY DETERMINISTIC MODEL OF FISHER AND FORD
PROGRAM BY P.M. SHEPPARD AND R.J. WHITE (BISHOP AND SHEPPARD, 1972)

POLYOMMATUS ICARUS TEAN ISLES OF SCILLY 1938

RECAPTURES ANALYZED BY DAY OF RELEASE :

DATE	NUMBER RELEASED	SUBSEQUENTLY RECOVERED	DAYS SURVIVED
26	40	24	130
27	40	40	194
29	12	18	60
30	50	50	170
31	51	24	79
33	52	30	93
34	50	14	52
37	15	6	6
38	20	4	4

SURVIVAL ESTIMATION

DAY	MARKS-AT-LARGE RELEASED	TOTAL	RECAPS	AGE-FROM-RELEASE EXPECTED	ACTUAL	--- DAYS SURVIVED --- EXPECTED	ACTUAL	DIFF.
26	40							
27	40	33	5	1·00	1·00	5·00	5	0·00
28		61		1·45				
29	12	50	3	2·45	2·00	7·36	6	-1·36
30	50	52	16	2·98	2·56	47·67	41	-6·67
31	51	84	39	2·51	2·69	97·96	105	+7·04
32		112		2·56				
33	52	93	47	3·56	3·74	167·48	176	+8·52
34	50	120	40	3·28	3·62	131·31	145	+13·69
35		141		3·32				
36		117		4·32				
37	15	97	17	5·32	5·35	90·38	91	+0·62
38	20	92	30	5·60	5·07	168·03	152	-16·03
39		93	13	5·60	5·15	72·85	67	-5·85

POPULATION ESTIMATES FOR   14 DAYS T.D.S. =   788·00
SURVIVAL RATE = 0·82831   ESTIMATED T.D.S. =   788·04

POPULATION ESTIMATION

DAY	TOTAL CAPTURES	RECAPS	MARKS-IN-POP NO.	PERCENT	POPULATION SIZE RAW	SMOOTH	GAINS	LOSSES
27	43	5	33	13·6	243	204		
28			61			193	25	35
29	13	3	50	28·6	176	170	10	33
30	52	16	52	32·1	161	148	7	29
31	56	39	84	70·2	120	123	1	25
32			112			108	5	21
33	52	47	93	90·6	102	114	25	18
34	50	40	120	80·4	149	120	25	20
35			141			112	12	21
36			117			94	2	19
37	15	17	97	112·5	86	76	-2	16
38	20	30	92	147·6	63	65	1	13
39	7	13	93	175·0	53	57	4	11

TOTAL GAINS =   116
TOTAL LOSSES =   262

14. The Estimation of Parameters from Epidemic Models

NORMAN T. J. BAILEY

World Health Organization, Geneva, Switzerland

1. Introduction

Theoretical investigations of epidemic models are expected to provide "insight" into the structure of transmission phenomena. What is this "insight", and does it have any objective existence? To what extent is any real knowledge gained a purely private vision, and to what extent can it supply public information, susceptible to scientific investigation. In my opinion, real insight means a better grasp of what is going on; a more useful understanding of the logical relationships between the epidemiological events taking place; a revelation of the underlying mechanisms behind surface phenomena; a clearer appreciation of what consequences inevitably flow from any given set of assumptions, etc. But all this initially *presumed* understanding of the real world must be validated in due course by successful prediction, and, in many cases of importance, by actual *control* of the phenomena in question.

This means that it is not enough merely to analyze purely theoretical models, important though this is. The models must *first* be adequately fitted to actual data, and this involves the special subject of this presentation, namely the estimation of parameters. And, *secondly*, the models must ultimately be able, directly or indirectly, to guide successful intervention and control.

2. General Considerations

Concepts such as contact-rate, recovery-rate, latent period, infectious period, incubation period, serial interval, etc. are commonplace ingredients of epidemiological discussion. There are, however, some difficulties, first in defining some of the concepts precisely, and secondly in measuring relevant indices for any given disease. For many of the better understood diseases, such as measles and smallpox, the case-to-case transmission interval, or serial

253

interval, is known with some accuracy. Moreover, the incubation period, measured from the receipt of infection to the appearance of symptoms, can be gauged quite reliably from situations in which primary cases can be identified with near certainty. Of course it is accepted that appreciable statistical variation may be involved, but this can be handled routinely if sufficient data are available.

Unfortunately in practice the data may not be very extensive, and may in any case have been assembled from highly heterogeneous sources. Again, items like the latent period are not in general observable, and are therefore much harder to assess in a sufficiently accurate way.

One method of dealing with these difficulties is to develop a well-defined quantitative model of the disease in question, and to use it to analyze appropriate data in a mathematically and statistically valid manner. There are many advantages in such an approach. First, the basic assumptions and hypotheses are set out in an explicit quantitative manner. This clarifies thought and helps to reach agreement that the model proposed is neither too oversimplified to have practical value, nor too complex to be satisfactorily analyzed and interpreted. Secondly, the implications inherent in the assumptions can be deduced by using known mathematical and computing techniques. Even though the intricacies of these techniques may be difficult for some non-mathematicians to follow, the results can usually be interpreted in straightforward scientific language, using the epidemiological concepts already agreed on in the original model. Thirdly, we obtain numerical estimates of the basic parameters of the model, together with indications of their probable accuracy. Thus, given the assumptions, we have some measure of the reliability of the conclusions, and this facilitates discussion and interpretation. Fourthly, it is frequently possible to test statistically whether the model chosen, and the hypotheses it implies, are a sufficiently good description of the observed data to be at least provisionally acceptable, or whether the divergence between the model and reality is so great as to require rejection or modification of some of the basic assumptions.

This paper describes the application of the foregoing approach to the problem of elucidating the finer detail of an epidemic process, paying special attention to the measurement of epidemiologically significant features such as latent, infectious and incubation periods. For this type of investigation the best material comes from the careful observation of a large number of small families. However, one approximate method of analysis can also be usefully applied to observations on a single sufficiently large epidemic in a well-defined population of moderate size.

It should be noted in passing that the type of analysis to be described below is distinct from, though complementary to, the less detailed mathematical models available for describing the broad features of epidemic processes in large populations. In the latter case it is usual to concentrate on such aspects as average infection rates and average symptom-recognition rates, while ignoring the finer details altogether. But such methods are usually inappropriate for the analysis of family data or of epidemics in only moderate sized populations. This kind of data needs a model which can represent in a fairly realistic way the biological mechanisms involved, and thus adequately account for the great statistical variations normally observed. It is only in very large populations that the finer structure can safely be averaged out and effectively ignored.

3. Epidemiological Data

Let us now consider some of the types of epidemiological data that are available in certain instances from the observations made by general practitioners or other investigators of communicable disease. Since we hope to be able to unravel some of the intricacies of the fine structure of the mechanisms involved, it is useless in general simply to look at statistics that have been aggregated over large groups of individuals. Weekly or monthly totals of new cases may be adequate for describing general population behaviour. But for the measurement of anything as specific as an infectious or latent period it is essential to have reliable information on individuals.

Now, so far as the mechanism of the spread of infection from one person to another is concerned, it is unfortunate that in general we cannot observe the event of infection itself. At best we only have an accurate record of when cases are recognized by the appearance of symptoms. By this time such cases have already played their major role in possibly passing on their infection to others.

This means that in good data on a disease like measles in families of two susceptible children, both of whom were infected, we may be able to get an adequate idea of the *serial interval*, or case-to-case transmission interval. Although in such families this interval is the basic epidemiological unit observed, it is not so easy to deduce directly the incubation period or infectious period. Moreover, the picture may easily be confused by the occurrence of double primaries, both of whom may well have been independently infected by a common source outside the home.

Consider for example the now classic data on measles collected by R. E. Hope Simpson in the Cirencester area during the years 1946–1952 (quoted in Bailey, 1957). Table 1 shows the distribution of the observed time interval between two cases of measles in 219 families with two children under the age of 15. There are also 45 families with one case only. The second column shows the actual number of families seen for each time interval. The latter varies between 0 and 21 days, with an obvious peak at about $10\frac{1}{2}$ days.

TABLE 1. Observed time-interval distribution between two cases of measles in families of two (Hope Simpson's data from Cirencester area, 1946–52)

Time interval between the two cases in days	Total number of families observed	Presumed double primaries	Presumed case-to-case transmission	Number of families predicted
0	5	5	.	4·7
1	13	13	.	8·6
2	5	5	.	6·7
3	4	4	.	4·5
4	3	2	1	2·8
5	2	.	2	2·2
6	4	.	4	4·0
7	11	.	11	8·8
8	5	.	5	16·6
9	25	.	25	24·7
10	37	.	37	29·4
11	38	.	38	29·3
12	26	.	26	25·4
13	12	.	12	20·0
14	15	.	15	14·3
15	6	.	6	9·0
16	3	.	3	4·8
17	1	.	1	2·1
18	3	.	3	1·0
19	.	.	.	0·2
20	.	.	.	0·0
21	1	.	1	0·0
Subtotals	219	29	190	219·1
Number of families with one case only	45	—	—	44·1
Grand total	264			

It can be presumed that the bulk of the observations represent case-to-case transmission within the family. These show a considerable amount of variation, although they constitute a fairly symmetrical distribution. However a small number of families, where the observed serial interval is only a few days, give a marked peak at the beginning of the distribution. It may be surmised that these are really double primaries, simultaneously infected from some outside source. The third and fourth columns show the two parts of the distribution separately. These parts overlap slightly but there is little difficulty with these data in making a sufficiently accurate separation intuitively. When the overlap is more marked special statistical methods can be used.

In order to explain how the model finally chosen was arrived at, we first look at the simplest possible explanation of the observed data, and then consider what modifications are required. As long ago as 1928, L. J. Reed and W. H. Frost (unpublished, but see Wilson and Burke, 1942) were using so-called chain-binomial models in the United States. The essence of this is to assume that, following the receipt of infection, there is a constant incubation period, which is terminated by a very short interval of high infectiousness. Symptoms then appear, and the case is promptly removed from circulation. Susceptibles in contact with the case at the time of infectiousness have a certain probability of themselves becoming infected. This leads to successive crops of cases in a group of susceptibles, the crops being separated in time by the incubation period. The variable number of cases actually observed at any given stage can be shown to have a binomial distribution. Hence the name "chain-binomial model". Greenwood independently developed a somewhat similar model in England in 1931.

Such models have been fairly successful in explaining the broad features of the spread of measles in families. But it is evident from Table 1 that the model breaks down as soon as we look at the details of the serial interval distribution. For if the incubation period were constant or nearly so, not only would the double primaries all be clumped very closely together, but the case-to-case intervals would also all be approximately the same and equal to the incubation period.

Since the double primaries show an appreciable degree of variation in their time of appearance it is clear that we must at least include some variation in the incubation period. Suppose therefore that we assume a variable incubation period, but still retain the very short period of high infectiousness. It is easily shown by simple statistical reasoning that the variance of the observed case-to-case distribution would be a direct measure of the theoretical variance of the incubation period, while the mean square of the double primary distribution about the origin would be *twice* the variance of the incubation period. Examination of the figures in Table 1 shows that in fact the first measure of variation is observed to be nearly twice the second, and not conversely. An exact statistical test reveals that the model is significantly contradicted by the data.

The next obvious modification of the model is to introduce an infectious period, extended in time instead of being virtually contracted to a point as above. Of course it would perhaps be more realistic to consider a variable infectious period. But it is normal scientific practice not to introduce more complications than are strictly necessary to explain phenomena. So let us first assume an extended infectious period of constant length. In any case this already introduces a new source of variation into the model since susceptibles at risk may be infected at any point of the infectious period. This modified model was first introduced by Bailey in 1954, and analysed in greater detail later.

Although the model can be described fairly easily, the appropriate statistical analysis is rather more difficult. Before going on to discuss the results and interpretations let us define the model quite specifically in elementary quantitative terms in the next section.

4. The Mathematical Model

Let us define rather more precisely the mathematical model arrived at above, after rejecting two simpler versions as being obviously inadequate to explain the observed facts.

We assume that, after the receipt of infection by a susceptible individual, there ensues a *latent period*, during which the disease develops clandestinely. The latent period is taken to have approximately "normal" statistical frequency distribution with mean m and standard deviation s.

This latent period is followed by an *infectious period* of length a. During this period any exposed susceptible may become infected in a purely chance way. Specifically, we assume that the probability of a given susceptible becoming infected in a short interval of time dt is $\lambda\, dt$. Thus λ is an *infection-rate*.

Finally, we suppose the infectious period is effectively terminated by the appearance of symptoms, when the individual in question is removed from circulation amongst the group of susceptibles and isolated from further contacts. The new case could of course continue to be infectious for some time after removal, but this would not be significant for further spread of the disease if isolation were reasonably effective. In this event the "infectious period" is really the time elapsing between the onset of infectiousness and the appearance of symptoms.

Although this model is a relatively simple one it does contain the essential ingredients of latent period, infectious period, appearance of symptoms, removal, and infection-rate, quite explicitly, with clear mathematical definitions of these features. This enables us to examine the implications of the model in appropriate circumstances to see how successful it is in accounting for observed phenomena. As we shall see in the following section quite promising applications have been made to family data on measles and infectious hepatitis, with a tentative extension to a smallpox outbreak in a single large group.

The model in its present form is only relevant to diseases where the type of transmission is a direct case-to-case one. If environmental factors play a major role, or if animal vectors or reservoirs are involved, then these aspects would have to be modelled explicitly. However, the present discussion illustrates what can be done in a certain class of diseases when the right kind of data is available or can be collected.

5. Applications

First, let it be emphasized that in this paper we are primarily concerned (1) with the kind of data available on certain types of infectious disease, (2) with exhibiting and discussing an appropriate quantitative model, and (3) with presenting, reviewing and interpreting the biological and epidemiological implications of various kinds of mathematical analysis and computerized handling. The technical mathematical details can if desired be studied in the theoretically oriented research papers that are referenced below.

The statistical approach adopted in each case has been to estimate the parameters in the model, i.e. the average length and standard deviation of the latent period, the length of the infectious period, and the infection-rate, using the well-known principle of maximum likelihood. This means that we choose those values of the parameters that make the observations most probable. This simple principle is now well-attested, both by commonsense and by a vast technical literature which shows that maximum-likelihood estimates are usually best in the sense of being at least as accurate as those produced by any other method of analysis. The method also supplies standard errors for each estimate, so that we have a *measure* of how accurate the estimate is likely to be. In some cases we can go a step further and ask how satisfactory is the composite scientific hypothesis entailed by the assumed mathematical model and the actual set of estimates chosen. This is the so-called "goodness-of-fit" test. If the fit, or agreement, between the data and the explanatory model is not close enough, i.e. if there is a statistically significant divergence, then the model must be modified or abandoned. Conversely, if no such alarm is sounded, then we can continue to have at least some provisional confidence in the model.

Let us now look at some practical applications to three specific diseases, namely measles, infectious hepatitis and smallpox.

(a) Measles

We have already examined the excellent data of Hope Simpson, presented in Table 1, on measles in family groups of size two. This material was first analysed by Bailey in two papers in 1956 (see Bailey, 1957, Ch. 7 for full discussion and references) using a very considerable amount of mathematical analysis, as well as necessitating several weeks' work with a desk calculating machine. These circumstances have probably prevented the approach being adopted by epidemiologists to any great extent.

However in 1970, Bailey and Alff-Steinberger described a complete computerization of the whole method. The extensive use of automatic computing actually modified the mathematical approach as well. It became necessary only to be able to write down the probability of the observations in terms of the four parameters to be estimated. Another standard maximization programme could then be used to calculate exactly which values of the four parameters made the probability of the observations a maximum, and also to calculate the relevant standard errors. All this took only a *few minutes* on an IBM 360 Model 40 computer, and would be even faster on the Model 65 now available.

The result obtained for the data in Table 1 were as shown in Table 2, where the figures following the "$\pm$" signs are the standard errors.

TABLE 2. Estimated parameters for Measles data

Average latent period	$m = 8\cdot6 \pm 0\cdot4$ days
Standard deviation of latent period	$s = 1\cdot8 \pm 0\cdot2$ days
Length of infectious period	$a = 6\cdot6 \pm 1\cdot0$ days
Infection-rate	$\lambda = 0\cdot26 \pm 0\cdot04$
Average incubation period	$m + a = 15\cdot2 \pm 0\cdot5$ days
Standard deviation of incubation period	$s = 1\cdot8$ days

It appears immediately that a striking finding is the relatively long infectious period, about $6\frac{1}{2}$ days. This contrasts strongly with the assumption, made in many earlier epidemiological discussions, of infectiousness being concentrated almost at a single point of time (as in the chain-binomial models of Reed, Frost and Greenwood). The question of how well the model fits the data is therefore a crucial one. The last column of Table 1 shows the number expected on the basis of the assumptions and the results of the calculations. Comparison with the observed numbers shows that the model does not fit spectacularly well. A straightforward application of a chi-square goodness-of-fit test gives a result at about the 5 % significance point—the point beyond which one would normally consider rejecting a hypothesis.

However, further careful inspection of Table 1 indicates an obvious irregularity in the data. There appear to be local peaks at 7 days and 14 days which could hardly be explained by any reasonable biological hypothesis. It is more likely that these peaks are due to unconscious bias in the data collection towards a whole number of weeks. The isolated observation at exactly three weeks tends to support this conclusion. We can make some allowance for such a possible bias, by amalgamating the classes for 6, 7 and 8 days in one group, and the classes for 13, 14 and 15 days in another group, before carrying out the significance test. When this is done we get a chi-square well within the 5 % significance point, so that we can assume that the model fits and explains the data reasonably well.

This kind of analysis can be extended to families of three individuals, though it is then much more complicated. First of all the basic epidemiological data consist of three main parts, namely families with one, two or three cases. The families with only two cases present much the same appearance as Table 1, with a distinction having to be made between double primaries and case-to-case transmission. The families with three cases have a fourfold classification: triple primaries; double primaries and secondary case; single primary with two secondary cases derived from it; and a single primary with one secondary case followed by a third, tertiary, case infected from the secondary case. Plotting the pairs of time intervals between the three successive cases in a two-way table often enables a separation into the four classes to be made intuitively with sufficient accuracy.

Working as before, we use the basic mathematical model to arrive at a formula giving the probability of the observed types of infection patterns and the observed time distributions between successive cases. We then employ the standard computer maximization programme to find those estimates of the parameters that make the probability of the observations a maximum. In Hope Simpson's data there were only 78 families of three individuals, and the estimates obtained for the four basic parameters were less accurate than those derived from families of two given above. However, it is interesting that in no case did the estimates of any parameter differ significantly between the two different sizes of a family. Moreover, a satisfactory chi-square goodness-of-fit was obtained when the data peaks at 7 and 14 days were smoothed out as before. It should perhaps be mentioned here that with families of size three (but not of size two) it is also feasible to study the possibility of the infection rate varying from one family to another. It turns out that there is in fact a significant amount of such variation in the measles data which the analysis allows us to disentangle.

(b) *Infectious hepatitis*

Since measles data have been quite extensively studied over the last 40 years it seemed worth applying the model discussed in this paper to some other diseases. Dr. K. Petersen (1970) had accumulated data on infectious hepatitis in Hamburg during 1959–1967 for outbreaks in 224 households containing three children under the age of 16. No parents were affected and they were therefore assumed not to be involved. It was of course accepted that there were many difficulties and complications involved in making a satisfactory clinical diagnosis of infectious hepatitis. However, great care had been taken in checking these data. All patients had been admitted to hospital and subjected to extensive clinical and laboratory investigation. It was accordingly considered that the data could be regarded as highly reliable and subject only to minor inaccuracies.

It is interesting to record that it was Dr. Petersen's suggestion that Bailey's model might be applied to his data that led to a re-examination of the original rather cumbersome methods of analysis, and hence to the more streamlined computerized technique of Bailey and Alff-Steinberger (1970) already referred to.

Of the total of 224 families there were 157 with one case only, 41 with two cases and 26 with three cases. The distribution of the time interval in days between the 41 pairs of cases in families with only two cases is shown in Table 3. And the distribution of the 26 pairs of intervals, each pair consisting of the interval between the first and second cases, and the interval between the second and third cases, is displayed in Table 4.

TABLE 3. Distribution of time interval, measured in days, between two cases of infectious hepatitis in 41 families of three individuals each having just two cases

Interval (days)	0	1	2	3	4	5	6	7	8	9	10	11	12	13	
No of cases	2	1	2	.	2	1	1	.	.	.	1	.	1	.	
Interval (days)	14	15	16	17	18	19	20	21	22	23	24	25	26	27	
No of cases	2	1	.	1	1	.	.	1	1	.	2	4	2	.	
Interval (days)	28	29	30	31	32	33	34	35	36	37	38	39	40	41	42
No of cases	1	.	1	2	3	.	1	1	1	2	1	.	1	.	1

TABLE 4. Distribution of pairs of time intervals, measured in days, between three successive cases of infectious hepatitis in 26 families of three, all of whom were infected

Type of transmission	Pairs of intervals (in days) observed
Triple primary	(1, 0) (1, 2)
Double primary with one secondary	(1, 24) (7, 22) (9, 18)
Single primary with two secondaries	(13, 1) (16, 0) (16, 6) (19, 9) (21, 5) (23, 3) (23, 11) (24, 1) (25, 1) (25, 5) (27, 1) (28, 3) (34, 2)
Single primary with single secondary and single tertiary	(16, 15) (22, 37) (25, 36) (28, 31) (32, 33) (34, 14) (36, 21) (36, 30)

These are all the data required for analysis. There is a small problem about where the cut-off should occur in making the distinction between the overlapping distributions. We decided in the first instance to make an arbitrary break between 12 and 13 days, after careful consideration of all the figures in Tables 3 and 4, in order to distinguish between cases derived from a common source and those involved in case-to-case transmission. This led to the groupings actually exhibited in Table 4 for the families with three cases.

Implementation of the computerized analysis already described then led to the results appearing in Table 5.

TABLE 5. Estimated parameters for hepatitis data

Average latent period	$m = 15.6 \pm 1.5$ days
Standard deviation of latent period	$s = 4.0 \pm 0.6$ days
Length of infectious period	$a = 21.7 \pm 2.5$ days
Infection rate	$\lambda = 0.007 \pm 0.001$
Average incubation period	$m + a = 37$ days
Standard deviation of incubation period	$s = 4$ days

The chi-square goodness-of-fit turned out to be entirely satisfactory, so that the model and the estimated parametric values appeared to explain the data rather well.

The calculations were actually reworked making a slightly different choice of cut-off point between overlapping distributions. Taking this to be between 9 and 10 days gave results that were a little different from before, though the probability of the observations was then a bit smaller, and the goodness-of-fit marginally less good.

Overall conclusions are that the best quantitative interpretation of the Hamburg data is that the infectious hepatitis outbreaks observed there entailed a latent period of about 16 days, having a standard deviation of 4 days, followed by an infectious period of 22 days. The incubation period from the receipt of infection to the time of isolation was approximately 37 days with a standard deviation of 4 days.

This standard deviation suggests a rather smaller range of variation in the incubation period than has been proposed in the literature. In fact, our results imply that about 95% of incubation periods should lie in the range between 29 and 45 days. It will be interesting to compare these findings with others that can be made when suitable data become available from sufficiently large outbreaks elsewhere.

(c) *Smallpox*

The modelling approach described so far can be used and evaluated for families of size two or three. For large families the same kind of relatively precise analysis has so far proved intractable because of the complications arising when we try to describe the transmission patterns in bigger groups. Nevertheless, data are sometimes available for a small number of quite large households, and occasionally are presented for a single outbreak in a much bigger population such as school, classroom or a village. It is challenging therefore to try to develop some way of extracting the maximum amount of information from such data when they arise.

In discussing a more theoretical though mathematically exact approach to the problem of large groups, Bailey and Thomas (1971) illustrated their analysis by application to some data kindly made available by Dr. David M.

Thompson and Dr. William H. Foege, taken from an outbreak of smallpox that occurred in 1967 in a well defined, socially isolated community in Abakaliki in southeastern Nigeria. Bailey and Thomas noted however that a model making allowance for an appreciable latent period following infection would be more suitable for the treatment of smallpox data, although their mathematically oversimplified approach provided some useful results.

The data in question came from a religious group belonging to the Faith Tabernacle that in principle objects to vaccination, as well as to other preventive and curative health measures. There was a total of 30 cases of smallpox in a community of size 120. The basic 29 successive interremoval times are shown in Table 6. If cases occur on the same day this fact is indicated by the appearance of zeros. Thus the fourth case was observed two days after the third case, and was followed by three cases three days later.

TABLE 6. Interremoval times for 30 observed cases of smallpox in a population of 120 (taken from Thompson and Foege)

17, 7, 2, 3, 0, 0, 1, 4, 5, 3, 2, 0, 2, 0, 5,
3, 1, 4, 0, 1, 1, 1, 2, 0, 1, 5, 0, 5, 5.

A finer analysis of the original data showed a certain amount of recognizable detailed structure. The people lived, for example, in several different compounds, and about a quarter of them bore some evidence of previous vaccination. Earlier cases in the epidemic were isolated only at home, but some later cases were sent to hospital. Nevertheless, because of the existence of regular religious meetings, it seemed reasonable as a first approximation to treat the whole group as homogeneous.

A method of analysis developed by Bailey and Alff-Steinberger (1972, unpublished) does not attempt to identify specific patterns of case-to-case transmission, but works systematically through the set of interremoval times, computing the probability of each observed interval *given* the previously observed pattern. The product of all such conditional probabilities is then taken to give the probability of the whole set of interremoval times for any chosen parametric values. Using the computerized maximization technique referred to previously thus automatically supplies maximum-likelihood estimates of the parameters.

Although the mathematical method actually used by Bailey and Steinberger is intuitively appealing, it makes certain mathematical conjectures that are not entirely correct. Thus the frequency distribution of the length of any interval depends not merely on the past configuration, but is correlated to some extent with future events. The process described is not in fact Markovian, nor is it easy to discern any simple associated process which has this property. It is, however, feasible to use the method simply as an empirical technique, and then check the results from data actually observed by careful investigation of extensive simulated data of the same type.

Applying the technique in question to the data of Table 6 yields the preliminary results shown in Table 7.

TABLE 7. Estimated parameters for smallpox data

Average latent period	$m = 9 \cdot 8 \pm 2 \cdot 2$ days
Standard deviation of latent period	$s = 4 \cdot 2 \pm 1 \cdot 0$ days
Length of infection period	$a = 1 \cdot 0 \pm 0 \cdot 4$ days
Infection-rate	$\lambda = 0 \cdot 010 \pm 0 \cdot 004$
Average incubation period	$m + \alpha = 10 \cdot 8$ days
Standard deviation of incubation period	$s = 4 \cdot 2$ days

The immediate problem is to get some indication of how reliable these results are, using an independent criterion. Now, the assumptions of the model are precisely specified. So our question can be rephrased as "How intrinsically reliable are the results, given the assumptions?" One way of answering this question is in principle as follows. We first construct a large number of computer-based simulations of epidemics for a known set of parametric values, i.e. we construct typical, though artificial data sets, for which the true answers are of course known in advance. The proposed method of analysis is then used to obtain *estimates* of the parameters, which are then compared with the known true values.

We therefore run off a large number, e.g. 50 or 100 artificial epidemics of about the size actually observed, using parametric values estimated from the real data. Estimates are then calculated for each artificial epidemic and averaged over the whole set of epidemics. Next, the average biases could be computed and used to adjust the original estimates to obtain more accurate revised estimates. The figures obtained for a series of 50 simulations, starting from the actual estimates in Table 7, are given in Table 8.

TABLE 8. Average estimates for simulated data

Average latent period	$\bar{m} = 10 \cdot 4$
Standard deviation of latent period	$\bar{s} = 3 \cdot 7$
Length of infectious period	$\bar{a} = 0 \cdot 8$
Infection-rate	$\bar{\lambda} = 0 \cdot 013$

It can be shown that the average estimates appearing in Table 8 are significantly different from the "true" values, i.e. the estimates of Table 7, having regard for the information available from samples of 50. But the differences between corresponding figures in Tables 7 and 8 are not excessively large compared with the estimated standard errors of the estimates based on a single epidemic. This matter obviously requires further investigation, especially in the context of larger samples containing more information.

The above findings should of course be treated with caution since they are based on a very small amount of data. The average incubation period of about 11 days agrees well with the commonly accepted figure of 12 days. The calculated standard deviation of the latent period or incubation period seems somewhat high at 4.2 days, though a range of values between 7 and 17 is usually regarded as realistic. The estimated infection period of about one day is very short, though it must be remembered that we are measuring the end of the latent period to the point of time where the case is removed from circulation: we are not concerned in this model with infectiousness that may well exist after isolation. Moreover, there are some epidemiological reasons for thinking that the Faith Tabernacle outbreak might be atypical.

6. Conclusion

The results discussed in this paper indicate a number of steps already taken in the direction of constructing more realistic, though tractable, models of epidemic outbreaks. Much remains to be done, though it is already possible to see the kind of practical public health applications that could be made. Thus, after instituting various disease control measures it would be useful to know to what extent the basic biology of the disease had been changed, whether for example the infection rate had been significantly lowered, or whether an appreciable shortening of the infectious period had occurred. Such information would have obvious relevance for the planning of public health control measures.

Finally, it should be remarked that as yet data on most outbreaks of disease are insufficiently detailed to permit the estimation of parameters. But as the availability of acceptable transmission models improves, together with the tractable forms of analysis, we may find it easier to persuade busy field epidemiologists to collect more data in the form required for mathematical investigation.

REFERENCES

BAILEY, N. T. J. (1954). A statistical method of estimating the periods of incubation and infection of an infectious disease. *Nature* **174**, 139–140.

BAILEY, N. T. J. (1957). "The Mathematical Theory of Epidemics." Griffin, London.

BAILEY, N. T. J. AND CYNTHIA ALFF-STEINBERGER (1970). Improvements in the estimation of the latent and infectious periods of a contagious disease. *Biometrika* **57**, 141–153.

BAILEY, N. T. J. AND CYNTHIA ALFF-STEINBERGER (1972). The estimation of latent and infectious periods from a single epidemic outbreak (unpublished).

BAILEY, N. T. J. AND THOMAS, A. S. (1971). The estimation of parameters from population data on the general stochastic epidemic. *Theor. Popul. Biol.* **2**, 253–270.

GREENWOOD, M. (1931). On the statistical measure of infectiousness. *J. Hyg. Camb.* **31**, 336–351.
PETERSEN, K. (1970). Die Bedeutung der seuchenhygienischen Dokumentation durch den Amtsarzt. *Das öffentliche Gesundheitswesen* **32**, 86–95.
WILSON, E. B. AND BURKE, M. H. (1942). The epidemic curve. *Proc. nat. Acad. Sci., Wash.* **28**, 361–267.

15. Some Problems Associated with Computer Simulation of an Ecological System*

RICHARD A. PARKER

Departments of Zoology and Computer Science
Washington State University
Pullman, Washington
U.S.A.

1. Introduction

Accelerating modification of our physical environment has prompted expanded interest in quantitative analysis of ecological activity with the hope of controlling biological consequences. Mathematical models have played a useful role in this analytic process (Mann, 1969; Patten, 1971), yet the complexity of large-scale, real-life situations has acted as a significant deterrent. Certain failures can be attributed to deliberate oversimplification, others to lack of understanding. Theoretical problems encountered frequently arise from the underlying model structure (including statistical variation), quantification of parameters, and numerical techniques employed when exact solutions are not available. Recent studies have helped to bridge the gap between mathematical formalism and species characteristics (Sinko and Streifer, 1969; Goel et al., 1971). Nevertheless, a one-to-one correspondence between model output and observed system behaviour is not possible; success can only be judged in terms of how well the results meet their intended purpose, whether it be for general insight or for system management.

The ocean and its plankton, as well as lakes, often serve as subjects for analysis. These ecological systems are reasonably well defined, although they are obviously difficult to model (Parker, 1968; Patten, 1968; DiToro et al., 1970). The growth and death processes involved typically lead to exponential response, thus small errors in formulation or in associated parameters will

* This project has been financed in part with Federal funds from the United States Environmental Protection Agency under grant number R-800430, and by the Environmental Research Center at Washington State University.

produce unsatisfactory solutions to the nonlinear systems of coupled differential equations frequently utilized. Efforts by Parker (1972) to optimize the fit of model output (by parameter estimation) to observed data from Kootenay Lake in southeastern British Columbia, Canada have been partially successful. The searching techniques applied (Powell, 1968), were satisfactory only after a nearly "acceptable" solution was obtained by relaxation. Further improvement in performance by this model has been limited, thereby suggesting the need for structural change. Since the entire plankton community is extremely sensitive to fluctuating algal components, an attempt has been made in what follows to focus more precisely on photosynthesis and to emphasize the timing of population peaks.

2. Photosynthesis

Light intensity
The photosynthetic process depends both on light intensity and quality. In view of the fact that the physiological range for quality is approximately 400–700 nannometers, incoming solar radiation at the earth's surface for this range was calculated according to McCullough and Porter (1971) with the following exceptions: no distinction was made between true and apparent zenith angles, and the diffuse component D was assumed related to the direct component I at any zenith angle Z by $D(Z) = f(Z)I(Z)$. McCullough and Porter's values for $f(Z)$ were approximated by $(0.054 + 0.07\ \text{albedo})$ $\times \exp(0.226\ Z^{4.72})$, adopting a specific albedo of 0.1 over water.

Since the wavelength of maximum absorption by chlorophyll is highly variable (French, 1968), no species differential dependence within the range 400–700 nannometers was included. The basic equation chosen to relate relative (i.e. fraction of maximum) photosynthesis $\theta(z, t)$ at time t and depth z was one used by several investigators including Steele (1965):

$$\theta(z, t) = (I(z, t)/I_s) \exp(1 - I(z, t)/I_s) \tag{1}$$

where I_s is the intensity at which photosynthesis is maximum; perhaps the best support for this function is found in the data of Ryther (1956). Now recall that in a homogeneous column of water ("mixed"), light intensity decreases with depth in the following way:

$$I(z, t) = I(0, t) \exp(-\eta z),$$

where η is termed the extinction coefficient (here a mean over 400–700 nannometers). Therefore the total relative photosynthesis P occurring during one day between sunrise sr and sunset ss to a depth H is

$$P = \int_{sr}^{ss} \int_0^H \theta(z, t)\,\mathrm{d}z\,\mathrm{d}t$$

and the mean $\bar{P}$ during a 24-hour period per unit depth is

$$\bar{P} = P/(24H).$$

The general form of equation (1) is illustrated in Fig. 1 for $I_s = 10$ together with θ^3, θ^2, and $\theta^{0.5}$. Note that the power shapes the response over the range

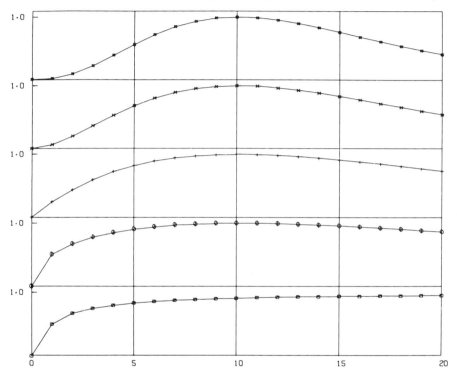

FIG. 1. θ^3, θ^2, θ, $\theta^{0.5}$ and Φ (top to bottom) with $I_s = 10$ and $K = 1$.

of the independent variable and provides the potential for approximating photosynthetic rates predicted by Vollenweider (1970) using a more elaborate expression (also see Fee, 1969).

Temperature
Thus far focus has been on the effect of light intensity, but it is equally obvious that temperature must play a significant role in photosynthesis and the associated enzyme kinetics. Although there is a general tendency for the rate to rise with intermediate temperatures, there must eventually be a decrease as temperature increases. The properties of heat-sensitive enzymes in micro-organisms have been discussed extensively by Stokes (1967). Furthermore,

Jorgensen and Steeman Nielsen (1965) demonstrated an influence of temperature on photosynthetic carbon fixation by the planktonic alga *Skeletonema*. Postulating that the overall temperature response is similar to that induced by light (and that temperature and light act independently), a function of the form $c\theta_I^{\alpha_I}\theta_T^{\alpha_T}$ was fit to Jorgensen and Steeman Nielsen's limited data using the approach of Powell (1964). Since α_I appeared close to 1 and had little effect on the value obtained for α_T, it was set equal to 1 for the remainder of this study. Fixation rates based on the other four parameters are given in Table 1 ($I_s = 11.7$, $T_s = 11.4$, $\alpha_T = 1.52$, $c = 1.62$).

TABLE 1. Predicted rate of carbon fixation by *Skeletonema* as a function of temperature and light intensity (data from Jorgensen and Steeman Nielsen, 1965)

Light intensity (klux)	Temperature (C)	Carbon fixation (μg C/10^6 cells/hour)	
		Observed	Predicted[a]
3	2	0.4	0.22
3	8	0.6	0.81
3	14	0.7	0.85
3	20	0.7	0.65
10	2	0.3	0.40
10	8	1.8	1.47
10	14	1.5	1.55
10	20	1.2	1.20
20	2	0.4	0.34
20	8	0.9	1.25
20	14	1.6	1.31
20	20	0.9	1.01

[a] $F = 1.62\,[(I/11.7)\exp(1 - I/11.7)]\,[(T/11.4)\exp(1 - T/11.4)]^{1.52}$.

If these results are representative, what is the general effect of temperature on daily values of $\bar{P}$ throughout the year? In anticipation of subsequent incorporation in the Kootenay Lake Model, computations were carried out with $H = 10$ metres at 50° latitude. Temperature change was assumed proportional to the difference between incoming solar radiation and current temperature (simulated by $dT/dt = 0.011$ daily radiation $-0.11\,T$). Figures 2 and 3 show that for $I_s = 4$ and 1 cal/sq cm/hr, respectively, numerical integration (10-point Gaussian quadrature) of the function $\theta_I\theta_T^2$ ($T_s = 10°$) often produced photosynthetic maxima where they did not exist using θ_I alone (depending on η). This coincident influence of light and temperature may provide the mechanism for development of spring and fall peaks in naturally occurring phytoplankton populations. If individual algal species

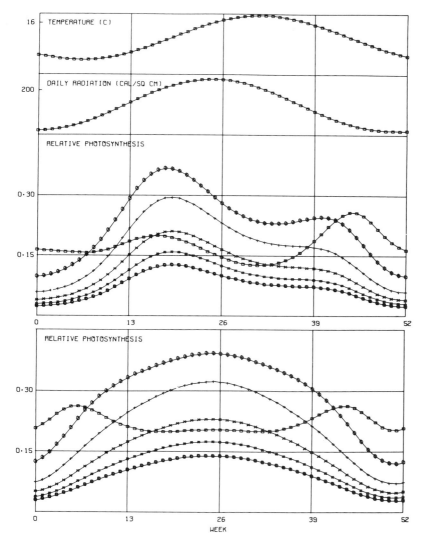

FIG. 2. $\theta_T^2 \int \theta_I$ (upper) and $\int \theta_I$ (lower) with $I_s = 4$ and $T_s = 10$ for $\eta = 0\cdot0$. ($\square$), $0\cdot2$ (Φ), $0\cdot4$ (+), $0\cdot6$, ($\times$), $0\cdot8$ (#), $1\cdot0$ ($\bigcirc$).

exhibit an "evolutionary strategy" which tends to increase growth rate subject to other environmental constraints, what value of I_s ("optimum") is required to maximize $\bar{P}$ on any given day of the year? Again using Powell's method (1964), I_s was determined for one day each week and plotted in Fig. 4.

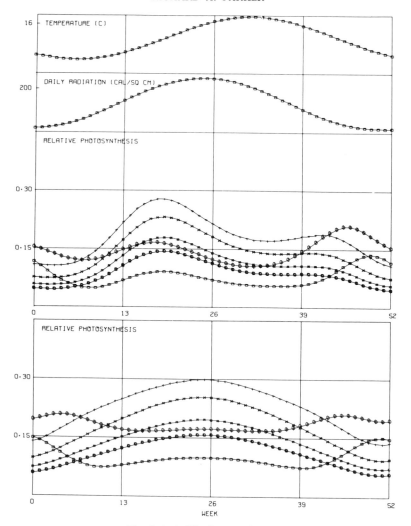

FIG. 3. As in Fig. 2 except $I_s = 1$.

Nutrients

Photosynthesis is usually considered to be affected by a nutrient N in a Michaelis–Menton manner; that is, proportional to

$$\Phi = N/(K + N).$$

Although it is apparent that θ^x might be more useful in accounting for the inevitable inhibitory effect of high concentrations, Φ has been employed here

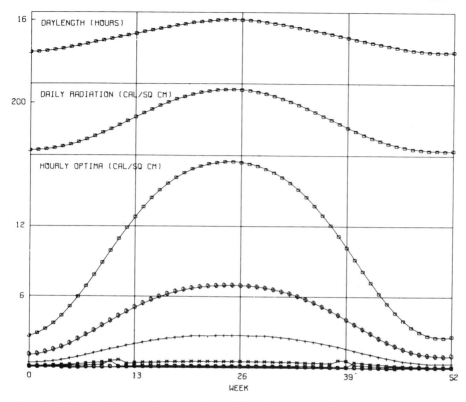

FIG. 4. I_s ("optimal") to produce maximum daily photosynthesis for $\eta = 0{\cdot}0$ ($\square$), $0{\cdot}2$ (Φ), $0{\cdot}4$ ($+$), $0{\cdot}6$ ($\times$), $0{\cdot}8$ ($\#$), $1{\cdot}0$ ($\bigcirc$).

due to the extensive supporting literature. For information, however, $\Phi = N/(1 + N)$ can be compared with $\theta^{0{\cdot}5}$ ($I_s = 10$) in Fig. 1.

Phosphate (N_1), nitrate (N_2), and ammonium (N_3) are inorganic nutrients which, when scarce, can limit phytoplankton photosynthesis. The measured values for K vary widely among species (for examples, see Thomas and Dodson 1968; MacIssac and Dugdale 1969; Eppley *et al.*, 1969). Although phosphate may be regarded as acting in a relatively independent fashion, it has been difficult to determine whether nitrate or ammonium is absorbed preferentially. This uncertainty prompts the assumption here of no preference, hence

$$\Phi_{N1} = N_1/(K_1 + N_1)$$

and

$$\Phi_{N2+N3} = (N_2 + N_3)/(K_2 + N_2 + N_3).$$

In summary, then, photosynthetic rate will be considered proportional to the following function of light intensity, temperature, and three primary nutrients:

$$p' = \theta_I \theta_T^z \Phi_{N1} \Phi_{N2+N3}.$$

One must bear in mind that we are dealing only with the mixed layer where it is reasonable to expect that, at any instant, temperature and nutrient concentrations are uniform throughout.

3. Kootenay Lake Model

Selection of organisms for inclusion is a difficult task since it is virtually impossible to construct a model in which all species are represented. Grouping of species according to general environmental requirements seems to be an acceptable alternative. In Kootenay Lake two algal groups are common: one, prevalent in August and September, consists largely of a phacotid and the second, frequent in April and May, is dominated by *Cryptomonas* and *Stephanodiscus*. The zooplankton can be loosely divided into cladocerans (*Daphnia, Bosmina, Diaphanosma*) and copepods (*Diaptomus, Cyclops*). Bacterial and fish components have been omitted for lack of data.

The general form for "growth" and "mortality" rates used in the model is presented in Table 2. "Growth" may be thought of as including births,

TABLE 2. Growth and mortality rates used in model

	Growth rate (G)	Natural (M_1)	Predation (M_2)
Algal Group 1 (A_1)	$a_{11}e_1(I)f_1(T)g_1(N_1)h_1(N_2,N_3)$	$a_{21}T$	$TB(c_{11}C_1 + c_{21}\mathcal{G}_2)$
Algal Group 2 (A_2)	$a_{21}e_2(I)f_2(T)g_2(N_1)h_2(N_2,N_3)$	$a_{22}T$	$TB(c_{12}C_1 + c_{22}C_2)$
Cladocera (C_1)	$b_1TB(c_{11}A_1 + c_{12}A_2)$	$c_{13}T$	$c_{14}TC_1$
Copepoda (C_2)	$b_2TB(c_{21}A_1 + c_{22}A_2)$	$c_{23}T$	$c_{24}TC_2$

I = light intensity; $\quad$ $f_i(T) = [(T/T_{si})\exp(1 - T/T_{si})]^{z_i}$;
T = temperature; $\quad$ $g_i(N_1) = N_1/(K_{1i} + N_1)$;
N_i = nutrients; $\quad$ $h_i(N_2,N_3) = (N_2 + N_3)/(K_{2i} + N_2 + N_3)$;
$e_i(I) = (I/I_{si})\exp(1 - I/I_{si})$; $\quad$ $B = \exp[-k(A_1 + A_2)]$;

increase in biomass, and even that portion of respiration directly dependent on growth. "Natural" mortality includes sinking and the balance of respiration as well as death from natural causes. The relative importance of each segment varies between zoo- and phytoplankton. "Predation" refers to grazing by the zooplankton and predation by fish (mimicked by making larger populations more vulnerable rather than assuming a constant fractional loss). Note that every process has incorporated a temperature effect. Also, photosynthesis is reduced by algal self-shading and depends here on $\eta = \eta_a + \eta_n$, where η_a represents the extinction coefficient due to phytoplankton and η_n that due to other causes (water, silt, detritus). For computation η_a was set equal to $0 \cdot 1 (A_1 + A_2)$ and η_n was that observed less the contribution by the observed algal concentration.

It is clear that food habits vary among zooplankton forms in Kootenay Lake; *Cyclops* is typically predaceous as an adult whereas the others may be regarded generally as filter feeders. Nevertheless all genera were treated similarly because of the grouping and, in the model, were allowed to feed selectively on the two algal groups.

TABLE 3. Differential Equation Model

$$A_1' = A_1(G_{A_1} - M_{1A_1} - M_{2A_1}) - v\, \partial A_1/\partial x$$

$$A_2' = A_2(G_{A_2} - M_{1A_2} - M_{2A_2}) - v\, \partial A_2/\partial x$$

$$C_1' = C_1(G_{C_1} - M_{1C_1} - M_{2C_1}) - v\, \partial C_1/\partial x$$

$$C_2' = C_2(G_{C_1} - M_{1C_2} - M_{2C_2}) - v\, \partial C_2/\partial x$$

$$N_1' = p(A_1 M_{2A_1} + A_2 M_{2A_2} - C_1 G_{C_1} - C_2 G_{C_2}) - p(A_1 G_{A_1} + A_2 G_{A_2}) + m\, \partial N_1/\partial z$$
$$- v\, \partial N_1/\partial x$$

$$N_2' = -n_1(A_1 G_{A_1} + A_2 G_{A_2})N_2/(N_2 + N_3) + m\, \partial N_2/\partial z - v\, \partial N_2/\partial x$$

$$N_3' = n_1(A_1 M_{2A_1} + A_2 M_{2A_2}) - n_2(C_1 G_{C_1} + C_2 G_{C_2}) - n_1(A_1 G_{A_1} + A_2 G_{A_2})$$
$$\cdot N_3/(N_2 + N_3) + m\, \partial N_3/\partial z - v\, \partial N_3/\partial x$$

Chemical components incorporated were phosphate, nitrate, and ammonium. They entered the model as indicated in Table 3, which summarizes the seven differential equations defining the simulated system (for four sampling locations). Regeneration by zooplankton, eddy diffusion, and upwelling ordinarily supply nutrients to the mixed layer (see, for example, Larrance 1971); in addition, Kootenay Lake receives large quantities via the Kootenay River. The organic pool of dissolved nutrients was neglected completely, and regeneration of inorganic phosphate as well as ammonium comprised the total contribution by the zooplankton. Horizontal gradients $(\partial/\partial x)$ at each sampling station were estimated from observed data assuming a linear function of distance. The associated average velocity was based on inflow records coupled with lake volume, multiplied by a function of density differences between stations due to temperature $\exp[D$ (density at station i − density at station $i - 1$)]. Vertical gradients $(\partial/\partial z)$ were calculated using the difference between the concentration in the mixed layer (defined here as the upper 10 metres) and the concentration in the lower waters at the time of fall overturn (about week 46 each year).

Finding suitable values for the large array of growth and mortality parameters has been an immense problem. Initially some were obtained from previous studies (Parker, 1968; 1972), and others (particularly those affecting photosynthesis) were fixed at levels consistent with observed environmental variables at times of maximum growth rates. For example, although the two algal groups seemed to peak at times when the temperatures were quite different (20, 11°), the value of I_s needed to maximize photosynthesis was approximately the same (1·6 for $\eta \simeq 0·5$). Final fitting of the model to data was based on a subjective evaluation of simulation patterns as well as local minimization of an objective function defined by the squared deviation relative to the predicted quantity summed over all observations, variates, and stations. Values used in the simulation runs presented here were:

$$a_{11} = 35 \qquad a_{21} = 0·02 \qquad k = 0·35$$

$$a_{21} = 8 \qquad a_{22} = 0·02 \qquad b_1 = b_2 = 0·47$$

$$c_{11} = 0·7 \qquad c_{12} = 0·10 \qquad c_{13} = 0·030 \qquad c_{14} = 1·5$$

$$c_{21} = 0·3 \qquad c_{22} = 0·15 \qquad c_{23} = 0·015 \qquad c_{24} = 0·5$$

$$I_{s1} = I_{s2} = 1·6$$

$$T_{s1} = 20 \qquad \alpha_1 = 4·80 \qquad T_{s2} = 11 \qquad \alpha_2 = 1·45$$

$$K_{11} = K_{12} = 1 \qquad\qquad K_{21} = K_{22} = 6$$

$$p = 0·3 \qquad n_1 = 2·0 \qquad n_2 = 5·0 \qquad m = 0·15$$

$$D = 2000$$

4. Observed Data and Simulation Results

Kootenay Lake has a surface area of nearly 500 km² and a mean depth of well over 100 metres; being fjord-like, it is narrow and about 100 km along its north–south axis. The Kootenay River enters the lake from the south and water leaves through a long shallow arm extending westward from a point mid-way. Acara (unpublished) indicates that Kootenay River water may reach the north end of the lake before returning to the exit point.

Five stations were utilized; STA 1 at the river mouth, STA 2–4 spaced over the southern half and STA 5 immediately north of the west arm. Multiple samples were collected at several depths (usually 1, 5, and 10 m at the four lake stations) at time intervals varying from 1 week to 1 month during the period 1966–1969. Station 5 was not established until mid-1967 nor were ammonium determinations made before then. The mean values of each variate

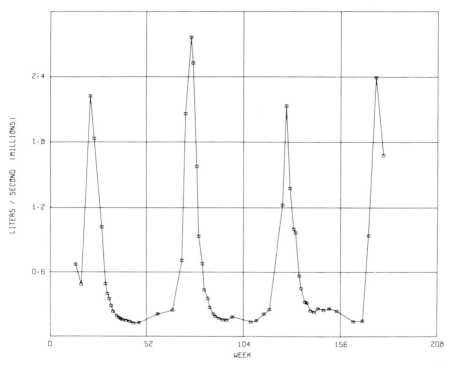

FIG. 5. Observed inflow (Kootenay River).

in the mixed layer will be used throughout the analysis here. Figure 5 illustrates the large contribution of water by the Kootenay River, particularly during May and June. Its effect on the extinction coefficient at all stations is readily apparent as is the obvious contribution to seasonal temperature pattern (Fig. 6).

Solutions to the system of differential equations were computed numerically using Hamming's modified predictor-corrector method as described in the IBM System/360 Scientific Subroutine Package. Time was expressed in weeks, and the interval during computation was automatically adjusted by the program for error control. It should be pointed out that integral photosynthesis for a given day was calculated holding all variables constant except light intensity.

Observed values and simulated results for phytoplankton, phosphate, nitrate, ammonium, copepods, and cladocera are compared in Figs. 7–12, respectively. Note that both algal groups are plotted in Fig. 7 since only total phytoplankton was observed in terms of chlorophyll a concentration converted to mg dry weight per litre (0·6% chlorophyll). Infrequent algal cell

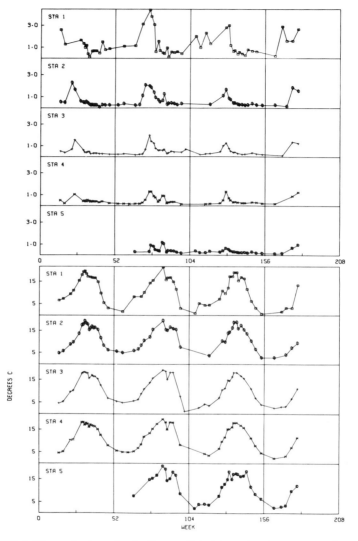

FIG. 6. Observed values of extinction coefficient (upper) and temperature (lower).

counts were adequate to establish general population trends for each group but could not be used for fitting purposes. Also, missing ammonium data was replaced by a seasonally sinusoidal approximation. In addition, the observed cladoceran and copepod quantities in mg/1 were calculated by multiplying numerical densities by 3·5 and 2·5 micrograms per individual,

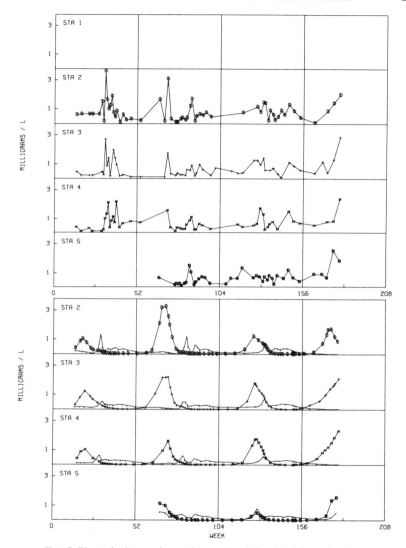

FIG. 7. Phytoplankton : observed (upper) and simulated (lower) values.

respectively (this procedure probably overestimated total weight during peak population periods due to the presence of more young animals).

The results speak for themselves. Insisting (idealistically) that the same model with the same parameter values applies at each of the four lake stations decreased the fitted accuracy. For example, the phytoplankton peaks which

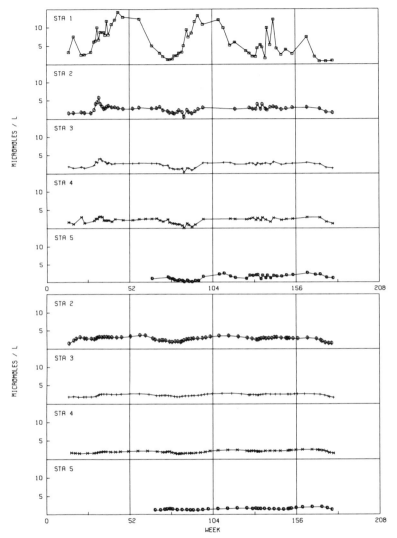

FIG. 8. Phosphate: observed (upper) and simulated (lower) values.

occur during late summer are reproduced in part at station 2 but not elsewhere. Some of this difficulty stems from the fact that predicted horizontal
transport between successive stations depends on estimates of velocity. All
too often the Kootenay River enters the lake and, depending on temperature
(hence density) differences, will not mix well with the lake water but rather

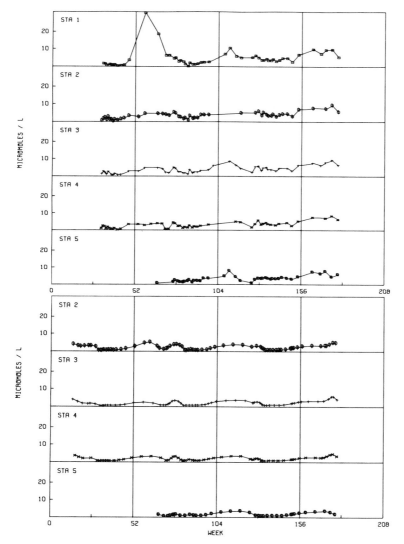

FIG. 9. Nitrate: observed (upper) and simulated (lower) values.

travels along the bottom or at the surface. One such occurrence may be seen in the phosphate data (Fig. 8) during August 1966. At that time the river water was 2° warmer than the water in the mixed layer at station 2, which in spite of relatively low river flow caused a large influx of this nutrient. Although velocity was modified by a function of density differences between stations, it is

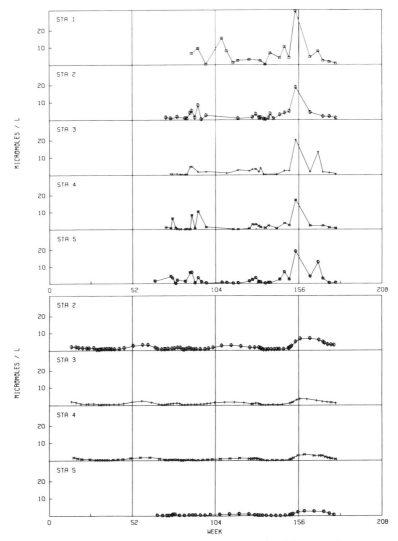

FIG. 10. Ammonium: observed (upper) and simulated (lower) values.

apparent that the modification was not sufficient. This inadequacy was of course partially responsible for the unsatisfactory model fit to nitrate and ammonium data (Figs. 9 and 10). Some room for optimism exists, however, in that predicted copepod (Fig. 11) and cladoceran (Fig. 12) population sizes are quite close to field observations.

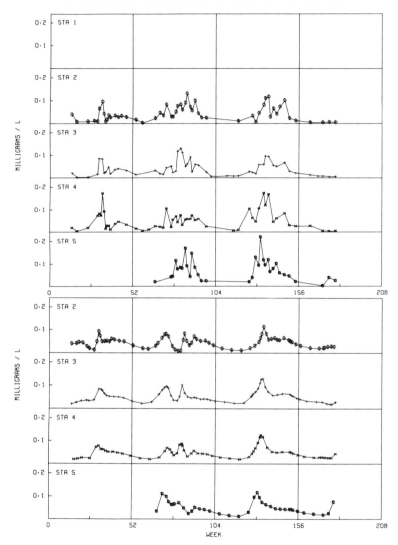

FIG. 11. Copepods: observed (upper) and simulated (lower) values.

It has become increasingly clear during the course of this investigation that those components which change most rapidly with time are the most difficult to model. One might avoid the problem by working with smoothed data; however, a more satisfying approach would increase sampling frequency and the number of samples collected at one time, thus reducing

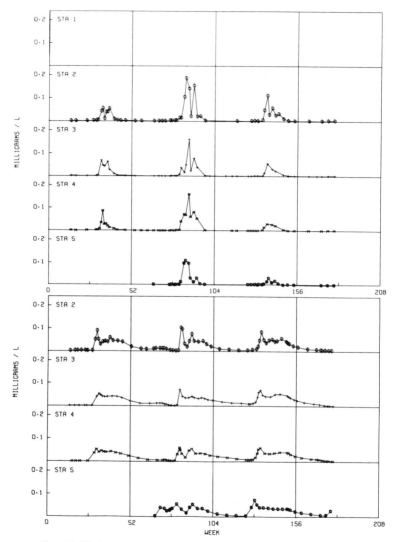

Fɪɢ. 12. Cladocerans: observed (upper) and simulated (lower) values.

measurement "noise". The resulting picture surely would provide greater clarity in model definition. Nevertheless, further improvement in simulating the system represented by Kootenay Lake is not likely to occur until there is a better understanding of its hydrodynamic properties, together with more knowledge about those components which affect nutrient cycling.

Acknowledgments

I wish to thank M. J. D. Powell for supplying computer programs based on his 1964 and 1968 papers. Field data were collected by K. G. Taylor and algal identifications are from D. Fillion (unpublished).

REFERENCES

DiToro, D. M., O'Connor, D. J. and Thomann, R. V. (1970). A dynamic model of phytoplankton populations in natural waters. Environmental Engineering and Science Program, Manhattan College.

Eppley, R. W., Rogers, J. N. and McCarthy, J. J. (1969). Half-saturation constants for uptake of nitrate and ammonium by marine phytoplankton. *Limnol. Oceanogr.* **14** (6), 912–920.

Fee, E. J. (1969). A numerical model for the estimation of photosynthetic production, integrated over time and depth, in natural waters. *Limnol. Oceanogr.* **14** (6), 906–911.

French, C. S. (1968). The absorption spectra of chlorophyll *a* in algae. *Carnegie Inst. Wash. Year B.* **66**, 177–186.

Goel, N. S., Maitra, S. C. and Montroll, E. W. (1971). On the Volterra and other nonlinear models of interacting populations. *Rev. Mod. Physics* **43** (2), 231–276.

Jorgensen, E. G. and Steeman Nielsen, E. (1965). Adaptation in plankton algae. *In* "Primary Productivity in Aquatic Environments." (C. R. Goldman, ed.), pp. 38–46. Mem. Ist. Ital. Idrobiol., 18 Suppl., Univ. Calif. Press, Berkeley.

Larrance, J. D. (1971). Primary production in the mid-subarctic Pacific Region, 1966–68. *Fish. Bull.* **69** (3), 595–613.

MacIsaac, J. J. and Dugdale, R. C. (1969). The kinetics of nitrate and ammonia uptake by natural populations of marine phytoplankton. *Deep-Sea Res.* **16** (1), 45–57.

Mann, K. H. 1969. The dynamics of aquatic ecosystems. *In* "Advances in Ecological Research." (J. B. Cragg, ed.), pp. 1–81. Academic Press, New York and London.

McCullough, E. C. and Porter, W. P. (1971). Computing clear day solar radiation spectra for the terrestrial ecological environment. *Ecology* **52** (6), 1008–1015.

Parker, R. A. (1968). Simulation of an aquatic ecosystem. *Biometrics* **24** (4), 803–821.

Parker, R. A. (1972). Estimation of aquatic ecosystem parameters. *Verh. Internat. Verein. Limmol.* 18, 257–263.

Patten, B. C. (1968). Mathematical models of plankton production. *Int. Rev. Ges. Hydrobiol.* **53** (3), 357–408.

Patten, B. C. (1971). "Systems Analysis and Simulation in Ecology." Vol. I. Academic Press, New York.

Powell, M. J. D. (1964). An efficient method for finding the minimum of a function of several variables without calculating derivatives. *Computer Journal* **7**, 155–162.

Powell, M. J. D. (1968). A FORTRAN subroutine for solving systems of non-linear algebraic equations. *UK Atomic Energy Research Establishment Report* R 5947, Harwell.

Ryther, J. H. (1956). Photosynthesis in the ocean as a function of light intensity. *Limnol. Oceanogr.* **1**, 61–70.

Sinko, J. W. and Streifer, W. (1969). Applying models incorporating age-size structure of a population to *Daphnia*. *Ecology* **50**, 608–615.

STEELE, J. H. (1965). Notes on some theoretical problems in production ecology. *In* "Primary Productivity in Aquatic Environments." (C. R. Goldman, ed.), pp. 385–398. Mem. Ist. Ital., Idrobiol., 18 Suppl., Univ. Calif. Press, Berkeley.

STOKES, J. L. (1967). Heat sensitive enzymes and enzyme synthesis in psychrophilic microorganisms. *In* "Molecular Mechanisms of Temperature Adaptation." (C. L. Prosser, ed.), pp. 311–323. AAAS Publ. No. 84.

THOMAS, W. H. and DODSON, A. N. (1968). Effects of phosphate concentration on cell division rates and yield of a tropical oceanic diatom. *Biol. Bull.* **134** (1), 199–208.

VOLLENWEIDER, R. A. (1970). Models for calculating integral photosynthesis and some implications regarding structural properties of the community metabolism of aquatic systems. *In* "Prediction and Measurement of Photosynthetic Productivity." pp. 455–472. Proc. IBP/PP Technical meetings, Trebon 14–21 September 1969.

Part V

Population Distribution and
Community Structure

16. Stochastic Formulations for Life Tables, Age Distributions and Mortality Curves

J. Gani

Department of Probability and Statistics,
University of Sheffield, Sheffield, England

1. Life Tables and Markov Chains

In the construction of life tables, the practice of listing numbers of survivors of age x for successive years ($x = 0, 1, \ldots, w$, where w is the maximum age at death) is most appropriate when recording the surviving individuals of an ageing cohort. For a current life table, such a procedure may be misleading; here, as Chiang (1968) and Keyfitz (1968) have pointed out, it is not the survivors l_x, but rather the proportions $\hat{p}_x(\hat{q}_x = 1 - \hat{p}_x)$ of individuals surviving (dying) during the age interval $(x, x + 1)$ which are basic. It is from these estimated quantities that the remaining columns of the current life tables are calculated synthetically.

To illustrate the two types of tables as simply as possible, let us suppose that over a 6-year period, we observe 6 artificially limited populations A, B, C, D, E, F of 100 pets (mice, say). Each population is assumed born on January 1 of every year, and all individuals die during their sixth year of life. Let the raw cohort data for these be:

TABLE 1. Data for 6 cohorts A, B, C, D, E, F

Age interval x to $x + 1$ years	Number l_x living at age x					
	A $x = 0$ in 1961	B $x = 0$ in 1962	C $x = 0$ in 1963	D $x = 0$ in 1964	E $x = 0$ in 1965	F $x = 0$ in 1966
0–1	100	100	100	100	100	100
1–2	68	70	74	77	80	83
2–3	49	53	57	62	69	73
3–4	39	42	46	51	61	64
4–5	31	34	36	41	50	52
5 and over	14	16	20	24	30	33

For each of these populations, we can readily obtain $\hat{p}'_x = l'_{x+1}/l'_x$, the proportion surviving in $(x, x+1)$, as well as the more commonly used proportions $\hat{q}'_x = (l'_x - l'_{x+1})/l'_x$ dying in $(x, x+1)$. Primes are used here to denote quantities based on cohort data. For convenience, these $\hat{p}'_x$ are listed below.

TABLE 2. Survival rates for 6 cohorts

	A $x = 0$ in 1961	B $x = 0$ in 1962	C $x = 0$ in 1963	D $x = 0$ in 1964	E $x = 0$ in 1965	F $x = 0$ in 1966
$\hat{p}'_0$	0·68	0·70	0·74	0·77	0·80	0·83
$\hat{p}'_1$	0·7206	0·7571	0·7703	0·8052	0·8625	0·8795
$\hat{p}'_2$	0·7959	0·7925	0·8070	0·8226	0·8841	0·8767
$\hat{p}'_3$	0·7949	0·8095	0·7826	0·8039	0·8197	0·8125
$\hat{p}'_4$	0·4516	0·4706	0·5556	0·5854	0·6000	0·6346
$\hat{p}'_5$	0	0	0	0	0	0

Note that these proportions have been deliberately chosen as time-dependent; this they are also in human populations where improvements in hygiene and medical care have caused continuing increases in survival rates. We should, in effect, write for the first line $\hat{p}'_0$, the values

$$\hat{p}'_0(1961) = 0·68, \quad \hat{p}'_0(1962) = 0·70, \quad \hat{p}'_0(1963) = 0·74,$$

$$\hat{p}'_0(1964) = 0·77, \quad \hat{p}'_0(1965) = 0·80, \quad \hat{p}'_0(1966) = 0·83,$$

to indicate that these are survival rates from ages 0 to 1 current for the years 1961–1962 to 1966–1967 respectively. Values $\hat{p}'_x(t)$ in the subsequent lines should be similarly distinguished for the year in which the age x is reached.

Suppose we now consider the relevant data for a current life table, using 1966 as our base year. What we require are the numbers of pets of age x in 1966 which will be alive and aged $x + 1$ in 1967. Our raw data, selected from the previous cohort figures will now be:

TABLE 3. Data and survival rates current for 1966–1967

Age interval x to $x + 1$ years	Number l_x living at age x in 1966	Number l_{x+1} living at age $x + 1$ in 1967	Proportion $\hat{p}_x = l_{x+1}/l_x$ surviving from x to $x + 1$
0–1	100	83	$\hat{p}_0(1966) = 0·83$
1–2	80	69	$\hat{p}_1(1966) = 0·8625$
2–3	62	51	$\hat{p}_2(1966) = 0·8226$
3–4	46	36	$\hat{p}_3(1966) = 0·7826$
4–5	34	16	$\hat{p}_4(1966) = 0·4706$
5 and over	14	0	$\hat{p}_5(1966) = 0$

The relevant proportions $\hat{p}_x(1966)$ of survival are identical with the north east diagonal $\hat{p}'_x(1966)$ of Table 2. Primes are now omitted to indicate that we are dealing with current data. It is these $\hat{p}_x$ which will be used to construct the typical current life table in the standard form given on p. 294, with $l_{x+1} = l_x \hat{p}_x$ ($x = 0, 1, \ldots, 5$), starting from the base $l_0 = 100$. While this layout is designed to resemble that of the cohort table, the figures l_x are not themselves raw data (as were the l'_x) but are synthetically calculated from the survival rates $\hat{p}_x$ for 1966. In human life tables, these one year survival rates are computed from age specific death rates of the current population (see Chiang (1968) for details).

Possibly the simplest way of exhibiting the interrelation between the current and cohort processes is by using a Markov chain formulation. It is clear from the derivation of the current survival rates $\hat{p}_x(t)$, where $t = 1966$ in Tables 3 and 4, that we could readily write the transition matrix $P(t) = \{p_{ij}(t)\}$ of survival probabilities from the year $t = 1966$ to $t + 1 = 1967$ as

$$P(1966) =$$

	0	1	2	3	4	5	6
0	$\hat{p}_0 = 0.83$						$\hat{q}_0 = 0.17$
1		$\hat{p}_1 = 0.8625$					$\hat{q}_1 = 0.1375$
2			$\hat{p}_2 = 0.8226$				$\hat{q}_2 = 0.1774$
3				$\hat{p}_3 = 0.7826$			$\hat{q}_3 = 0.2174$
4					$\hat{p}_4 = 0.4706$		$\hat{q}_4 = 0.5294$
5							$\hat{q}_5 = 1.0000$
6							1

$$(1.1)$$

Here, the states 0, 1, 2, 3, 4, 5, denote the ages x of survivors, while state 6 represents death. Similar matrices $P(t)$ could be written for other years t; their only non-zero elements would be $p_{x,x+1}(t) = \hat{p}_x(t)$ and $p_{x6}(t) = \hat{q}_x(t)$ ($x = 0, 1, \ldots, 6$). Note that the Markov chain will usually be non-homogeneous; entries from the consecutive matrices $P(1961), \ldots, P(1966)$ would thus be necessary to provide the transition probabilities

$$\hat{p}_0(1961), \hat{p}_1(1962), \hat{p}_2(1963), \hat{p}_3(1964), \hat{p}_4(1965), \hat{p}_5(1966) = 0$$

for the reconstruction of the cohort A in Table 1, say.

TABLE 4. Typical current life table for 1966–1967

Age interval x to x + 1 years	Number[a] l_x living at age x	Number[a] d_x dying in $(x, x + 1)$	Proportion $\hat{q}_x = \dfrac{d_x}{l_x}$ dying in interval $(x, x + 1)$	Proportion $\hat{p}_x = 1 - \hat{q}_x$ surviving in interval $(x, x + 1)$
0–1	100	17	0·1700	0·8300
1–2	83	11	0·1375	0·8625
2–3	72	13	0·1774	0·8226
3–4	59	13	0·2174	0·7826
4–5	46	24	0·5294	0·4706
5 and over	22	22	1·0000	0

[a] Rounded off to nearest digit.

Clearly, when the process is stationary the distinction between the survival probabilities $\hat{p}_x$ for the current and $\hat{p}'_x$ for the cohort life tables becomes unnecessary, and the notation $\hat{p}_x$ for $\hat{p}_x(t) = \hat{p}'_x(t)$ is adequate for both. While stationarity is rarely the case, $\hat{p}_x, \hat{p}'_x$ are sometimes taken as equal to a first approximation; it seems important, however, to insist on a general notation such as $\hat{p}_x(t), \hat{p}'_x(t)$ in order to highlight the time-dependence of the basic population process.

2. Chain Binomial Methods for Cohorts in Discrete Time

Probability distributions for the number of survivors (cf. Chiang (1968), Chapter 10) and the extinction time of a cohort in discrete time, based on empirical survival probabilities $\hat{p}'_x(t)$, may be derived very simply by chain binomial methods.

Suppose for the moment that in a cohort of initial size l'_0, the probability of survival of an individual from age x to age $x + 1$ is p, independent of age or time. Starting with l'_x individuals alive at age x, the probability of the number l'_{x+1} of survivors after the interval $(x, x + 1)$ is given by

$$\Pr\{l'_{x+1}|l'_x\} = \binom{l'_x}{l'_{x+1}} p^{l'_{x+1}}(1 - p)^{l'_x - l'_{x+1}}. \tag{2.1}$$

These are the transition probabilities governing the Markov chain $\{l'_x\}_{x=0}^{w}$; note that the states of this chain are the numbers of survivors $0, 1, \ldots, l'_0$, whereas, the states of (1.1) denoted the ages of a single survivor. The survival process characterized by this chain can be treated by methods similar to those used by Gani and Jerwood (1971) in an epidemic context; we shall, however, use simpler direct methods in our derivations.

It is clear that the joint distribution of survivors $l'_1, \ldots, l'_k$ ($k < w$) is given by the chain binomial

$$\Pr\{l'_1, \ldots, l'_k | l'_0\} = \prod_{x=0}^{k-1} \binom{l'_x}{l'_{x+1}} p^{l'_{x+1}} (1-p)^{l'_x - l'_{x+1}}. \tag{2.2}$$

The distribution of l'_k, irrespective of $l'_1, \ldots, l'_{k-1}$ is readily obtained as

$$\Pr\{l'_k | l'_0\} = \binom{l'_0}{l'_k} (p^k)^{l'_k} (1 - p^k)^{l'_0 - l'_k}, \tag{2.3}$$

since the k-step survival probability will now be p^k. The extinction time T of the cohort will thus be given by

$$\Pr\{T = \tau\} = \Pr\{l'_\tau = 0\} - \Pr\{l'_{\tau-1} = 0\}$$
$$= (1 - p^\tau)^{l'_0} - (1 - p^{\tau-1})^{l'_0}; \tag{2.4}$$

while this result may be expressed in terms of the transition probability matrix of the Markov chain $\{l'_x\}$, the form (2.4) and its variant (2.8) suggested by P. Armitage are far simpler.

Analogous results apply to the non-homogeneous Markov chain $\{l'_x\}_{x=0}^w$, in which the survival probabilities $\hat{p}'_x(t)$, dependent on both age and time, now replace the constant p. Equation (2.1) for this case remains identical in form, except for the substitution of $\hat{p}'_x(t)$ for p. The joint distribution of survivors $l'_1, \ldots, l'_k$ ($k \leq w$) is given by the chain binomial

$$\Pr\{l'_1, \ldots, l'_k | l'_0\} = \prod_{x=0}^{k-1} \binom{l'_x}{l'_{x+1}} \hat{p}'_x(t)^{l'_{x+1}} \hat{q}'_x(t)^{l'_x - l'_{x+1}}. \tag{2.5}$$

If, for $k = w$, one writes $d'_0 = l'_0 - l'_1, \ldots, d'_w = l'_w - l'_{w+1} = l'_w$ for the deaths in consecutive age intervals, then (2.5) leads to the multinomial distribution

$$\Pr\{d'_0, \ldots, d'_w | l'_0\} = \frac{l'_0}{d'_0! \ldots d'_w!} \hat{q}'_0(t)^{d'_0} \{\hat{p}'_0(t)\hat{q}'_1(t + 1)\}^{d'_1} \ldots \qquad (2.6)$$

$$\{\hat{p}'_0(t)\hat{p}'_1(t + 1) \ldots \hat{p}'_{w-1}(t + w - 1)\hat{q}'_w(t + w)\}^{d'_w},$$

where $\hat{q}'_w(t + w) = 1$, and the probabilities $\hat{p}'_0(t) \ldots \hat{p}'_{k-1}(t + k - 1)\hat{q}'_k(t + k)$
of death at age $1 \leq k \leq w$ satisfy the equation

$$\hat{q}'_0(t) + \hat{p}'_0(t)\hat{q}'_1(t + 1) + \ldots + \hat{p}'_0(t)\hat{p}'_1(t + 1) \ldots \hat{p}'_{w-1}(t + w - 1) = 1.$$

Equation (2.6) is a slight variation of the multinomial result of Chiang (1968; p. 225) to account for time dependence.

The distribution of survivors l'_k ($1 \leq k \leq w$) given l'_0 is now written as

$$\Pr\{l'_k | l'_0\} = \binom{l'_0}{l'_k} \{\hat{p}'_0(t) \ldots \hat{p}'_{k-1}(t + k - 1)\}^{l'_k} \qquad (2.7)$$

$$\{1 - \hat{p}'_0(t) \ldots \hat{p}'_{k-1}(t + k - 1)\}^{l'_0 - l'_k}$$

where $\{\hat{p}'_0(t) \ldots \hat{p}'_{k-1}(t + k - 1)\}$ is the survival probability from age 0 to age k. The extinction time probability of the cohort, similar in form to (2.4), will now take the form

$$\Pr\{T = \tau\} = \Pr\{l'_\tau = 0\} - \Pr\{l'_{\tau-1} = 0\}$$

$$= \left\{1 - \prod_{k=0}^{\tau-1} \hat{p}'_k(t + k)\right\}^{l'_0} - \left\{1 - \prod_{k=0}^{\tau-2} \hat{p}'_k(t + k)\right\}^{l'_0}.$$

This last formula could prove useful in actuarial problems of group life insurance, when the distribution of time to extinction of one or more cohort groups is of interest.

3. Age Distribution and Mortality Curves

Mortality or survival probability curves such as that in Fig. 1 have long been used in actuarial literature to specify probabilities of an individual's survival to age t; they are in fact, the theoretical continuous-time equivalents of our previous cohort life tables. We shall assume that such continuous time processes always underlie those in discrete time considered above. We note that if $\bar{L}_0(t)$ is the distribution function of the age $T > 0$ at death of an individual born at time 0, then the mortality curve is given by $L_0(t) = 1 - \bar{L}_0(t)$. To illustrate the relationship between the data for cohort life shown in Table 1 and the (crude) estimate $\hat{L}_\tau(t)$ of their associated mortality curves $L_\tau(t)$,

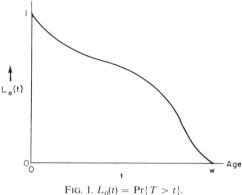

FIG. 1. $L_0(t) = \Pr\{T > t\}$.

assumed continuous and differentiable, we have drawn these in Fig. 2. Note that we have taken $\hat{L}_\tau(x) = [l'_x(\tau + x)/l'_0(\tau)]$ $(x = 0, 1, \ldots, 5)$ and $\hat{L}_\tau(6) = 0$, where $\tau = 1961, \ldots, 1966$, denotes the year of birth of the relevant cohort; thus

$$\hat{p}_x(\tau + x) = \frac{l'_{x+1}(\tau + x + 1)}{l'_x(\tau + x)} = \frac{\hat{L}_\tau(x + 1)}{\hat{L}_\tau(x)}$$

represents the estimated survival rate for those aged x in the year $\tau + x$.

To describe fully the relationship between the current age distribution for a closed population (without immigration or emigration) and its survival curve, we require a three-dimensional probabilistic representation with

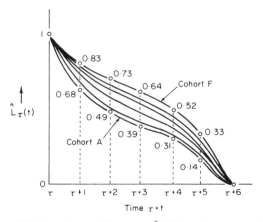

FIG. 2. Estimated mortality curves $\hat{L}_\tau(t)$ for cohorts $A, \ldots, F$.

orthogonal age and time axes in the horizontal plane. Following Bartlett (1970) let us define the expected age distribution density

$$f(x, t) = E(dN(x, t))/dx \qquad (3.1)$$

such that there are $f(x, t)\,\delta t\,\delta x$ expected individuals of age $x < X(t) < x + \delta x$ in the time interval $(t, t + \delta t)$. We denote by $\bar{F}(x, t) = \int_x^\infty f(u, t)\,du$ the expected cumulative age distribution function that there are $\bar{F}(x, t)\,\delta t$ expected individuals of age $X(t) > x$ in $t, t + \delta t$. Diagrams of the function $f(x, t)$, $\bar{F}(x, t)$ might be of the type illustrated in Fig. 3.

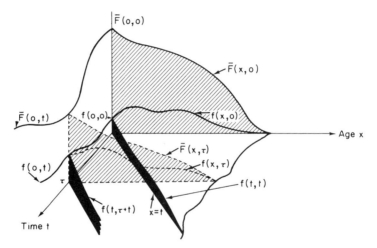

FIG. 3. The surfaces $f(x, t)$ and $\bar{F}(x, t)$.

For t fixed $\bar{F}(x, t)/F(0, t)$ will give the distribution

$$\Pr\{X(t) > x\} = P(x, t) = \bar{F}(x, t)/\bar{F}(0, t) \qquad (0 < x \le w) \qquad (3.2)$$

of the age $X(t)$ of a population at time t (or more strictly in the interval $t, t + \delta t$). This function, though shaped like a mortality curve, will in fact be quite distinct from it; $P(x, t)$ simply describes the age structure of a population at a fixed time t. On the other hand $f(t, \tau + t)/f(0, \tau)$ for fixed τ will give the mortality curve $L_\tau(t)$ for a population born at time τ (or strictly in the interval $\tau, \tau + \delta t$). A similar geometrical representation of the age-time structure of a population was given in 1965 and again in subsequent work by Skellam (1967).

In fact, it is neither of these which arises in practice, but rather their averaged values over a year, or some other period of time. For example, the distribution of age $x = 0, 1, \ldots, w$ in a population during the year τ may in practice be defined by

$$P'(x, \tau) = \frac{\int_\tau^{\tau+1} \bar{F}(x, t)\, dt}{\int_\tau^{\tau+1} \bar{F}(0, t)\, dt}, \tag{3.3}$$

where the age distribution functions $\bar{F}(x, t)$, $\bar{F}(0, t)$ are averaged over the year $(\tau, \tau + 1)$. Similarly, the mortality curve $L_\tau(t)$ may in practice be defined by

$$L_\tau(t) = \frac{\int_\tau^{\tau+1} f(t, v + t)\, dv}{\int_\tau^{\tau+1} f(0, v)\, dv}, \tag{3.4}$$

It is this which $\hat{L}_\tau(t)$ actually estimates in Fig. 2.

For every population subject to given birth and death (as well as immigration and emigration) rates, $f(x, t)$ will satisfy certain state equations. Following Bartlett (1970), we note that for a simple population, with birth and death rates $\lambda(x, t)$, $\mu(x, t)$ respectively, the density function $f(x, t)$ of individuals of age $(x, x + \delta x)$ in time $(t, t + \delta t)$ satisfies the partial differential equation

$$\frac{\partial f}{\partial t} + \frac{\partial f}{\partial x} = -\mu(x, t) f(x, t) \qquad (0 < x < t) \tag{3.5}$$

with

$$f(0, t) = \int_0^\infty \lambda(x, t) f(x, t)\, dx.$$

It is readily verified from probabilistic considerations that

$$f(x, t) = f(0, t - x)\, e^{-\int_0^x \mu(v, t - x + v)\, dv} \qquad (0 < x < t) \tag{3.6}$$

where $f(0, t)$ is the solution of the integral equation

$$f(0, t) = \int_0^t \lambda(x, t) f(0, t - x)\, e^{-\int_0^x \mu(v, t - x + v)\, dv}\, dx + f_0(t) \tag{3.7}$$

with $f_0(t)$ depending on the initial conditions at $t = 0$. This equation does not, in general, yield a simple solution. However, in the case where the birth and death rates $\lambda(x, t) = \lambda(x)$, $\mu(x, t) = \mu(x)$ depend on age only, Bartlett notes that if $L(s)$, $L_0(s)$, $M(s)$ refer to the Laplace transforms of $f(0, t)$, $f_0(t)$ and $\lambda(t) \exp\left[-\int_0^t \mu(v)\, dv\right]$ respectively, then (3.7) leads to

$$L(s) = \frac{L_0(s)}{1 - M(s)}. \tag{3.8}$$

As $t \to \infty$, the asymptotic solution is thus given by

$$f(0, t) \sim C\, e^{Kt} \tag{3.9}$$

where for $\lambda(x) > \mu(x)$ for all $x > 0$, $K > 0$ is the dominant root of the equation

$$M(s) = \int_0^\infty e^{-st} \lambda(t) e^{-\int_0^t \mu(v)\,dv}\,dt = 1, \tag{3.10}$$

and the constant C is given by

$$C = \frac{L_0(K)}{-M'(K)}.$$

Results of this type in the context of renewal theory have been previously derived by Feller (1941).

That K must be positive is obvious. For if $\lambda_{max} > \lambda(x) > \mu(x) > \mu_{min} > 0$ then assuming $s > 0$ real for simplicity, we have that

$$\int_0^\infty e^{-st} \lambda(t) e^{-\int_0^t \lambda(v)\,dv}\,dt < M(s) < \int_0^\infty e^{-st} \lambda_{max} e^{-\mu_{min} t}\,dt. \tag{3.11}$$

Integration by parts of the left bound, together with the inequality

$$-\int_0^\infty s\,e^{-st-\lambda_{min} t}\,dt < -\int_0^\infty s\,e^{-st-\int_0^t \lambda(v)\,dv}\,dt,$$

and ordinary integration of the right bound lead to

$$\frac{\lambda_{min}}{s + \lambda_{min}} = 1 - \int_0^\infty s\,e^{-st-\lambda_{min} t}\,dt < M(s) < \frac{\lambda_{max}}{s + \mu_{min}}. \tag{3.12}$$

Both bounding functions are monotonic decreasing in s; it follows that the solution K of $M(s) = 1$ will be contained between the roots $s_0 = 0$ of $\lambda_{min}(s + \lambda_{min})^{-1} = 1$ and $s_1 = \lambda_{max} - \mu_{min} > 0$ of $\lambda_{max}(s + \mu_{min})^{-1} = 1$. Hence

$$0 < K < \lambda_{max} - \mu_{min}.$$

Thus we see that for parameters depending on age only, as $t \to \infty$,

$$f(x, t) \sim C\,e^{K(t-x)} e^{-\int_0^x \mu(v)\,dv}. \tag{3.13}$$

A similar more complex result holds also for populations with two sexes.

4. Age and Time Dependence and Stability

The previous analysis throws light on the remarks of Hoem (1971) concerning formulae for period mortality rates. If, following Hoem, we define by $_n\mathcal{D}_x$ the number of deaths in $(0, T)$ with age between x and $x + n$ at death, then this is effectively written as

$$_n\mathcal{D}_x = \int_{u=x-T}^{x+n} \int_{t=\max(0,x-u)}^{\min(T,x+n-u)} f(u + t, t)\mu(u + t, t)\,du\,dt \tag{4.1}$$

where Hoem sets $f(x + t, t) = p(x)[l(x + t)/l(x)]$ and $\mu(x + t, t) = \mu(x + t)$ both age dependent only. In this, $p(x)\,\delta x$ denotes the expected number of individuals between ages $(x, x + \delta x)$ in the population, and $l(t)$ is a mortality function.

The gist of Hoem's argument is that unless the population is stable, with age distribution density

$$p(x) = \sum_{k=0}^{\infty} A_k\, e^{-r_k x} l(x) \tag{4.2}$$

and a birth frequency function

$$b(t) = \sum_{k=0}^{\infty} A_k\, e^{r_k t}, \tag{4.3}$$

then the simplification usually introduced in the weighted average

$$\int_0^n w(x + t)\mu(x + t)\,dt \Big/ \int_0^n w(x + t)\,dt$$

for the n-year death rate by Keyfitz (1970) and others, who set $w(x + t) = p(x + t)$, will not hold. What we show is that any formulation of the problem in a simple population of the type considered in Section 3 in which the birth and death rates $\lambda(x)$, $\mu(x)$ are purely age dependent is asymptotically of the type (4.2–4.3) for large t. The same applies to the analogous two-sex population. Where Hoem's warning remains valid is for more general birth and death rates $\lambda(x, t)$, $\mu(x, t)$ dependent on both time and age. For an illustration of the relevance of such age and time dependent rates to platelet kinetics, the reader is referred to Breny (1971).

For from (3.13) we see that in the population considered, as $t \to \infty$, we may write

$$f(x + t, t) \sim C\, e^{-Kx - \int_0^x \mu(v)\,dv}\, \frac{e^{-\int_0^{x+t}\mu(v)\,dv}}{e^{-\int_0^x \mu(v)\,dv}} = p(x)\frac{l(x + t)}{l(x)}. \tag{4.4}$$

Thus, asymptotically,

$$p(x) = C\, e^{-Kx} l(x)$$
$$b(t) = f(0, t) = C\, e^{Kt}; \tag{4.5}$$

these are precisely conditions of the type (4.2–3) necessary for the simplifications in weighted death rates to hold. It would thus appear that to avoid confusion in both discrete and continuous time population processes, it is essential to indicate the time-dependence of parameters governing them explicitly.

302   J. GANI

REFERENCES

BARTLETT, M. S. (1970). Age distributions. *Biometrics* **26**, 377–385.
BRENY, H. (1971). Non-stationary models. *In* "Platelet Kinetics." (J. M. Paulus, ed.), pp. 92–116. North Holland, Amsterdam.
CHIANG, CHIN LONG (1968). "Introduction to Stochastic Processes in Biostatistics." John Wiley, New York.
FELLER, W. (1941). On the integral equation of renewal theory. *Ann. Math. Statist.* **12**, 243–267.
GANI, J. AND JERWOOD, D. (1971). Markov chain methods in chain binomial epidemic models. *Biometrics* **27**, 591–603.
HOEM, J. M. (1971). On the interpretation of certain vital rates as averages of underlying forces of transition. *Theoret. Pop. Biol.* **2**, 454–468.
KEYFITZ, N. (1968). "Introduction to the Mathematics of Population." Addison-Wesley, Reading, Massachusetts.
KEYFITZ, N. (1970). Finding probabilities from observed rates, or how to make a life table. *Amer. Statist.* **24**, 28–33.
SKELLAM, J. G. (1967). Seasonal periodicity in theoretical population ecology. *In* "Proceedings of the Fifth Berkeley Symposium on Mathematical Statistics and Probability," Vol. 4, pp. 179–205. University of California Press, Berkeley.

17. On the Statistics of Cell Proliferation

P. D. M. MACDONALD

McMaster University, Hamilton, Ontario, Canada

1. Introduction

In studying biological populations it frequently happens that we are interested in learning about the life cycles of the individuals but, for one reason or another, there is no way to observe individuals over sufficiently long periods, and so inferences must be based on samples drawn from the population from time to time. Actuaries often have independent data on cohort survival, age-specific death rates and the current population age distribution. Biologists are rarely so fortunate.

Before population data can be used to draw inferences about individual life cycles it is necessary to construct a mathematical model to describe the life cycle, the possible interactions between individuals, the growth of the population and the method of sampling. This model can be used to deduce probability distributions for the population, and statistical inferences are based on these distributions and the sampling data. A good example in ecology is the study of the grasshopper life cycle described by Read and Ashford (1968). Another, mentioned by Macdonald (1971), concerns the feeding cycle of the tsetse, where flies were sampled more or less randomly from the population, and the amount of haematin remaining in the gut of a fly was taken to indicate the time since its last blood meal. The common feature in both these examples and in the present work is that, when an individual is drawn from the population, we are interested in the time that has elapsed since some past event in its life cycle or the time to some future event.

There are several practical difficulties with this approach. It might be impossible to obtain unbiased samples from the population, fish and winged insects being especially notorious in this respect. The population might exist in an inconstant environment, invalidating the use of "steady state" models. Often, the system is too complex to model mathematically. If the population is at all heterogeneous, this heterogeneity will be inextricably confounded with any natural variability within the life cycle.

These difficulties are minimized in the case of cell populations. This is particularly true for genetically uniform strains enjoying unrestrained proliferation in laboratory culture but much useful work has also been done by way of *in vivo* studies, notably in the areas of tumour growth and plant and animal morphogenesis (Cleaver, 1967; Baserga, 1971). In these latter cases the situation is much more complex and assumptions such as independence of cells must be viewed with some care.

The life cycle, or mitotic cycle, of a cell can be divided into four phases called G_1, S, G_2 and M, where S is a discrete period of DNA synthesis distinguished by the uptake of H^3 Tdr (tritiated thymidine, a radioactively labelled DNA precursor) and M is mitosis, the visible act of cell division. Each mitosis results in two daughter cells; however, allowing for the possibility of cell loss at mitosis, we let $A \leq 2$ denote the mean number of daughters remaining in the population. It follows that so long as the cells may be assumed to proceed independently of one another the cell population may be modelled by a multiphase branching process, and if $A > 1$ and the population has continued to proliferate under stable conditions for a sufficiently long time the population will increase exponentially and the distribution of cell ages will approach a limiting form.

The quantity of interest is then the joint distribution for the durations of the four phases, denoted by the probability density function (p.d.f.) $\phi_{1,2,3,4}(u_1, u_2, u_3, u_4)$. The object of this paper is to show how the limiting age and phase structure of the population can be derived in terms of $\phi_{1,2,3,4}$ and illustrate how these distributions can be applied to the statistical analysis of an experiment described by Puck and Steffen (1963) where a cell population, initially in a steady state of exponential growth, is exposed simultaneously to colcemide (which causes cells to accumulate in the mitotic phase) and H^3 Tdr. The approach used here is quite general and can be applied to the analysis of other, similar, experiments. Earlier work of this nature is reviewed in Macdonald (1970) and Brockwell *et al.* (1970).

2. Limiting Distributions for a Multiphase Branching Process

In this section we derive expressions for the statistical distributions of age and age-in-phase for a cell population which has reached a steady state of exponential growth. Some of these expressions have been derived more rigorously by Harris (1963, chapter VI) and Brockwell and Kuo (1972). It is hoped that the heuristic proofs given here will make these results accessible to readers less mathematically inclined.

Let the number of phases be p, and let $U_i \geq 0$ be the time that a cell just starting its cycle will spend in phase i $(i = 1, \ldots, p)$. Denote the joint p.d.f.

of $U_1, \ldots, U_p$ by $\phi_{1,\ldots,p}(u_1, \ldots, u_p)$ and its Laplace transform by

$$\phi^*_{1,\ldots,p}(s_1, \ldots, s_p) = \int_0^\infty \cdots \int_0^\infty e^{-(s_1 u_1 + \ldots + s_p u_p)} \phi_{1,\ldots,p}(u_1, \ldots, u_p)\, du_1 \ldots du_p.$$

Laplace transforms will provide a useful notation in what follows.

Successive phases may be combined and regarded as a single phase. Let the phases $1, 2, \ldots, i$ combined be indicated by the subscript $12 \ldots i$; $U_{12\ldots i}$, the time spent in phases 1 to i, has p.d.f. $\phi_{12\ldots i}(u)$ with Laplace transform

$$\phi^*_{12\ldots i}(s) = \phi^*_{1,2,\ldots,i,i+1,\ldots,p}(s, s, \ldots, s, 0, \ldots, 0)$$

and cumulative distribution function (c.d.f.) $\Phi_{12\ldots i}(u)$.

Total cycle time has a p.d.f. denoted by $\phi(u)$ with Laplace transform $\phi^*(s) = \phi^*_{1,\ldots,p}(s, \ldots, s)$ and c.d.f. $\Phi(u)$.

It should be clear that the rate of proliferation of the cell population will depend only on A and $\phi(u)$, and not otherwise on the phase structure of the cell cycle. If we ignore the phase structure for the moment we have a conventional age-dependent branching process for which many well-known results are available. For example, provided that $\int_0^\infty \{\phi(u)\}^p\, du < \infty$ for some $p > 1$, then for some $\varepsilon > 0$ the expected population size at time t, which we shall denote by $Z(t)$, approaches the limiting form

$$Z(t) = N\, e^{kt}\{1 + 0(e^{-\varepsilon t})\} \tag{1}$$

as $t \to \infty$, where N is a constant depending on the initial age distribution and k is the positive root of the equation

$$A\phi^*(k) = 1 \tag{2}$$

(Harris, 1963, Theorem 17.2, p. 143). Since $\{dZ(t)/dt\}/Z(t) = k$, k is the average rate of increase per cell per unit time; k is sometimes called the Malthusian parameter. A population that has continued to proliferate under stable conditions for a time sufficiently long that the asymptotic form

$$Z(t) = N\, e^{kt} \tag{3}$$

holds is said to be in a steady state of exponential growth.

A population in a steady state of exponential growth at time t will increase by an average amount

$$\{dZ(t)/dt\}\, dt = kN\, e^{kt}\, dt \tag{4}$$

between times t and $t + dt$. Since each cell division results in an average of A daughter cells, but increases the population by only $A - 1$ cells, it follows that the mean number of daughters born between times t and $t + \mathrm{d}t$ is given by

$$\{A/(A - 1)\}kN\,e^{kt}\,\mathrm{d}t. \tag{5}$$

A cell aged u at time t must have been born at time $t - u$ and have a cycle duration of u or longer. Hence

$$\{A/(A - 1)\}kN\,e^{k(t-u)}\{1 - \Phi(u)\}\,\mathrm{d}u \tag{6}$$

is the expected number of cells which are between ages u and $u + \mathrm{d}u$ at time t, and dividing by $N\,e^{kt}$, the expected total number of cells, we deduce that the probability that a randomly chosen cell is between ages u and $u + \mathrm{d}u$ is

$$\{A/(A - 1)\}k\,e^{-ku}\{1 - \Phi(u)\}\,\mathrm{d}u. \tag{7}$$

It can be shown that under very weak conditions the actual population age distribution approaches this form with probability 1 as $t \to \infty$ (Harris, 1963, Theorem 25.1, p. 154).

Now let the cycle be divided into p phases and consider a cell born at time $t - u$. If it has not divided by time t, it will be in phase i with probability

$$\frac{\Phi_{12\ldots(i-1)}(u) - \Phi_{12\ldots i}(u)}{1 - \Phi(u)}, \tag{8}$$

where $\Phi_{12\ldots(i-1)}(u)$ is interpreted as 1 when $i = 1$. Furthermore, given that that cell will be in phase i at time t, the joint p.d.f. for the phase durations $U_1, \ldots, U_p$ is given by

$$\begin{cases} \dfrac{\phi_{1,\ldots,p}(u_1,\ldots,u_p)}{\Phi_{12\ldots(i-1)}(u) - \Phi_{12\ldots i}(u)} & \left(\displaystyle\sum_{j=1}^{i-1} u_j < u \le \sum_{j=1}^{i} u_j\right) \\ 0 & \text{(otherwise)}. \end{cases} \tag{9}$$

Multiplying (7) and (8) gives the probability that a randomly chosen cell is in phase i and between ages u and $u + \mathrm{d}u$, viz.

$$\{A/(A - 1)\}k\,e^{-ku}\{\Phi_{12\ldots(i-1)}(u) - \Phi_{12\ldots i}(u)\}\,\mathrm{d}u; \tag{10}$$

while integrating out u gives the fraction of the population in phase i:

$$\begin{aligned} P_i &= \{A/(A - 1)\}\,\{\phi^*_{12\ldots(i-1)}(k) - \phi^*_{12\ldots i}(k)\} \\ &= \frac{\phi^*_{12\ldots(i-1)}(k) - \phi^*_{12\ldots i}(k)}{1 - \phi^*(k)}. \end{aligned} \tag{11}$$

Finally, multiplying (9) and (10) and changing variables give

$$\frac{A}{A-1} k \, e^{-k(u_1 + \ldots + u_{i-1} + x)} \phi_{1,\ldots,p}(u_1, \ldots, u_{i-1}, x + y, u_{i+1}, \ldots, u_p)$$
$$\times \prod_{\substack{j=1 \\ j \neq i}}^{p} du_j \, dx \, dy \tag{12}$$

which is the joint probability that a randomly chosen cell is in phase i, having spent times $u_1, \ldots u_{i-1}$ in the previous phases and attained age x in phase i, and is to remain an additional time y in phase i then spend times $u_{i+1}, \ldots, u_p$ in the remaining phases. All of the required population distributions can be derived directly from (11) and (12); most can be found quite simply by manipulating the Laplace transform of (12) which is

$$\frac{A}{A-1} \frac{k}{q-r+k} \{\phi^*_{1,\ldots,p}(s_1 + k, \ldots, s_{i-1} + k, r, s_{i+1}, \ldots, s_p)$$
$$- \phi^*_{1,\ldots,p}(s_1 + k, \ldots, s_{i-1} + k, q + k, s_{i+1}, \ldots, s_p)\} \tag{13}$$

where $s_1, \ldots, s_{i-1}, q, r, s_{i+1}, \ldots, s_p$ correspond to $u_1, \ldots, u_{i-1}, x, y, u_{i+1}, \ldots, u_p$, respectively. In the case $p = 4$ and $i = 1$, (13) reduces to

$$\frac{A}{A-1} \frac{k}{q-r+k} \{\phi^*_{1,2,3,4}(r, s_2, s_3, s_4) - \phi^*_{1,2,3,4}(q + k, s_2, s_3, s_4)\}, \tag{13a}$$

while if $p = 4$ and $i = 2$, (13) becomes

$$\frac{A}{A-1} \frac{k}{q-r+k} \{\phi^*_{1,2,3,4}(s_1 + k, r, s_3, s_4)$$
$$- \phi^*_{1,2,3,4}(s_1 + k, q + k, s_3, s_4)\}. \tag{13b}$$

The manipulations involve assigning appropriate values to the transform variables in (13), then inverting the resulting transform by inspection.

Some examples are:

(a) Duration of phase i, for a cell sampled just as it enters phase i:

p.d.f.: $h_i(u) = \displaystyle\int_0^\infty e^{-kv} \phi_{12\ldots(i-1),i}(v, u) \, dv / \phi^*_{12\ldots(i-1)}(k)$

c.d.f.: $H_i(u) = \displaystyle\int_0^u h_i(v) \, dv$

Laplace transform: $h_i^*(s) = \phi^*_{12\ldots(i-1),i}(k, s) / \phi^*_{12\ldots(i-1)}(k)$.

Clearly, $h_1 \equiv \phi_1$; also, if phase i is independent of phase $12 \ldots (i-1)$, or if $k = 0$, then $h_i \equiv \phi_i$.

If, for instance, $\phi_{1,2}$ is bivariate normal with mean vector $\begin{pmatrix} \mu_1 \\ \mu_2 \end{pmatrix}$ and co-variance matrix $\begin{pmatrix} \sigma_1^2 & \rho\sigma_1\sigma_2 \\ \rho\sigma_1\sigma_2 & \sigma_2^2 \end{pmatrix}$, where it is assumed that the means are sufficiently large that the negative tails may be ignored, then ϕ_2 is normal with mean μ_2 and variance σ_2^2 but h_2 is normal with mean $\mu_2 - k\rho\sigma_1\sigma_2$ and variance σ_2^2. More generally, using the relation

$$\log h_2^*(s) = \log \phi_{1,2}^*(k, s) - \log \phi_1^*(k)$$

and letting κ_{ij} denote the i, jth cumulant of $\phi_{1,2}$, the jth cumulant of ϕ_2 is just κ_{0j} but the jth cumulant of h_2 is given by $\kappa_{0j} - k\kappa_{1j} + \frac{1}{2}k^2\kappa_{2j} - \ldots$; hence h_2 has mean $u_2 - k\rho\sigma_1\sigma_2 + 0(k^2)$ and variance $\sigma_2^2 - k\kappa_{12} + 0(k^2)$.

In examples (b) to (e) below it is worth noting that a cell cycle with p phases correlated and distributed according to the p-variate p.d.f. $\phi_{1,\ldots,p}$ will give rise to the same population structure as a cell cycle with p phases statistically independent and distributed according to the p.d.f.'s $h_1, \ldots, h_p$.

(b) Duration of phase i, for a cell sampled just as it leaves phase i:

p.d.f.: $f_i(u) = e^{-ku}h_i(u)/h_i^*(k)$

c.d.f.: $F_i(u) = \int_0^u f_i(v)\,dv$

Laplace transform: $f_i^*(s) = h_i^*(s+k)/h_i^*(k)$.

If h_i is normal with mean μ_i and variance σ_i^2 it can then be shown that f_i is normal with mean $\mu_i - k\sigma_i^2$ and variance σ_i^2. Also, if h_i is a gamma distribution with shape parameter α_i and scale parameter λ_i then f_i is a gamma distribution with shape parameter α_i and scale parameter $\lambda_i + k$. More generally, if h_i has mean μ_i and variance σ_i^2 and the skewness is not too large, then f_i has mean $\mu_i - k\sigma_i^2 + 0(k^2)$ and variance $\sigma_i^2 + 0(k)$.

(c) Age in phase i, for a cell sampled from phase i:

p.d.f.: $g_i(x) = k\,e^{-kx}\dfrac{1 - H_i(x)}{1 - h_i^*(k)}$

c.d.f.: $G_i(x) = 1 - e^{-kx}\dfrac{1 - B_i(x)}{1 - h_i^*(k)}$

where

$$B_i(x) = \int_0^\infty \exp\{-k \max(0, u - x)\}h_i(u)\,du$$

$$= H_i(x) + e^{kx}h_i^*(k)\{1 - F_i(x)\}.$$

(d) Time to end of phase $i + 1$, for a cell sampled from phase i:

p.d.f.: $\gamma_{i:(i+1)}(z) = k\, e^{kz} h^*_{i(i+1)}(k)\dfrac{F_{i+1}(z) - F_{i(i+1)}(z)}{1 - h^*_i(k)}$

c.d.f.: $\Gamma_{i:(i+1)}(z) = \dfrac{B_{i(i+1)}(z) - h^*_i(k)B_{i+1}(z)}{1 - h^*_i(k)}.$

(e) Time to end of phase i, for a cell sampled from phase i:

p.d.f.: $\gamma_i(y) = k\, e^{ky} h^*_i(k)\dfrac{1 - F_i(y)}{1 - h^*_i(k)}$

c.d.f.: $\Gamma_i(y) = \dfrac{B_i(y) - h^*_i(k)}{1 - h^*_i(k)}.$

3. Puck and Steffen's Experiment

Puck and Steffen (1963) describe an experiment where colcemide and H^3 Tdr were added simultaneously to a cell culture which was in a steady state of exponential growth. It was assumed that a cell in S at time $t > 0$ will take up H^3 Tdr and become labelled, a cell in M at $t = 0$ will proceed to normal division but a cell entering M at $t > 0$ will be blocked in M. It has been shown that these assumptions may be satisfied if the correct concentrations of colcemide and H^3 Tdr are used and the experiment is not continued too long. A microscope slide can be prepared from a sample of cells; mitotic cells are then recognized by the configuration of their chromosomes and labelled cells can be detected by autoradiography (Cleaver, 1967). The data record the numbers of mitotic, labelled and labelled mitotic cells in independent samples drawn at successive times after $t = 0$.

Let N be the total population size at time zero, let $N(t)$ be the expected number of cells at time t, and let $M(t)$ be the expected number of cells in mitosis at time t. It follows that

$$N(t) = N + (A - 1)NP_4\Gamma_4(t)$$
$$= NAh^*_{123}(k)B_4(t)$$

and

$$M(t) = NP_{123}\Gamma_{123}(t) + NP_4\{1 - \Gamma_4(t)\} + NAP_4\Gamma_{4:123}(t)$$
$$= NAh^*_{123}(k)\{A/(A - 1)\}\{B_{4123}(t) - h^*(k)B_4(t)\};$$

hence the fraction of the population in M at time t is

$$m(t) = M(t)/N(t)$$
$$= \{A/(A - 1)\}\{B_{4123}(t) - h^*(k)B_4(t)\}/B_4(t).$$

Similarly, the fraction labelled at time t is

$$l(t) = \{A/(A - 1)\} \{B_{41}(t) - h^*_{412}(k)\}/B_4(t)$$

and the fraction labelled and in M at time t is given by

$$lm(t) = \{A/(A - 1)\} \{B_{4123}(t) - h^*_{412}(k)B_3(t)\}/B_4(t).$$

Since $h^*(k) = A^{-1}$ and $h^*_{412}(k) = h^*_{4123}(k)/h^*_3(k)$, and since each B_i is defined solely in terms of k and h_i, the three theoretical curves $m(t)$, $l(t)$ and $lm(t)$ are completely specified when the four distributions h_{4123}, h_4, h_{41} and h_3 are given, along with the values of A and k. Statistical analysis of the experimental data can therefore lead to inferences about parameters of h_i ($i = 4123, 4, 41, 3$) and, as explained in Section 2a, the mean and variance of h_i differ from the mean and variance of ϕ_i by terms of $0(k)$. Typically, k is small; a useful approximate solution to (2) is

$$k \simeq \{\mu - \sqrt{(\mu^2 - 2\sigma^2 \log A)}\}/\sigma^2, \tag{14}$$

obtained by writing

$$\phi^*(k) = \exp \{-k\mu + \tfrac{1}{2}k^2\sigma^2 + 0(k^3)\}. \tag{15}$$

In applications where σ^2 is small, k is best computed by the limit $k \to (\log A)/\mu$ as $\sigma^2 \to 0$.

In most applications the length of time spent in mitosis is relatively short and not appreciably affected by the length of time spent in interphase. It is then possible to simplify the analysis by assuming that phase 4 is independent of phase 123, with the result that $h_{4123} \equiv \phi$. Then, if A is given and if the mean μ and variance σ^2 of the cycle duration are estimated from the experimental data, k can be computed from (2) or (14).

The statistical analysis is done by partitioning the counts at each time into mutually exclusive categories such as: labelled mitotic, unlabelled mitotic, labelled interphase and unlabelled interphase; or, at a time when labelled cells were not scored: mitotic and interphase. The theoretical curves $m(t)$, $l(t)$ and $lm(t)$ can be used to compute the probability associated with each category and the situation is essentially that of parametric multinomial estimation (Rao, 1965, Section 5e). A likelihood function for the complete experiment is obtained by summing the log-likelihoods over all sampling times, and maximum likelihood estimates can be found by "scoring" if the appropriate derivatives of $m(t)$, $l(t)$ and $lm(t)$ are available, otherwise by some nonlinear optimization method.

The choice of parameters for estimation is crucial to the success of the estimation procedure. Let μ_i, σ_i^2, denote the mean and variance, respectively, of the distribution h_i. Inspection of the curves $m(t)$, $l(t)$ and $lm(t)$, plotted in Figs. 1, 2 and 3, suggests that information for μ comes mainly from the slopes

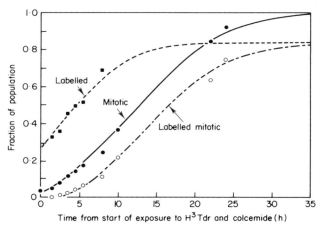

FIG. 1. Experimental points and fitted curves for the data of Puck and Steffen (1963). Parameter estimates are given in Table 1.

of the three curves. The initial levels $m(0)$ and $l(0)$ are determined mainly by μ_4 and μ_{41}, respectively, while the apparent displacement in time between $m(t)$ and $lm(t)$ is a rough estimate of μ_3. Comparison of Fig. 1 and Fig. 2, where the curves were computed with the same means as those in Fig. 1 but with all variances set to zero, shows that information about the variability of phase duration comes from the degree to which the "corners" of the curves, abrupt in the deterministic case, are rounded off; in particular, the approach of $m(t)$ to the asymptote gives σ^2, the initial rise of $lm(t)$ gives σ_3^2, and the very early

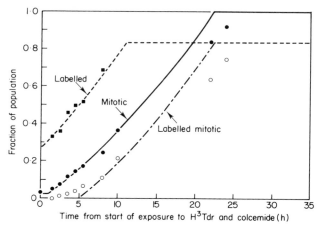

FIG. 2. The theoretical curves of Fig. 1 were recomputed with the variances set to zero and all other parameters unchanged.

behaviour of $m(t)$ gives σ_4^2. However, it will not be possible to estimate these variances unless there are sufficient experimental points to define these features; the experimental points in Figs. 1 and 2, for example, convey almost no information about σ_4^2. If the experiment is carried on long enough to define the approach to the asymptote there is the danger that cells previously blocked in mitosis may lose their mitotic configuration and be no longer recognizable as such. The proportions of mitotic and labelled mitotic cells will then be underestimated, and hence the variances will be overestimated. In any event, the variances may be difficult to interpret biologically as the apparent variances may have been inflated by such factors as heterogeneity within the population or indistinct demarcations between phases. The curves are relatively insensitive to the value of A in the range $1 < A \le 2$, and so it is not possible to estimate A from this experiment.

We have assumed throughout that the population consists entirely of proliferating cells. If the population includes a fixed proportion of quiescent cells prior to the experiment the distributions of Section 2 still apply, but to the proliferating cells only, and it is straightforward to modify the derivations of $m(t)$, $l(t)$ and $lm(t)$ accordingly.

4. Examples

Computations were done using four gamma distributions with a common scale parameter to represent the distributions ϕ, h_4, h_{41} and h_3. This allowed estimation of the four means (for the durations of the cycle, M, $M + G_1$ and G_2, respectively) and an overall measure of the variability, which was expressed as the standard deviation of the cycle duration.

TABLE 1. Parameters estimated by maximum likelihood from Puck and Steffen's data are given with their covariance matrix and compared with the estimates originally given by Puck and Steffen. We have taken $A = 2$. The fitted curves are shown in Fig. 1. Goodness-of-fit: $\chi^2(20\text{D.F.}) = 83$.

Cycle duration (h)		Mean phase durations (h)		
Mean	Std. dev.	M	G_2	$M + G_1$
$22\cdot43 \pm 0\cdot31$	$6\cdot50 \pm 0\cdot30$	$0\cdot91 \pm 0\cdot14$	$4\cdot91 \pm 0\cdot14$	$10\cdot92 \pm 0\cdot21$
Covariance matrix of the estimates:				
$0\cdot0934$				
$-0\cdot0237$	$0\cdot0894$			
$0\cdot0064$	$-0\cdot0050$	$0\cdot0190$		
$-0\cdot0057$	$0\cdot0210$	$-0\cdot0014$	$0\cdot0197$	
$0\cdot0466$	$-0\cdot0300$	$0\cdot0056$	$-0\cdot0140$	$0\cdot0452$
Estimates given by Puck and Steffen:				
$20\cdot1$	$1\cdot85$	$1\cdot10$	$4\cdot56$	$9\cdot50$

Curves were fitted to the original data of Puck and Steffen (1963). These are shown in Fig. 1 and the estimates appear in Table 1. Puck and Steffen (1963) had devised a graphical method for fitting these curves, and their estimates (also given in Table 1), are very close to ours. Their method assumed that all variances were zero; the estimate of σ they quoted was determined by monitoring individual cells by time-lapse cinemicrophotography, and this is apt to be an underestimate due to cells with very long cycle times not being noticed. As mentioned in Section 3, our estimate of σ is apt to be an overestimate.

Figure 3 shows curves fitted to experimental data provided by P. W. Barlow of the Agricultural Research Council, Unit of Developmental Botany,

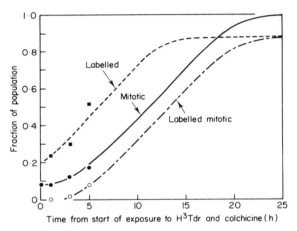

FIG. 3. Experimental points and fitted curves for P. W. Barlow's data. Parameter estimates are given in Table 2.

TABLE 2. Parameters estimated by maximum likelihood from P. W. Barlow's data are given with their covariance matrix. We have taken $A = 2$. The fitted curves are shown in Fig. 3. Goodness-of-fit: $\chi^2(5\text{D.F.}) = 16.9$.

Cycle duration (h)		Mean phase durations (h)		
Mean	Std. dev.	M	G_2	$M + G_1$
19.6 ± 1.2	3.0 ± 1.2	2.1 ± 0.2	3.2 ± 0.3	12.1 ± 0.7

Covariance matrix of the estimates:

1·531				
0·594	1·347			
0·153	0·026	0·056		
0·129	0·324	0·005	0·116	
0·830	0·161	0·094	0·012	0·522

Cambridge. The cells were the initial cells of the cap columella in the root tip of *Zea mays*, and colchicine was used in place of colcemide, being the more effective with plant cells. The estimates appear in Table 2. The roots were about 100 cm long and the mitotic cycle appears to proceed more slowly than in the younger roots described by Barlow and Macdonald (1973), the difference being mainly in the duration of G_1. The example illustrates how powerful this experimental technique can be for getting good parameter estimates from one small-scale experiment.

REFERENCES

BASERGA, R. (1971). "The Cell Cycle and Cancer." Marcel Dekker, New York.

BARLOW, P. W. AND MACDONALD, P. D. M. (1973). An analysis of the mitotic cell cycle in the root meristem of *Zea mays*. *Proc. Roy. Soc.* B. (In press).

BROCKWELL, P. J. AND KUO, W. H. (1972). Generalized asymptotic age distributions for a multiphase branching process. Michigan State University, Statistical Laboratory Publication 38.

BROCKWELL, P. J., MACLAREN, M. D. AND TRUCCO, E. (1970). Monte Carlo simulation of PLM-curves and collection functions. *Bull. Math. Biophys.* **32**, 429–443.

CLEAVER, J. E. (1967). "Thymidine Metabolism and Cell Kinetics." North-Holland, Amsterdam.

HARRIS, T. E. (1963). "The Theory of Branching Processes." Springer-Verlag, Berlin.

MACDONALD, P. D. M. (1970). Statistical inference from the fraction labelled mitoses curve. *Biometrika* **57**, 489–503.

MACDONALD, P. D. M. (1971). Discussion of a paper by Boneva, Kendall and Stefanov. *J. R. Statist. Soc.* B **33**, 43–44.

PUCK, T. T. AND STEFFEN, J. (1963). Life cycle analysis of mammalian cells. I. A method of localizing metabolic events within the life cycle, and its application to the action of colcemide and sublethal doses of X-irradiation. *Biophys. J.* **3**, 379–397.

RAO, C. R. (1965). "Linear Statistical Inference." Wiley, New York.

READ, K. L. Q. AND ASHFORD, J. R. (1968). A system of models for the life cycle of a biological organism. *Biometrika* **55**, 211–221.

18. The Significance of Clutch-Size

M. D. Mountford

The Nature Conservancy, Belgrave Square, London, England

1. Introduction

Each bird species has a characteristic frequency distribution of clutch-sizes. The Wood-Pigeon invariably lays a clutch of two eggs. The Swift, in the Oxford region, sometimes lays two eggs and sometimes three. The clutch-size of the Great Tit is much more variable; in the same season and locality clutches ranging from one to fifteen have been observed.

Two conflicting interpretations of the evolutionary significance of the clutch-size distribution have been proposed. First, Lack (1947) argues that the clutch-size has evolved through natural selection to that number which, *on average*, gives rise to the greatest number of offspring. He further argues that most bird populations are limited in numbers by their food supplies, and that this density-dependent regulation operates primarily on the mortality rate and not significantly on the clutch-size. The second theory, developed by Wynne-Edwards, rejects the argument that the clutch-size is such as to maximize the average number of surviving offspring: it proposes instead that populations have evolved mechanisms—his so-called epideictic displays—which prevent their number reaching a level at which any direct competition for food is necessary. In particular clutch-size has been evolved to balance the mortality rate and thus overpopulation is avoided. The clash between Lack (1964) and Wynne-Edwards (1964) over the population of Oxford Swifts revealed the serious inadequacy of Lack's hypothesis to explain the clutch-size data. The clutch of the Oxford Swifts is 2 or 3—the brood sizes are approximately in the ratio 2:1. The number of young that the parents are able to raise depends on the weather: being greater in fine summers when air borne insects are abundant than in cold, wet summers when the insects are few. In fine summers the broods of 3 produce more surviving offspring than broods of 2: in cold summers the broods of 2 are most productive. Lack argues, and no-one would disagree, that the Swift has evolved a clutch of 2 or 3 because 2 is more advantageous in some years and 3 more advantageous in other years. It all depends on what one means by advantageous. Lack's measure of advantage is the average number of surviving offspring. Wynne-Edwards pointed out that the Swift clutch-size distribution is well below the most efficient clutch-size, this being the size that on average has yielded the largest number of surviving offspring. Wynne-Edwards showed that the broods of 3, averaged over the observed sequence of fine and of cold summers, produced substantially more surviving offspring than broods of 2. Hence, according to Lack's criterion, the Swift should have a constant clutch of 3. Indeed if Lack's criterion did apply, then each species would have evolved a single fixed clutch-size. Any mixture of clutch sizes is, by his criterion, sub-optimal. It may be argued in defence, that with differences in the availability of food, timing of the breeding season, and in parental skill, selection does not always work to the same figure. But this line of defence crumbles when species with a wide range of clutch sizes are considered. The Great Tit lays, in the same year and locality, clutches of 1 to 15 eggs. This wide variation cannot be ascribed to differences in food supply and parental skill. There is no doubt that when averaged over good and bad years, the clutch sizes in the middle of the frequency distribution are most productive. Yet the extreme sizes are palpably there. Why? The extreme clutch-sizes are there to guard against or take advantage of extreme conditions. In catastrophically bad years only the nestlings of small clutches have any chance of survival. These clutches do not, on average, leave the maximum number of surviving offspring: they merely ensure that in times of crisis, the population hangs on in existence. It is yet another example of the evolutionary strategy of meeting variability in the environment with variability in the population. Nature does not put all her eggs in one clutch-size; she does not maximize the average number of offspring but prudently insures herself against all possible breakages. Only if the environment is unchanging would we expect a constant clutch size. The fact of variability in clutch size argues that natural selection operates not by maximizing the average productivity but rather by minimizing the possibility of extinction. In view of this the average growth rate is a somewhat irrelevant measure of the success or failure of a population. A more appropriate measure is the probability of extinction. This measure is the one adopted in the succeeding analysis.

2. Mathematical Model

In this section the argument is given a mathematical formulation. Though the model is highly-simplified, it retains the main structure of, and bears some resemblance to reality. The changes in population size are described by a branching process in a random environment. A species breeds once a year. For simplicity it is assumed that a parent breeds only once then dies, and that a nestling takes less than a year to mature. Only counts of females are considered. In year i let $p_i(r)$ be the probability that a female produces r females that survive to breed in the following season.

$\phi_i(s)$ is the corresponding probability generating function i.e.

$$\phi_i(s) = \sum_0^\infty p_i(r)s^r$$

Starting with 1 individual in year 0, the p.g.f. of females at the start of the nth breeding season is

$$\phi_1(\phi_2(\phi_3(\ldots \phi_n(s)\ldots)))$$

The ϕ_i vary randomly from year to year depending on the breeding conditions prevailing. For simplicity it is assumed that the environment varies between only three states, bad years, average years, and good years with equal probability $1/3$ and with corresponding p.g.f's $\phi_\alpha(s)$, $\phi_\beta(s)$, and $\phi_\gamma(s)$. If in fact the environment is unvarying, i.e. $\phi_\alpha(s) = \phi_\beta(s) = \phi_\gamma(s) = \phi(s)$ say, then the probability of the ultimate extinction of the population is that value q satisfying $q = \phi(q)$, a well-known result of the theory of the classical branching process.

This model was first proposed by Smith and Wilkinson (1969) who provided necessary and sufficient conditions for the probability of extinction to be less than 1. Wilkinson (1969) subsequently provided a method of calculating the extinction probability. More recently Athreya and Karlin (1971a and b) have produced a definitive theoretical analysis and extension of the model. The concern of the present paper is merely to apply the model to the problem of clutch-size.

Let us now assume that the breeding conditions in any one year are determined by the abundance of food.

In particular the survivorship of a nestling is assumed to be related to the food abundance by the sigmoid function

$$P = \frac{c}{1 + b\,e^{-F}}$$

where P is the probability of survival, F is the food abundance, and b and c are fixed parameters.

This relation describes the survival rate of a nestling from a brood of 1. The probability of survival increases with increasing food until the upper asymptote is approached. For clutches (or rather broods), of 2 the food available to each nestling is halved, with a corresponding reduction in the probability of survival; and in general for broods of size n, the food is equally divided between the n nestlings. This again is, of course, a simplification. Parents of large clutches work harder in providing food—but the main line of the argument is preserved in that there is certainly an inverse relation between clutch size and the rate of individual feeding.

F_α is the food available in a bad year. The probability of survival for a nestling belonging to a brood of size n is ${}_\alpha p_n = f(F_\alpha/n)$ where f is the sigmoid function. The p.g.f. of survivors is the binomial $({}_\alpha q_n + {}_\alpha p_n s)^n$. If λ_n is the probability that a parent produces a brood of size n, then in a bad year the overall p.g.f. is $\phi_\alpha(s) = \sum \lambda_n({}_\alpha q_n + {}_\alpha p_n s)^n$; $\phi_\beta(s)$ and $\phi_\gamma(s)$, the p.g.f.s of survivors in average and in good years, are similarly derived.

A numerical example was computed in which the parameters were

$$f(F) = \frac{0.6}{1 + 125\,e^{-F}} \qquad F_\alpha = 13.5;\ F_\beta = 17.5;\ F_\gamma = 21.5$$

Using Wilkinson's matrix inversion method to calculate the extinction probabilities, it was found that the combination of the λ_i producing the minimum probability of extinction was

$$\lambda_1 = 0; \qquad \lambda_2 = 0.88; \qquad \lambda_3 = 0.12; \qquad \lambda_4 = 0$$

The optimum mix is 88% broods of size 2 and 12% broods of size 3.

It is of comparative interest to note that in an unchanging environment in which the abundance of food equals $(F_\alpha + F_\beta + F_\gamma)/3$—the abundance averaged over years,—the probability of extinction is minimized by a constant clutch size of 3.

This analysis is rather labouring the obvious—but at least it does provide an explanation for the observed variability in clutch-size. When the clutch-size distributions of different species of birds are looked at in this light, the noticeable constancy of clutch size in birds of maritime behaviour may be explained by the relative constancy of their environment compared with inland birds. Similarly the dove and pigeon species—which invariably have clutches of two eggs—provide their young with a uniform supply of food in the form of pigeon-milk produced in the crop of both parents. The variability in the clutch size of the Great Tit may reflect their dependence on the uncertain supply of the caterpillars of the Winter Moth and on the unpredictable and widely varying beechmast crop.

3. Proportion of Non-Breeding Individuals

In an earlier paper, Mountford (1971), an allied problem was discussed—that of the non-breeding reserve of mature individuals which many populations maintain to insure themselves against periods of harsh environmental conditions. The non-breeding individuals, unweakened by parentage, are better able to withstand extreme situations. The model was similar to the one just discussed. The environment varied randomly between bad and good years—each with probability $\frac{1}{2}$—and with p.g.f.'s

$$\phi_\alpha(s) = (1 - R)f_\alpha(s) + Rg_\alpha(s)$$

$$\phi_\beta(s) = (1 - R)f_\beta(s) + Rg_\beta(s)$$

where R is the probability that a mature animal breeds, the $f(s)$ are the p.g.f.'s of the non-breeders and the $g(s)$ the p.g.f.'s of the breeding individuals. In bad years the breeders do less well than the non-breeders and in good years the converse holds. In this earlier paper I considered the example

$$\phi_\alpha(s) = (1 - R)(0{\cdot}2 + 0{\cdot}8s) + R(0{\cdot}9 + 0{\cdot}1s^3)$$

$$\phi_\beta(s) = (1 - R)(0{\cdot}1 + 0{\cdot}9s) + R(0{\cdot}2 + 0{\cdot}8s^6).$$

I showed—what indeed was perfectly clear from the outset—that the proportion of non-breeders that minimized the extinction problem was different from that proportion that maximized the relative growth rate. The minimal probability of extinction obtains at $R = 0{\cdot}55$ and the relative growth rate is maximized at $R = 0{\cdot}68$. I went on to surmise that if the population had a mechanism that allows the proportion of non-breeders to vary with density, then as the population changes from low to high densities the proportion of breeders would increase from $0{\cdot}55$ to $0{\cdot}68$. I now think that this argument is incorrect. The fact is that those bird species whose clutch sizes do vary in response to changes in density, show a decreasing clutch size with increasing density. For example this inverse relation holds for both the Great and the Coal Tit. Lack (1966) ascribes the reduction in clutch-size to the increased competition between parents for the limited food supply. Be that as it may, let us consider the application of the criterion, the minimizing of the probability of extinction—on a population that allows the proportion of non-breeders to vary with density. It is important to note that the branching process model used here, is based on the assumption of unlimited food supplies—the p.g.f.'s $f_\alpha, f_\beta, g_\alpha, g_\beta$ do not alter with changing density. The analysis applies to population sizes below the level at which any direct competition for food is necessary. Let R_i be the proportion of breeders when the population size is i. The p.g.f. of survivors is

$$\tfrac{1}{2}\{[_i\phi_\alpha(s)]^i + [_i\phi_\beta(s)]^i\}$$

where

$$_i\phi_\alpha(s) = (1 - R_i)f_\alpha(s) + R_i g_\alpha(s)$$

$$_i\phi_\beta(s) = (1 - R_i)f_\beta(s) + R_i g_\beta(s).$$

What are the proportions R_i such that the probability of extinction is minimized? An analytical treatment of this question has not as yet produced an answer. The only idea of the nature of the solution so far acquired is contained in the computed results of two particular numerical examples. These were

$$f_\alpha(s) = 0.1 + 0.9s; g_\alpha(s) = 0.7 + 0.3s^2; f_\beta(s) = s; g_\beta(s) = 0.1 + 0.9s^2 \quad (1)$$

$$f_\alpha(s) = 0.1 + 0.9s; g_\alpha(s) = 0.85 + 0.15s^3; f_\beta(s) = 0.05 + 0.95s;$$
$$g_\beta(s) = 0.2 + 0.8s^3 \quad (2)$$

The values of the proportions R_i corresponding to the minimum extinction problem were calculated by the Nelder–Mead Simplex Minimization procedure combined with the Wilkinson inverse matrix method. In order to limit the cost of computer hire and also to guard against numerical instability, the values of R_i for $i \geq 7$ were constrained to be equal to the same value. The results of the two runs are given in Table 1.

TABLE 1. Proportions of Breeders minimizing probability of ultimate extinction

Population size	$f_1(s) = 0.1 + 0.9\,s$ $g_1(s) = 0.7 + 0.3\,s^2$ $f_2(s) = s$ $g_2(s) = 0.1 + 0.9\,s^2$	$f_1(s) = 0.1 + 0.9\,s$ $g_1(s) = 0.85 + 0.15\,s^3$ $f_2(s) = 0.05 + 0.95\,s$ $g_2(s) = 0.2 + 0.8\,s^3$
1	0.92	0.70
2	0.88	0.60
3	0.85	0.55
4	0.75	0.54
5	0.66	0.49
6	0.60	0.47
>6	0.42	0.37

The proportion of breeders decreases with increasing density in both cases. This slight evidence—though when one considers the structure-preserving properties of convex functions, the results can no doubt be extended beyond these two particular examples—is something of a paradox.

Remembering that the model we have used to obtain these results postulates a pair of environments which are unaffected by changes in the population size—the p.g.f.'s $f_\alpha, f_\beta, g_\alpha, g_\beta$ do not alter with changing density—then the results demonstrate that population regulation can be brought about, not by direct density-dependent factors, such as competition for a limited food-supply, but for the excellent reason that by ever reducing its reproductive rate with increasing density, the population thereby enhances its chance of survival.

The argument gains some marginal support from a reconsideration of the earlier problem of determining the minimum extinction probability for a population in which the proportion of non-breeders is unvarying over all densities. We now consider the determination of the minimum probability for different initial population sizes.

As before

$$\phi_\alpha(s) = (1 - R)f_\alpha(s) + Rg_\alpha(s)$$

and

$$\phi_\beta(s) = (1 - R)f_\beta(s) + Rg_\beta(s)$$

are the p.g.f's in bad and good years respectively.

The probability of ultimate extinction of a population starting with K individuals is

$$q_K = \operatorname*{Lt}_{n \to \infty} \frac{1}{2^n} \sum_{i,j,\ldots} [\phi_i(\phi_j(\ldots(0)\ldots))]^K$$

If $f_\alpha(s) \leq g_\alpha(s)$ and $f_\beta(s) \geq g_\beta(s)$ for s in $(s_\beta, 1)$ where $\phi_\beta(s_\beta) = s_\beta$ i.e. breeders do less well than non-breeders in bad years and better than in good years, then $\phi_\alpha(s)$ increases and $\phi_\beta(s)$ decreases with increasing R for all values of $s \geq s_\beta$. So, arguing informally, $\phi_i(\phi_j(\ldots(0)\ldots))$ increases or decreases with increasing R depending on the preponderance of ϕ_α's or ϕ_β's—especially in the functions corresponding to the later generations. A similar preponderance of ϕ_α's or of ϕ_β's associates with respectively high and low values of $\phi_i(\phi_j(\ldots(0)\ldots))$. It follows that at that value of R minimizing q_1, i.e. where

$$\operatorname*{Lt}_{n \to \infty} \frac{1}{2^n} \sum \frac{\partial}{\partial R}(\phi_i(\phi_j(\ldots(0)\ldots))) = 0$$

the value of

$$\frac{\partial q_2}{\partial R} = \operatorname*{Lt}_{n \to \infty} \frac{1}{2^n} \sum \phi_i(\phi_j(\ldots(0)\ldots)) \frac{\partial}{\partial R} \phi_i(\phi_j(\ldots(0)\ldots))$$

is positive. Hence as q_2 is a convex function of R then the value of R minimizing q_2 is less than that minimizing q. Similarly the value of R minimizing q_K is less than that minimizing q_{K-1} for all K. The argument leads to the conclusion that the larger the initial population size, the larger is that proportion of non-breeding individuals minimizing the probability of extinction.

4. Discussion

The preceding analysis shows that in certain cases a population minimizes its probability of extinction by maintaining a non-breeding reserve which increases proportionately with increasing size of population. Clearly the argument can be elaborated to provide an explanation both of the range of variability of clutch-size and also of the reduction in clutch-size with increasing density. Lack's hypothesis is unable to explain either of these types of behaviour. The variability in clutch-size runs counter to his criterion that the clutch-size is such as to give rise, on average, to the greatest number of off-spring. The inadequacies of his thesis stem from its use of the concept of an average. However defined, an average may be a misleading measure of a population's success in a variable environment.

As an extreme example it is easy to specify a population which has an infinitely large eventual expected or average size and which is also doomed to certain extinction. In the circumstances of a variable environment the fate of a population is more appropriately measured by its probability of extinction; it is a measure that is unambiguous and biologically meaningful.

The model of clutch-size proposed in this paper has been criticized for lacking a sound genetic basis. It has been argued that if the difference between breeders and non-breeders is genetically determined, then breeders will eventually predominate. I have two comments to make on this criticism.

First, extensive field observations have far from established the nature of the genetic bases for clutch-size. Indeed Kluyver (1963), who has made a detailed study of the problem, concluded that clutch-size variability had no apparent genetic component. Secondly, and of more importance, the usual measure of the fitness of a genotype ill-describes the evolutionary movement of a population in a varying environment. In current usage the fitness of an individual is defined as the average number of gametes contributed to the next generation. The emergence of the fittest genotype is thus in exact corre-spondence to the evolution of the most efficient clutch-size in the sense defined by Lack (1947). Hence the criticism in this paper of Lack's criterion applies equally to the conventional measure of fitness; in a varying environ-ment both measures are misleading and inappropriate.

5. Summary

Lack's theory of clutch-size is criticized for its failure to provide an explana-tion of the observed variability in clutch-sizes. An alternative measure, the probability of extinction in a random environment, is shown to provide an explanation of the variability in clutch-sizes and also of the reduction in clutch-size with increasing density. For populations that maintain a reserve of non-breeding individuals, Monte Carlo studies demonstrate that the proba-bility of extinction is minimized by a mechanism that increases the proportion of non-breeders with increasing population size.

It is argued that the conventional measure of fitness is inapplicable to a variable environment.

REFERENCES

ATHREYA, K. B. AND KARLIN, S. (1971a). Branching processes with random environments, I: Extinction probabilities. *Ann. Math. Statist.* **42**, 1499–1520.
ATHREYA, K. B. AND KARLIN, S. (1971b). Branching processes with random environments, II: Limit theorems. *Ann. Math. Statist.* **42**, 1843–1858.
KLUYVER, H. N. (1963). The determination of reproductive rates in Paridae. Proc. 13th Int. Orn. Congr. Ithaca, 706–716.
LACK, D. (1947). The significance of clutch-size. *Ibis* **87**, 302–352.
LACK, D. (1964). Significance of clutch-size in Swift and Grouse. *Nature* **203**, 98–99.
LACK, D. (1966). "Population Studies of Birds." Oxford University Press, London.
MOUNTFORD, M. D. (1971). Population survival in a variable environment. *J. theor. Biol.* **32**, 75–79.
SMITH, W. L. and WILKINSON, W. (1969). On branching processes in random environments. *Ann. Math. Statist.* **40**, 814–827.
WILKINSON, W. (1969). On calculating extinction probabilities for branching processes in random environments. *J. Appl. Prob.* **6**, 478–492.
WYNNE-EDWARDS, V. C. (1964). Significance of clutch-size in Swift and Grouse. *Nature* **203**, 99.

19. Species Diversity in Ecological Communities

MARK WILLIAMSON

Department of Biology, University of York, York, England

The study of diversity is the study of the variation in the numbers of different species under different ecological circumstances. It is, for instance, well known that there tend to be more species in the tropics than in temperate regions, and fewer again in the Arctic. Arnold (1972) gives diagrams showing the change in the numbers of sympatric species of snakes, lizards and some other groups in the Americas. As these figures are for species found together, it seems that the gradient in diversity is not just a matter of a greater range of habitats in warmer climates. The study of the nature and causes of variations in species diversity has become extremely popular over the last ten years, and there has been a major symposium on the topic (Woodwell and Smith, 1969). In this paper I can do no more than outline some of the studies that have been made and show the relation of this study to the science of population dynamics.

It has frequently been stressed that there are at least two important aspects of diversity. The first is the one which, *pace* Hurlbert (1971), is the original meaning of the word, namely the variation of total number of species. The other aspect is the distribution of the numbers of individuals amongst the species. If most individuals belong to only a few of the species represented, then it is conventional and natural to consider the assemblage as less diverse than if the numbers of the different species are more even. A concept of evenness is readily applied to an assemblage of species of trees, or birds or any other fairly homogeneous set of biological species with similar life habits. It is much less easy to apply it to the community as a whole, where there will be great variations in mode of life, and of size, so that the relative abundance would in any case be expected to differ by several orders of magnitude in many cases. There are few studies that have attempted to deal with this aspect of diversity, yet it must be tackled if the structure of communities is to be understood.

The changes in diversity from one community to another that can be seen are so great that certain generalizations can be made irrespective of the method used to quantify these changes. It is usually sufficient just to consider the number of species. Typically, diversity increases or has increased:

1. Towards the tropics, and an example of this is shown in Fig. 2 and is discussed more fully below.

2. Through geological time. Simpson (1969) gives some simple diagrams of this. In fossil assemblages too, tropical faunas are often more diverse (Stehli *et al.*, 1969).

3. Away from toxic pollution (Woodwell, 1970), and in some cases pollutants that are not clearly toxic have the same effect.

4. During ecological succession, that is, the change from pioneering communities to, eventually, stable climax communities (Goulden, 1969).

5. Towards the more stable environments, of which the most surprising example is the richness of the cold, deep-sea benthos. The fauna here, living in temperatures just above freezing point, in total darkness, under high pressure and with a very poor food supply, is almost as rich as that living on the floor of shallow tropical seas, and much more diverse than the faunas of shallow temperate seas (Sanders, 1968, especially his Fig. 3). As the different environmental factors are non-commensurate, it is difficult to gauge their relative stability except by their effect on biological organisms, and so there are dangers of circular arguments about stability and biological diversity.

Although these rules are frequently observed, there are also very frequent exceptions. For instance, it is well known that there are fewer species of penguins at the equator than in the sub-antarctic and there are many other less well known groups for which this is true. There have been many occasions in the geological records in which diversity has sharply decreased, as with the dinosaurs at the end of the Cretaceous. It's a fairly common occurrence for diversity to increase steadily through most of the successional series, and then to decrease somewhat to the climax, and for some organisms there may be no relation between diversity and the successional state reached. There are important variations in diversity not encompassed by any of the generalizations above. For instance the Amazonian jungle is much more diverse than the African jungle, and it is possible that the most diverse plant communities are not jungles at all but sclerophyllous scrub (Richards, 1969).

One important consequence of the changes in diversity in geological time is that all modern patterns of diversity have an important historical and evolutionary element to them. There are, for instance, no bears in Africa though there are in tropical America and tropical Asia. Great variations in the available diversity are readily shown in the British insects (Fig. 1). Any insect community is likely to have large numbers of species of moths (Lepidoptera), flies (Diptera), beetles (Coleoptera) and wasps (Hymenoptera), and the predominance of these groups in any species list reflects their evolutionary success. At a lower taxonomic level Vuilleumier (1972) gives an example

Number of species in the orders etc. of British Insects

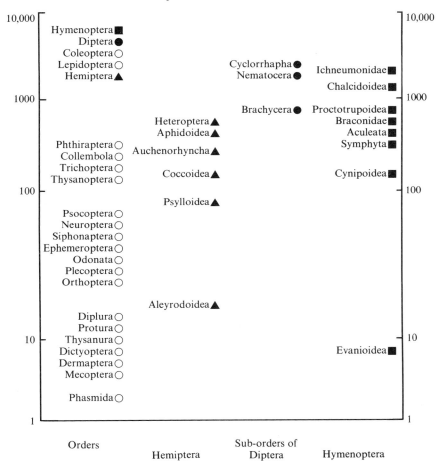

FIG. 1. Numbers of species in different orders of insects recorded from the British Isles. The numbers in the sub-orders of three of the orders are also shown. Data from Kloet and Hincks (1945 and 1964), and Kerrich *et al.* (1967).

where the diversity of birds in different vegetational types in South America differs from the expectation derived from the study of North American faunas. The explanation probably lies in the availability of bird species, and so is a historical and evolutionary one rather than an ecological one. For ecological studies, an attempt at the separation of these two aspects is desirable, but so far it has been avoided almost entirely by considering taxonomically homogeneous groups. Indeed, the doubts that may be felt about the possibility of making a clear separation of these aspects suggests that there may be a considerable limitation to the usefulness that diversity studies can have in ecological research.

The Measurement of Diversity

The simplest index of diversity is the total number of species found, but clearly this does not measure the evenness component and numerous other measures have been proposed, some devised specifically to measure evenness, others designed to balance both aspects, and yet others developed to fit one worker's particular concept of what is meant by diversity. The mathematical theory of some of the commonest of these is given by Pielou (1969). The commonest alternative to just counting the number of species is to use an index based in information theory, such as H which is defined as $-\Sigma p_i \log p_i$. One of the more satisfactory variants of this is to standardize by dividing it by H max, the maximum value the index could take for a given number of species. Such indices tend to be dominated by the relative proportions of the common species, and so to measure evenness, while indices based on the total number of species are dominated by the number of rare species. Even so, richness and evenness usually go together in ecological communities, and so all the indices tend to be correlated. Auclair and Goff (1971), and Johnson and Raven (1970) have demonstrated these correlations over a wide range of plant communities. There seems little point in using, or further studying, most of these indices, unless one particular index could be shown to illuminate some aspect of the population dynamics of a community which is ignored by the others. Mathematically, one could suggest that, if it were possible to get an indication of the richness which was not dependent on the size of the sample, then some other index which was orthogonal to that one might be useful in measuring the evenness of the community which is not correlated with the richness. Unfortunately all indices, including the count of the total number of species, vary with the sample size, and this will now be considered.

In terrestrial ecology, particularly when the worker is concerned only with his own data, it is easy to avoid problems of sample size by making all samples the same size. When one wants to compare different communities studied by different workers, and particularly when one is studying communities that are hard to sample, such as those in the depth of the ocean, it is necessary to take account of the variation in sample size. Sanders' (1968) suggestion was to reduce all samples to the size of the smallest by a process he called rarefaction. This obviously leads to a considerable loss of information, but the chief objection to his method is that it is mathematically unacceptable as it stands. Fortunately in his work the differences in diversity are so great that neither the loss of information, nor the inaccuracies caused by his rarefaction method, affect the validity of his conclusions. Simberloff (1972) has shown how to get a more acceptable rarefaction, based on computer sampling from the larger samples. Fager (1972) shows that even this is not sufficient, in that the rarefaction that should be applied depends on the grouping or pattern of the organisms. In very few species are the individuals distributed at random; normally they are clumped to a greater or lesser extent. Fager shows that Sanders' technique is roughly equivalent to assuming that the organisms are grouped in clumps of half an individual, or in other words that the individuals are somewhat more evenly spread than if they were randomly distributed. All this work is essentially an attempt to measure the number of species independently of the sample size: Sheldon (1969) has considered this problem with the other indices.

Patterns of Species Abundance

Biologists find indices of diversity attractive, because the properties of their samples are summed up in one single statistic. The number of indices that have been proposed, the correlations between them, and the problems arising from sampling, all suggest that diversity would be better measured by rather more complicated statistics than a single figure. Figure 2 gives some data for one of the most discussed comparisons: that between temperature and tropical forests. Curtis (1959) gives figures for the number of trees over 4 inches in diameter per acre per species for a number of forest types in Wisconsin: Black *et al.* (1950) and Murca Pires *et al.* (1953) do the same for three Amazonian forests, using trunks over 10 cm diameter at breast height, based in two cases on one hectare and in one on three hectares (1 hectare = 2·4 acres). Figure 2 was derived by taking the geometrical mean number of trees for the commonest, second and so on for the three tropical forests and for seven temperate forests listed by Curtis (including types dominated both by deciduous trees and by conifers). The mean numbers of trees of all species, and one standard deviation about it, is also shown for both sets of communities. All the data have been scaled to numbers per hectare. The plot used in Fig. 2, which is the logarithm of abundance against the logarithm of rank, is used here simply because it tends to give the straightest plot of a number that were tried. Of course, each forest community was also plotted separately, and although these fall reasonably about the mean lines for their set, some clearly are concave upwards and others concave downwards. It is clear that there is no one transformation which would produce a straight line from the data of each separate community.

Very roughly, there are the same number of trees per unit area in Wisconsin and in Brazil, and knowing this and the abundance of the few commonest

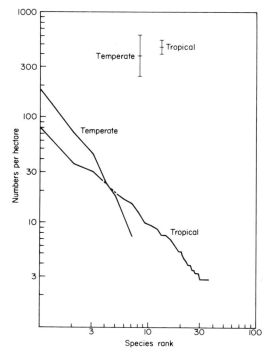

F<small>IG</small>. 2. Plots of the mean number of trees in standard plots in temperate and tropical forests. The steeper curve relates to seven one acre plots in Wisconsin (Curtis, 1959), the shallower to three plots in Brazil (Black *et al.*, 1950 and Murca *et al.*, 1953) on one and three hectare plots. The data have all been scaled to show log numbers per hectare. The abundances have been arranged in rank order and these ranks are given, also on a logarithmic scale, on the *x* axis. The mean numbers per plot ± one standard deviation are also shown. The shorter span, with the slightly higher mean refers to the Amazonian forests, the longer to the Wisconsin forests.

species, would allow one to say, again rather roughly, what the abundance of the rare species would be. These approximate relationships are quite sufficient to explain the correlations of the indices of diversity that have been found. As this sort of plot might suggest a return to the theoretical studies of the logarithmic curve (Williams, 1964), or the lognormal curve (Preston, 1962) or even to the morass of papers that were produced as a result of MacArthur's broken stick hypothesis (Pielou, 1969) it is perhaps well to emphasize not only that the individual forests do not follow the same line, but also that these curves are monotonic only because they have been plotted from samples taken at a single point of time. We will return to this point below. What Preston's and Williams' work does bring to mind, is that there is a relationship between the diversity in a sample of a given size, and what are known as species/area curves. These are plots for a number of species found in samples of successively larger area.

Species/Area Relationships

Recently, species/area relationships have been studied by better statistical methods than in the past. Johnson and Raven (1970) is only one study that shows the value of multiple regression in relating the number of species to environmental variables in the areas in which they are found.

Species/area curves have in fact been studied by botanists since the 1920's, and early on two relationships seemed to receive empirical support. The first was that there was a straight line relationship between the number of species and the logarithm of the area, the second that there was a straight line relationship between the logarithm of a number of species and the logarithm of the area. In view of this work it is a pity that some of the recent multiple regression studies have tried comparing the number of species and the logarithm of this directly to the area. Dony (1970, see also 1963) has reviewed 80 sets of data covering much of this old controversy, and one might say in passing how fortunate it was that the editors in those days allowed the data to be published in full. He concludes that in 47 studies it is reasonable to say that the log species to log area relationship is straight, in 5 this plot appears to produce two straight lines with a reasonably clear break between them, and the same situation less clearly in 2 more. However, 12 studies definitely do not produce a straight line on this sort of plot, but do produce one when it is the number of species that is plotted against the logarithm of the area. In addition ten studies produce data that is not straight on any of these plots, and a further four involve too few species for any statement to be made about them. The conclusion seems to be that although no one relationship will fit all communities, the log species to log area relationship is the one which is most usually found to be satisfactory. There is an obvious relationship between this and the type of plot shown in Fig. 2, even though Fig. 2 refers to fixed areas. The x axis, the rank order of species, is relatable to the number of species found, while the y axis, the logarithm of the abundance of individual species, is relatable to the area that one would expect to have to search to find a single individual of that species.

It would appear, then, that diversity might be better studied by making a pragmatic plot of the abundancies of the species. In the studies discussed so far, the individuals of different species are all reasonably comparable. When this is not so, for instance when both plants and animals are included, some more sophisticated measure of the species role in the community, such as its productivity, would probably be needed (Dickman, 1968; Hurlbert, 1971; Whittaker, 1969). It would appear sensible to consider the logarithm of the species abundance in the first instance, and this is scarcely surprising as it is almost invariably valuable to do so in population dynamics (Williamson, 1972).

Explanations of Diversity

As studies on diversity have so far mostly used the indices, which are all rather unsatisfactory and inadequate, it is natural that the explanations of the variations in diversity are also rather crude. Pianka (1966) surveyed the explanations that had been produced to that date of latitudinal gradients in diversity, though as was seen earlier there are many other systematic variations. Variations in environmental heterogeneity may partly explain the variations in diversity. For instance, Johnson and Raven (1970) show that the number of soil types had an important influence on the number of species of plants found in the Scottish islands. However, Arnold's (1972) study of the variation in the number of sympatric species of reptiles, and more particularly the diversity found in the very stable homogeneous and inhospitable deep-sea benthos seems to suggest that environmental heterogeneity is confusing the issue. Possibly it should be removed statistically, by standard covariance methods. This would presumably leave an important problem: the variations in diversity in homogeneous areas. Extra heterogeneity was one of the explanations for the increased diversity in the tropics reviewed by Pianka (1966). The other five were: time (which relates to the evolutionary component of diversity, and is also considered important by Sanders (1968)), competition, predation, stability and productivity.

As there is little if any critical evidence relating to any of these, the only one I want to consider here is that of predation. Since Pianka's paper, Janzen (1970) has argued that herbivores attacking seeds are an important factor contributing to tropical forest diversity, and Dayton and Hessler have argued (1972) that disturbance, which includes predation, is important in the maintenance of diversity in the deep sea. At the very least, it is only reasonable to suppose that food chain relationships affect diversity in some communities. The implication of this is that diversity cannot be measured simply by the numbers, or even as suggested above by the productivity of the individual species; the role of each species in the community must also be assessed and quantified. An important aspect of the theory of predator/prey systems is they take time, and the lags that may result can produce oscillations, in theory at least. Whether or not predator/prey interactions are a partial cause, in all communities that have been studied not only does the abundance of each individual species vary in time, but so do the relative abundances of the different species (Williamson, 1972). So if the species are ranked by their average abundance over a series of samples, the plots like those of Fig. 2

for the individual samples would not normally be monotonic. Conversely, if the species are plotted in strict rank order always in each sample, then different species would occupy different ranks in different samples, and this is unlikely to be helpful in producing an explanation of the shape of the plot from the study of the population dynamics of the community. In other words a realistic study of the diversity of communities, and of the variations of diversity between different communities, must consider not only the mean structure of that community, but also the variation about it, probably measured in the first place by the variance of some appropriate measure.

Some studies have already been made of data matrices of communities, in which the elements are the abundance of different species at different times. The variation in such matrices can be ascribed to some extent to measurable environmental factors (Williamson, 1972), but no real progress has been made towards explaining the size of the variation, let alone the mean structure of these communities, by population models. Such models would appear to require another matrix, a data matrix of the strength of interactions. The elements in this matrix would be the strength of the various interactions in a community at successive sampling times. This raises enormous problems both of definition and measurement. But until this approach is attempted, it is likely that the study of diversity will remain where it seems to be at the moment, a mixture of inadequate empirical studies and interesting speculations not checked by critical observation or experiment. My own view is that the direct study of diversity is less likely to be profitable in the long run than attempts to study the population dynamics of sets of species, but that for the time being studies of diversity can still be helpful to ecologists, provided that they remember the limitations imposed by past evolution, and provided they attempt to use more sophisticated statistics than the indices that have dominated the subject so far.

Acknowledgements

I am grateful to Dr. P. K. Dayton and the other members of his seminars on biological diversity at the Scripps Institution of Oceanography in 1970–1971 for introducing me to many aspects of the problem, and to Dr. J. H. Lawton for many useful discussions.

REFERENCES

ARNOLD, S. J. (1972). Species diversity of predators and their prey. *Am. Nat.* **106**, 220–236.

AUCLAIR, A. N. AND GOFF, F. G. (1971). Diversity relations of upland forests in the western Greak Lakes area. *Am. Nat.* **105**, 499–528.

BLACK, G. A., DOBZHANSKY, T. AND PAVAN, C. (1950). Some attempts to estimate species diversity in trees in Amazonian forests. *Bot. Gaz.* **111**, 413–425.

CURTIS, J. T. (1959). "The vegetation of Wisconsin: An ordination of plant communities." Univ. Wisconsin Press, Madison.

DAYTON, P. K. AND HESSLER, R. R. (1972). Role of biological disturbance in maintaining diversity in the deep sea. *Deep Sea Res.* **19**, 199–208.

DICKMAN, M. (1968). Some indices of diversity. *Ecology* **49**, 1191.

DONY, J. G. (1963). The expectation of plant records from prescribed areas. *Watsonia* **5**, 377–385.

DONY, J. G. (1970). Species-area relationships. Unpubl. m.s. report to the Natural Environment Research Council, London.

FAGER, E. W. (1972). Diversity: a sampling study. *Am. Nat.* **106**, 293–310.

GOULDEN, C. E. (1969). Temporal changes in diversity. *In* "Diversity and Stability in Ecological Systems." (G. M. Woodall and H. H. Smith, eds.), Brookhaven Symposium in Biology No. 22, pp. 178–195. Brookhaven National Laboratory.

HURLBERT, S. H. (1971). The non-concept of species diversity: a critique and alternative parameters. *Ecology* **52**, 577–586.

JANZEN, D. H. (1970). Herbivores and the number of tree species in tropical forests. *Am. Nat.* **104**, 501–528.

JOHNSON, M. P. AND RAVEN, P. H. (1970). Natural regulation of plant species diversity. *Evolutionary Biology* **4**, 127–162.

KERRICH, G. J., MEIKLE, R. D. AND TEBBLE, N. (1967). Bibliography of key works for the identification of the British fauna and flora (3rd Edn.). London, Systematics Association.

KLOET, G. S. AND HINCKS, W. D. (1945). "A Check List of British Insects." Kloet and Hincks, Stockport.

KLOET, G. S. AND HINCKS, W. D. (1964). "A Check List of British Insects." (2nd Edn., revised). Part I: Small orders and Hemiptera. R. Ent. Soc. London. Handb. Indent. Br. Insects 11 (1).

MURCA PIRES, J., DOBZHANSKY, T. AND BLACK, G. A. (1953). An estimate of the number of species of trees in an Amazonian forest community. *Bot. Gaz.* **114**, 467–477.

PIANKA, E. R. (1966). Latitudinal gradients in species diversity: a review of concepts. *Am. Nat.* **100**, 33–46.

PIELOU, E. C. (1969). "An Introduction to Mathematical Ecology." Wiley-Intersciences. New York and London.

PRESTON, F. W. (1962). The canonical distribution of commonness and rarity. *Ecology* **43**, 185–215 and 410–432.

RICHARDS, P. W. (1969). Speciation in the tropical rain forest and the concept of the niche. *Biol. J. Linn. Soc. Lond.* **1**, 149–153.

SANDERS, H. L. (1968). Marine benthic diversity: a comparative study. *Am. Nat.* **102**, 243–282.

SHELDON, A. W. (1969). Equitability indices: dependence on the species count. *Ecology* **50**, 466–467.

SIMBERLOFF, D. (1972). Properties of the rarefaction diversity measurements. *Am. Nat.* **106**, 414–415.

SIMPSON, G. G. (1969). The first three billion years of community evolution. *In* "Diversity and Stability in Ecological Systems." (G. M. Woodwell and H. H. Smith, eds.), Brookhaven Symposium in Biology No. 22, pp.162–176. Brookhaven National Laboratory.

STEHLI, F. G., DOUGLAS, R. G. AND NEWELL, N. D. (1969). Generation and maintenance of gradients in taxonomic diversity. *Science* **164**, 947–949.

VUILLEUMIER, F. (1972). Bird species diversity in Patagonia (Temperate South America). *Am. Nat.* **106**, 266–271.

WHITTAKER, R. H. (1969). Evolution of diversity in plant communities. *In* "Diversity and Stability in Ecological Systems." (G. M. Woodwell and H. H. Smith, eds.), Brookhaven Symposium in Biology No. 22, pp. 178–195. Brookhaven National Laboratory.

WILLIAMS, C. B. (1964). "Patterns in the Balance of Nature." Academic Press, London and New York.

WILLIAMSON, M. (1972). "The Analysis of Biological Populations." Edward Arnold, London.

WOODWELL, G. M. (1970). Effects of pollution on the structure and physiology of ecosystems. *Science* **168**, 429–433.

WOODWELL, G. M. AND SMITH, H. H. (eds.) (1969). "Diversity and Stability in Ecological Systems." Brookhaven Symposium in Biology. No. 22. Brookhaven National Laboratory.

Author Index

Subject Index